CRM Series in Mathematical Physics

Springer Science+Business Media, LLC

CRM Series in Mathematical Physics

Yvan Saint-Aubin
Luc Vinet
Editors

Algebraic Methods in Physics

A Symposium for the 60th
Birthdays of Jiří Patera and
Pavel Winternitz

With 17 Illustrations

 Springer

Yvan Saint-Aubin
Département de mathématiques
et de statistique
CRM, Université de Montréal
C.P. 6128, succ. Centre-ville
Montréal, Québec H3C 3J7
Canada
saint@math.ias.edu

Luc Vinet
Département de mathématiques
et de statistique
CRM, Université de Montréal
C.P. 6128, succ. Centre-ville
Montréal, Québec H3C 3J7
Canada
vinet@crm.umontreal.ca

Library of Congress Cataloging-in-Publication Data
Algebraic methods in physics : A symposium for the 60th
  birthdays of Jiří Patera and Pavel Winternitz
  Yvan Saint-Aubin, Luc Vinet [editors].
    p. cm.—(The CRM series in mathematical physics)
  Includes bibliographical references and index.
  ISBN 978-1-4612-6528-3   ISBN 978-1-4613-0119-6 (eBook)
  DOI 10.1007/978-1-4613-0119-6
    1. Mathematical physics—Congresses.   2. Algebra—
Congresses. I. Patera, Jiří. II. Winternitz, Pavel
III. Saint-Aubin, Yvan. IV. Vinet, Luc. V. Series: CRM series
in mathematical physics.
QC19 .2 .A44 2000
530 .15′2 21   00-61863

Printed on acid-free paper.

Production managed by A. Orrantia; manufacturing supervised by Jeffrey Taub.
Photocomposed copy prepared by the editors.

9 8 7 6 5 4 3 2 1

ISBN 978-1-4612-6528-3   SPIN 10778273

Series Preface

The Centre de recherches mathématiques (CRM) was created in 1968 by the Université de Montréal to promote research in the mathematical sciences. It is now a national institute that hosts several groups, holds special theme years, summer schools, workshops, and a postdoctoral program. The focus of its scientific activities ranges from pure to applied mathematics, and includes statistics, theoretical computer science, mathematical methods in biology and life sciences, and mathematical and theoretical physics. The CRM also promotes collaboration between mathematicians and industry. It is subsidized by the Natural Sciences and Engineering Research Council of Canada, the Fonds FCAR of the Province de Québec, the Canadian Institute for Advanced Research and has private endowments. Current activities, fellowships, and annual reports can be found on the CRM web page at `www.CRM.UMontreal.CA`.

The CRM Series in Mathematical Physics includes monographs, lecture notes, and proceedings based on research pursued and events held at the Centre de recherches mathématiques.

Yvan Saint-Aubin
Montréal

Preface

This book contains papers presented at the Symposium in honour of Jiří Patera and Pavel Winternitz, held in January 1997 at the Centre de recherches mathématiques (CRM).

Both were born in 1936, a mere 60 years ago, Pavel in July and Jiří in October. These 60th anniversaries provided a wonderful occasion to express our friendship and appreciation for their remarkable scientific achievements, the mammoth role that they have played in the life of the CRM, and what they have done for mathematical physics at the Université de Montréal and in Canada. This is extremely pleasing to us since Jiří was Yvan's supervisor and Pavel, Luc's.

We are most grateful for the participants' enthusiastic willingness to be part of this celebration, and for their contributions to this volume. This speaks for the high esteem in which Jiří and Pavel are held by their friends and colleagues. The extent of Jiří's and Pavel's network of collaborators, former students, postdocs, and associates is such that many other eminent scientists could and should have been invited to speak. Obvious limitations prevented us from inviting all those that we wished.

The theme of this symposium was *Algebraic Methods and Theoretical Physics*. It encompasses most of Jiří's and Pavel's work. They both did their graduate studies in the former Soviet Union: Pavel in Leningrad first and then in Dubna, Jiří in Moscow for the main part. They were trained as particle theoreticians. They began their formal academic career at the time of the Prague Spring. Then came August 21st and they elected to leave their country. Little did we suspect, watching the 1968 events in Prague unfold, that they would thus have such a direct and important impact on our lives and more generally on Montreal's scientific evolution.

You might be curious to know to what we owe the good fortune of their coming to Montreal. Interestingly, Jiří first came to the Université de Montréal in January 1965 with a postdoctoral fellowship from the National Research Council. He stayed here until the end of 1966. During these two years he continued his work in high energy physics but also started to collaborate with colleagues in Montreal on more mathematical projects bearing, for instance, on the representation theory of Lie algebras, an area for which he became a champion. He returned again to Montreal for four months in the winter of 1967–1968. He was in Prague during the ensuing famous spring and, in the aftermath of the invasion, managed to attend the High Energy Physics conference in Vienna at the end of August. He was then offered a position at Imperial College, which he took immediately and held until the summer of 1969.

Pavel had also received a number of offers. He declined those in the United States and also went to England, where he had spent some of his childhood during the war. This brought him to the Rutherford High Energy Physics Lab, where he stayed until 1969.

Because of the visits that Jiří had made to the Université de Montréal and the links that he had developed with this university settling in Montreal was natural for him.

At that time, in 1968 and 1969, with great vision, Roger Gaudry, Maurice L'Abbé, Jacques St-Pierre, André Aisenstadt, and others were setting up at the Université de Montréal a research institute in the mathematical sciences: le Centre de recherches mathématiques, commonly referred to as the CRM. They also showed great insight in hiring a number of highly talented Czechs, among them Jiří, who thus came to the nascent CRM in August of 1969.

Meanwhile Pavel left England for the United States and became an associate professor, first at the Carnegie-Mellon University and then at the University of Pittsburgh in 1970. In those years, his research focused mainly on two-variable expansions of scattering amplitudes and their expansions.

It was Jiří basically who, soon after, in 1972, convinced Pavel to join him in Montreal at the CRM. The rest, as is often said, is history: a tremendous research team in mathematical physics was thus created and it became one of the trademarks of the CRM.

The accomplishments of Jiří and Pavel since their arrival at the CRM are, to say the least, impressive.

Let us briefly recall their seminal work in—

• the theory and the application of two-variable expansions of scattering amplitudes;

• the connection between symmetries of partial differential equations and the separation of variables;

• the classification with Hans Zassenhaus of the subgroups of groups important in physics, which led to the formidable enterprise of classifying maximal Abelian subalgebras of complex and real simple Lie algebras. From 1973 to 1983 they jointly wrote more than 30 papers (in addition to conference proceedings) on these topics, many of which are now classics.

In 1979–1980 Pavel was on leave in Paris and Jiří spent a year at Caltech in 1983–1984. At that time they established new collaborations extending the team's expertise. Jiří explored new problems in Lie algebras: elements of finite order in groups, the relationship between grading and contractions, and more recently, cells of root lattices, quasicrystals, and aperiodic order. Pavel moved into the field of integrable systems then known as soliton theory. After work on matrix Riccati equations and superposition principles, he initiated a vast research program on Lie symmetries of differential equations and their Painlevé property in which he is still much involved and that has recently been extended to finite-difference equations.

Jiří is also world famous for his tables of "vital data" on Lie algebras. Every high energy physicist has a copy of them on his bookshelf, or more probably on his desk, and many pure mathematicians in representation theory and finite and Lie group theory use them regularly.

In 1976, the year the Summer Olympic Games were held there, Jiří and Pavel brought the International Colloquium on Group Theoretical Methods in Physics to Montreal. It was the first time this conference was held outside Marseilles and Nijmegen; since then, the colloquium has been hosted by 15 other cities in 12 different countries. They also organized a three-week summer school before the Colloquium, under the auspices of the Séminaire de Mathématiques Supérieures, a 30-year-old tradition of the Département de Mathématiques et de Statistique of the Université de Montréal. They organized many additional meetings since then and served on numerous committees and editorial boards.

Jiří and Pavel have always been much involved with the training of graduate students and supervised scores of Masters, Ph.D.'s, and postdocs. They have also made countless invited presentations at international conferences and have received various awards. We mention particulary the Killam fellowship that Jiří held in 1992–1993.

We conclude by again expressing to them our gratitude and admiration for the inspiration, energy and dedication. The hard work that they did has made Montreal one of the places where the action is in mathematical physics and helped the Centre de recherches mathématiques acquire its worldwide reputation. Their impact here will be long-lasting.

Yvan Saint-Aubin
Luc Vinet

Contents

14 Bargmann Representation for Some Deformed Harmonic Oscillators with Non-Fock Representation   205
Michèle Irac-Astaud and Guy Rideau

15 The Vector-Coherent-State Inducing Construction for Clebsch–Gordan Coefficients   219
D.J. Rowe

16 Highest-Weight Representations of Borcherds Algebras   233
Richard Slansky

17 Graded Contractions of Lie Algebras of Physical Interest   247
J. Tolar

List of Contributors

Michéle Irac-Astaud, Laboratoire de Physique Théorique de la matière condensée, Université Paris VII, 2 place Jussieu F-75251 Paris Cedex 05, France
mici@ccr.jussieu.fr

Michael Baake, Institut fur Theoretische Physik, Universität Tübingen, Auf der Morgenstelle 14, D-72076 Tübingen, Germany
michael.baake@uni-tuebingen.de

Mark Bodner, Department of Psychiatry, University of California, Los Angeles, CA 92004, USA
mbodner@ucla.edu

D.J. Britten, Department of Mathematics, University of Windsor, Windsor, On, N9B 3P4, Canada
britten@uwindsor.ca

Bernard Champagne, 5780 rue Duquesne, Montréal, Qué., H1M 2K4, Canada

Mark Fels, School of Mathematics, University of Minnesota, Minneapolis, MN 55455, USA
fels@math.umn.edu

Jean-Pierre Gazeau, LPTMC, Université Paris 7–Denis Diderot, 2 Place Jussieu, 75251 Paris Cedex 05, France
gazeau@ccr.jussieu.fr

B. Grammaticos, LPN, Université Paris VII, Tour 24-14, 5e étage, 75251 Paris, France
grammati@paris7.jussieu.fr

E.G. Kalnins, Department of Mathematics and Statistics, University of Waikato, Hamilton, New Zealand
math0236@waikato.ac.nz

N. Kamran, Department of Mathematics and Statistics, McGill University, Montréal, Qué., H3A 2K6, Canada
NKamran@math.mcgill.ca

Ronald C. King, Department of Mathematics, University of
Southampton, Southampton SO17 1BJ, England
rck@maths.soton.ac.uk

F. W. Lemire, Department of Mathematics, University of Windsor,
Windsor, On, N9B 3P4, Canada
lemire@server.uwindsor.c

D. Levi, Dipartimento di Fisica, Universitá di Roma 3 and INFN-Sezione
di Roma, Via della Vasca Navale 84, 00146 Roma, Italy
levi@roma1.infn.it

Louis Michel, IHES, 91440 Bures-sur-Yvette, France
michel@ihes.fr

W. Miller, Jr., School of Mathematics, University of Minnesota,
Minneapolis, MN 55455, USA
miller@ima.umn.edu

Robert V. Moody, Department of Mathematical Sciences, University of
Alberta, Edmonton, Alberta, T6G 2G1,Canada
rvm@miles.math.ualberta.ca

M. Moshinsky, Instituto de Física, UNAM, Apartado Postal 20-364,
01000 México, D.F. México
moshi@fenix.ifisicacu.unam.mx

Peter J. Olver, School of Mathematics, University of Minnesota,
Minneapolis, MN 55455, USA
olver@ima.umn.edu

L.O'Raifeartaigh, School of Theoretical Physics, Dublin Institute for
Advanced Studies, 10 Burlington Rd, Dublin 4, Ireland
lor@stp01.stp.dias.ie

G.S. Pogosyan, Laboratory of Theoretical Physics, Joint Institute for
Nuclear Research, Dubna, Moscow Region 141980, Russia
pogosyan@thsun1.jinr.dubna.su

A. Ramani, CPT, École Polytechnique, CNRS, UPR 14, 91128 Palaiseau,
France
ramani@orphee.polytechnique.fr

Guy Rideau, Laboratoire de Physique Théorique de la matière condensée,
Université Paris VII, 2 place Jussieu F-75251 Paris Cedex 05, France
rideau@ccr.jussieu.fr

D.J. Rowe, Department of Physics, University of Toronto, Toronto, On.,
M5S 1A7, Canada
rowe@physics.utoronto.ca

Gordon L. Shaw, Department of Physics and Center for the Neurobiology of Learning and Memory, University of California, Irvine, CA 92697-4575, USA
gshaw@uci.edu

Richard Slansky, Theoretical Division, MS B210, Los Alamos National Laboratory, Los Alamos, NM 87545, USA

K. Tenenblat, Departamento de Matemática, Universidade de Brasília, Brasília, DF 70910, Brasil
keti@Ipe.mat.unb.br

J. Tolar, Department of Physics and Doppler Institute, Faculty of Nuclear Sciences and Physical Engineering, Czech Technical University, Břehová 7, CZ-115 19 Prague 1, Czech Republic
tolar@br.fjfi.cvut.cz

R. Yamilov, Ufa Institute of Mathematics, Russian Academy of Sciences, 112 Chernyshevsky Street, Ufa 450000, Russia
yamilovimat.rb.ru

1

Self-Similarities and Invariant Densities for Model Sets

Michael Baake and Robert V. Moody

ABSTRACT Model sets (also called cut and project sets) are generalizations of lattices. Here we show how the self-similarities of model sets are a natural replacement for the group of translations of a lattice. This leads us to the concept of averaging operators and invariant densities on model sets. We prove that invariant densities exist and that they produce absolutely continuous invariant measures in internal space. We study the invariant densities and their relationships to diffraction, continuous refinement operators, and Hutchinson measures.

1 Model Sets and Self-Similarities

In this paper we introduce the notion of averaging operators on suitable spaces of functions on model sets. An averaging operator encodes information about the entire set of self-similarities with a given inflation factor for a given model set. It can be interpreted as a Hilbert–Schmidt operator on the space of continuous functions on the corresponding acceptance window and, remarkably, from this point of view is seen to be an example of the recently studied continuous refinement operators. Using this connection we can determine the spectrum and associated set of eigenfunctions for any inflation factor of any given model set. In particular, the leading eigenvalue 1 gives rise to an *invariant density* for the model set. We derive some properties of the Bragg spectrum of a model set that has been weighted by an invariant density. We also show that an invariant density leads to an absolutely continuous invariant measure on internal space and we relate this measure to a weakly converging sequence of Hutchinson measures. The full mathematical development of this work will appear in Ref. [2].

1.1   Cut and project schemes

We begin with the notion of a cut and project scheme. By definition, this consists of a collection of spaces and mappings:

$$\mathbb{R}^m \overset{\pi_1}{\longleftarrow} \mathbb{R}^m \times \mathbb{R}^n \overset{\pi_2}{\longrightarrow} \mathbb{R}^n, \tag{1.1}$$

$$\cup$$
$$\widetilde{L}$$

where $\mathbb{R}^m$ and $\mathbb{R}^n$ are two real spaces, π_1 and π_2 are the projection maps onto them, and $\widetilde{L} \subset \mathbb{R}^m \times \mathbb{R}^n$ is a lattice. We assume that $\pi_1|_{\widetilde{L}}$ is injective and that $\pi_2(\widetilde{L})$ is dense in $\mathbb{R}^n$. We call $\mathbb{R}^m$ (resp. $\mathbb{R}^n$) the physical (resp. internal) space. We will assume that $\mathbb{R}^m$ and $\mathbb{R}^n$ are equipped with Euclidean metrics and that $\mathbb{R}^m \times \mathbb{R}^n$ is the orthogonal sum of the two spaces. For x lying in any of these spaces, $|x|$ denotes its length.

A cut and project scheme involves, then, the projection of a lattice into a space of smaller dimension, but a lattice that is transversally located with respect to the projection maps involved.

1.2   Example

A simple, and very useful, example of such a scheme arises from a real quadratic irrationality q. We form the ring $\mathbb{Z}[q] \subset \mathbb{R}$ and let $*$ be the $\mathbb{Z}$-mapping that takes q into its algebraic (quadratic) conjugate. Then the set of points $\widetilde{\mathbb{Z}[q]} := \{(x, x^*) \mid x \in \mathbb{Z}[q]\}$ is a lattice in $\mathbb{R}^2$ and

$$\mathbb{R} \overset{\pi_1}{\longleftarrow} \mathbb{R}^2 \overset{\pi_2}{\longrightarrow} \mathbb{R}, \tag{1.2}$$

$$\cup$$
$$\widetilde{\mathbb{Z}[q]}$$

where we use the coordinate projections, is a cut a project scheme. An important case of this occurs when $q = \tau := (1 + \sqrt{5})/2$.

1.3   Model sets

Let $L := \pi_1(\widetilde{L})$ and let

$$(\ )^* : L \to \mathbb{R}^n \tag{1.3}$$

be the mapping $\pi_2 \circ (\pi_1|_{\widetilde{L}})^{-1}$. This mapping extends naturally to a mapping on the rational span $\mathbb{Q}L$ of L, also denoted by $(\ )^*$. Note that the lattice $\widetilde{L}$ can also be written as

$$\widetilde{L} = \{(x, x^*) \mid x \in L\}. \tag{1.4}$$

Now, let $\Omega \subset \mathbb{R}^n$. Define

$$\Lambda = \Lambda(\Omega) := \{x \in L \mid x^* \in \Omega\}. \tag{1.5}$$

We call such a set Λ a *model set* (or *cut and project set*) if the following three conditions are fulfilled:

W1: $\Omega \subset \mathbb{R}^n$ is compact.

W2: $\Omega = \overline{\text{int}(\Omega)}$.

W3: The boundary of Ω has Lebesgue measure 0.

The mathematical reasons for studying model sets are that they are very natural generalizations of lattices, they share many properties with them, and they allow symmetries that are otherwise unavailable in lattices of the corresponding dimensions. For example, the following properties are shared by all model sets Λ:

M1: Λ is *uniformly discrete:* that is to say, there is an $r > 0$ so that for all distinct $x, y \in \Lambda$, $|x - y| \geq r$.

M2: Λ is *relatively dense:* that is to say, there is an $R > 0$ so that for each $x \in \mathbb{R}^m$ the open ball of radius R around x contains a point of Λ.

M3: There is a finite set F so that $\Lambda - \Lambda \subset \Lambda - F$.

M4: $\Lambda - \Lambda$ is a Delone set (i.e., satisfies [**M1, M2**]).

M5: Λ has a well-defined density d, i.e.,

$$d = \lim_{s \to \infty} \frac{\#\Lambda_s}{\text{vol}(B_s(0))} = \lim_{s \to \infty} \frac{\#\Lambda_s}{c_m s^m}, \tag{1.6}$$

exists, where $B_s(0) := \{x \in \mathbb{R}^m \mid |x| \leq s\}$ and

$$c_m := \frac{\pi^{m/2}}{\Gamma(m/2 + 1)}, \tag{1.7}$$

is the volume of the unit sphere in $\mathbb{R}^m$.

M6: Λ diffracts. (See Section 3.1 for more on this.)

A set with the properties **M1** and **M2** is called a *Delone* set. A lattice is nothing else than a Delone set that is a group. If $F = \{0\}$, then **M3** states that Λ is a group; so **M3** is in fact a generalization of the group law. The limit in (1.6) is easily seen to be independent of the choice of origin for the Euclidean space. What is more, it even exists *uniformly* for sets. This means that for any subset S of Ω with boundary of measure 0, the relative frequency of the points of $(\Lambda_s)^*$ falling into S, as $s \to \infty$, is $\text{vol}(S)/\text{vol}(\Omega)$, and the convergence is uniform with respect to translation of the set S. For more on these properties one may consult [6, 13, 14, 17].

Model sets arise in situations in which one is looking for Delone structures with symmetries that are incompatible with lattices. The most famous example is that of the icosahedral group which cannot appear as the point symmetry of any lattice in 3-space. It is known [5, 15] that if G is any finite group acting irreducibly in $\mathbb{R}^m$ and X is any nontrivial orbit of G, then either G acts crystallographically in $\mathbb{R}^m$; that is to say, there is a G-stable lattice of $\mathbb{R}^m$; or there is a G-stable cut and project set in $\mathbb{R}^m$ that contains the set X. This is the origin of the interest in these sets in the theory of quasicrystals, see [8] for background material.

However, there is a serious price to be paid for moving from lattices to model sets. Lattices, by definition, have an entire lattice of translational symmetries. By comparison, in a model set Λ described by a cut and project scheme (1.1), the set of translational symmetries is the kernel of ( )* in (1.3), and in all the standard examples this is in fact $\{0\}$. Fortunately, in many cases of interest, there is nonetheless an abundance of symmetry, as long as one is prepared to consider self-similarities instead of group symmetries.

1.4   Self-similarities

Definition 1.1. A *self-similarity* of Λ is an affine linear mapping $t = t_{Q,v}$

$$t_{Q,v}\colon x \mapsto Qx + v \qquad (1.8)$$

on $\mathbb{R}^m$ that maps Λ into itself, where Q is a (linear) similarity and $v \in \mathbb{R}^m$. Thus $Q = qR$, i.e., it is made up of an orthogonal transformation R and an *inflation factor* q.

Let $t_{Q,v}$ be a self-similarity of Λ. If Λ is uniformly discrete, we must have $|q| \geq 1$. This is the reason we also call such a self-similarity an *affine inflation*. We are interested in the *entire* set of affine inflations with the same similarity factor Q. It is convenient to have $0 \in \Lambda$ and $0 = 0^* \in \mathrm{int}(\Omega)$. Using a translation of Λ by a vector $v_0 \in L$ and the corresponding translation of Ω by $v_0^* \in L^*$, we assume this to be the case. This makes no structural difference to the set of inflations of Λ, but simplifies the algebra: if $0 \in \Lambda$ and $t_{Q,v}$ is an affine inflation, then $v \in L$ and $Q(L) \subset L$. In fact, we are going to also assume that $QL = L$.

Let us then fix once and for all a (linear) similarity transformation Q on $\mathbb{R}^m$ such that $QL = L$. How do we describe the set of all self-similarities with similarity factor Q ? In preparation for answering this question, it is useful to note that there are three different ways of looking at the same cut and project set Λ: first as a Delone set in $\mathbb{R}^m$, which we may think of as the *discrete* picture; second as part of the lattice $\widetilde{L}$, which we may think of as the *arithmetic* picture; and finally as a dense subset of Ω via the mapping ( )*, which may be thought of as the *analytic* picture.

As an illustration of these ideas, note that Q naturally gives rise to an automorphism $\widetilde{Q}$ of the lattice $\widetilde{L}$, i.e., an element of $GL_{\mathbb{Z}}(\widetilde{L})$, and a linear mapping Q^* of $\mathbb{R}^n$ that maps Ω into itself. From the arithmetic nature of $\widetilde{Q}$ we deduce that the eigenvalues of Q and Q^* are algebraic integers and from the compactness of Ω that Q^* is contractive. Furthermore, one can deduce [2] that Q^* is diagonalizable from the corresponding property of Q. Strictly speaking we can only deduce that the eigenvalues of Q^* do not exceed 1 in absolute value, but we will always assume that in fact they are less than one in absolute value. In the sequel we will normally denote the contraction Q^* by A to match various sources we will refer to frequently.

Define

$$\Omega_Q := \{u \in \mathbb{R}^n \mid A\Omega + u \subset \Omega\}, \tag{1.9}$$

we say that Q is *compatible* with Λ if $\mathrm{int}(\Omega_Q) \neq \emptyset$. In this article, we shall always assume that not only Ω, but also Ω_Q is Riemann measurablel; i.e., $\partial\Omega_Q$ has zero Lebesgue measure. Interpreting (1.8) on the window side we obtain:

Proposition 1.1. *Let $\Lambda = \Lambda(\Omega)$ be a model set based on a window Ω that satisfies the window conditions* **W1–W3**. *Let Q be a similarity compatible with Λ. Then the set T_Q of affine inflations with the same similarity Q is the set of mappings $t_{Q,v}$ where v runs through the set*

$$T_Q = \{v \in L \mid v^* \in \Omega_Q\}. \tag{1.10}$$

In particular, T_Q is also a model set.

Let us pause to consider the special situation where Ω is convex. In this case, Ω_Q is also convex and hence satisfies the conditions that we need. If in addition $Q^* = \varepsilon \cdot \mathbf{1}$, $0 < \varepsilon < 1$, which actually is often the case in examples of physical relevance, one obtains

$$\Omega_Q = (1 - \varepsilon)\Omega. \tag{1.11}$$

If $-1 < \varepsilon < 0$, but $\Omega = -\Omega$, 1.11 is still true if ε is replaced by $|\varepsilon|$. This happens in our examples.

1.5 Example

In Example 1.2 above, take $q = \tau$. A simple model set is defined by

$$\Lambda = \{x \in \mathbb{Z}[\tau] \mid x' \in [-1, 1]\}. \tag{1.12}$$

Let us look at the inflation factor $q = \tau$. Multiplication by q determines the contraction $A = \tau' \cdot \mathbf{1}$, $\tau' = -1/\tau$, on the internal side and

$$\Omega_\tau = \left\{u \in \mathbb{R} \,\middle|\, -\frac{1}{\tau} \cdot [-1, 1] + u \subset [-1, 1]\right\} = \left[-1 + \frac{1}{\tau}, 1 - \frac{1}{\tau}\right]. \tag{1.13}$$

Thus, for all $v \in \mathbb{Z}[\tau]$ with $v' \in [-1/\tau^2, 1/\tau^2]$,

$$t_{\tau,v} \colon x \mapsto \tau x + v, \tag{1.14}$$

is an inflation of Λ, and

$$t^*_{\tau,v} : y \mapsto \frac{-y}{\tau} + v', \tag{1.15}$$

is the corresponding contraction in internal space.

2   Averaging Operators and Invariant Densities

One of the most familiar techniques in the theory of group representations is the use of group averages in order to produce invariants. Thus for a finite group G one typically invokes the process

$$F \mapsto \frac{1}{\#G} \sum_{g \in G} g \cdot F, \tag{2.1}$$

which averages the function F over the group, where $g \cdot F(x) := F(g^{-1}x)$. We intend to do exactly the same thing replacing G by the set of all self-similarities $\mathcal{T}_Q$ of a model set Λ. Since $\mathcal{T}_Q$ will be infinite we have to be a little careful in averaging.

For any subset $T \subset \mathbb{R}^m$ and for any $s \geq 0$, we thus define

$$T_s := \{x \in T \mid |x| \leq s\}, \tag{2.2}$$

where $|x| = (x \cdot x)^{1/2}$ is the standard Euclidean norm on $\mathbb{R}^m$.

2.1   Averaging operators

Definition 2.1. Let $\Lambda = \Lambda(\Omega)$ be a model set based on a window Ω that satisfies the window conditions **W1–W3**. Let Q be a similarity (with inflation factor q) that is compatible with Λ, let $\mathcal{T} = \mathcal{T}_Q$ be the set of all self-similarities of Λ with similarity Q, and let $T = T_Q$ be the corresponding set of translations. Let $p : L \to \mathbb{R}$ be any function on L that vanishes off Λ. Then the average of p over $\mathcal{T}$ is defined to be

$$(\mathcal{A}p)(x) = \lim_{s \to \infty} \frac{|\det(Q)|}{\#T_s} \sum_{v \in T_s} p(t_v^{-1}x), \tag{2.3}$$

provided this limit exists. We say that p is a Q-invariant density on Λ if

ID1: p is nonnegative on Λ,

ID2: $\mathcal{A}p = p$, and,

ID3: p is normalized, i.e.,

$$\lim_{s \to \infty} \frac{1}{\#\Lambda_s} \sum_{x \in \Lambda_s} p(x) = 1. \tag{2.4}$$

Note that for $x \in \Lambda$ and $v \in T$, $t_v^{-1}x \in L$, but $t_v^{-1}x$ does not in general lie in Λ. Thus in (2.3) one can expect that many of the summands on the right-hand side will be 0 because p vanishes off Λ. Note also that (2.4) is a normalization per point of Λ. This can be changed to a normalization per unit volume if necessary because the density d of points of Λ exists.

Let $\mathcal{C}(\Omega)$ be the space of all continuous complex-valued functions on $\mathbb{R}^n$ with support in Ω. Via the mapping $(\ )^*$ of (1.3) we obtain a space $\mathcal{C}(\Lambda)$ of functions on L, vanishing off Λ: for $f \in \mathcal{C}(\Omega)$, we define $p = p_f \in \mathcal{C}(\Lambda)$ by

$$p(x) = \text{vol}(\Omega) \cdot f(x^*), \tag{2.5}$$

for $x \in \Lambda$, and $p(x) = 0$ otherwise. Here, the normalization constant $\text{vol}(\Omega)$ is thrown in to make things more convenient later on.

Bearing in mind (2.5), we rewrite (2.3) as a new averaging operator on $\mathcal{C}(\Omega)$:

$$(\mathcal{A}f)(x^*) = \lim_{s \to \infty} \frac{|\det(Q)|}{\# T_s} \sum_{v \in T_s} f\big((t_v^*)^{-1}x^*\big), \tag{2.6}$$

for all $x^* \in \Lambda^*$.

Now, it is well known [6, 16, 17] that the points $\pi_2(\widetilde{L})$ are uniformly distributed in $\mathbb{R}^n$, in the sense described in Section 1.3. In particular, the points of T^* are uniformly distributed in Ω_Q, hence $\lim_{s \to \infty} \overline{(T_s)^*} = \overline{\Omega}_Q$. Using Weyl's theorem [12, 18], the continuity of f, and the fact that Ω_Q is Riemann integrable, we obtain

$$(\mathcal{A}f)(x^*) = \frac{|\det(Q)|}{\text{vol}(\Omega_Q)} \int_{\Omega_Q} f\big(A^{-1}(x^* - u)\big)\, du. \tag{2.7}$$

Using the fact that $\det(Q) = \det(A)^{-1}$ (remember that $A = Q^*$) and introducing the normalized indicator (or characteristic) functions

$$X_S := \frac{\mathbf{1}_S}{\text{vol}(S)}, \tag{2.8}$$

defined for all measurable subsets S of $\mathbb{R}^n$, we can rewrite (2.7) in a number of equivalent ways:

$$\begin{aligned}
(\mathcal{A}f)(x) &= \int_{\mathbb{R}^n} X_{Ay + \Omega_Q}(x) f(y)\, dy, \\
&= \frac{1}{|\det(A)|} \int_{\mathbb{R}^n} X_{A^{-1}\Omega_Q}(A^{-1}x - y) f(y)\, dy, \\
&= \frac{1}{|\det(A)|} \int_{\mathbb{R}^n} X_{\Omega_Q}(x - y) f(A^{-1}y)\, dy.
\end{aligned} \tag{2.9}$$

This shows that the averaging operator $\mathcal{A}$ is a *continuous refinement operator* in the sense of [9] in the one-dimensional case and [10] in the multidimensional case. In Fourier space, by application of the convolution theorem, (2.9) reads as

$$\widehat{(\mathcal{A}f)}(k) = \widehat{X_{\Omega_Q}}(k) \hat{f}(A^t k), \tag{2.10}$$

where $\hat{f}(k) := \int_{\mathbb{R}^n} e^{-2\pi i k \cdot x} f(x)\, dx$ and $f(x) = \int_{\mathbb{R}^n} e^{2\pi i k \cdot x} \hat{f}(k) dk$.

2.2 Invariant densities

At this point, we can determine a function $f = f_p$ corresponding to an *invariant density* p. We are looking for a 1-eigenfunction of the operator $\mathcal{A}$. The normalization condition **ID3**, seen on the window side, becomes, using Weyl again, $\int_{\mathbb{R}^n} f(u)\, du = 1$. Iteration of (2.10) leads to

$$\hat{f}(k) = \widehat{X}_{\Omega_Q}(k) \cdot \widehat{X}_{\Omega_Q}(A^t k) \cdot \cdots \cdot \widehat{X}_{\Omega_Q}\big((A^t)^N k\big) \cdot \hat{f}\big((A^t)^{N+1} k\big). \quad (2.11)$$

Since A is a contraction, so is A^t. Consequently, $(A^t)^{N+1} k \to 0$ as $N \to \infty$. Since $\hat{f}$ and $\widehat{X}_{\Omega_Q}$ are C^∞ and $\hat{f}(0) = 1$, we can take the limit and obtain

$$\hat{f}(k) = \prod_{N=0}^{\infty} \widehat{X}_{\Omega_Q}\big((A^t)^N k\big) = \prod_{N=0}^{\infty} \frac{\hat{1}_{\Omega_Q}\big((A^t)^N k\big)}{\mathrm{vol}(\Omega_Q)}. \quad (2.12)$$

This is an infinite product with compact convergence that solves our problem: $\hat{f}$ is an infinite product of C^∞-functions, and is itself C^∞. The function f is now the inverse Fourier transform of $\hat{f}$. By construction, it is the Radon-Nikodym derivative, and hence the density, of an absolutely continuous invariant (3.8) measure (with respect to Lebesgue measure) in internal space. Again applying the convolution theorem we arrive at—

Proposition 2.1. *Let $\Lambda = \Lambda(\Omega)$ be a model set based on a window Ω that satisfies the window conditions* **W1**–**W3**. *Let Q be a similarity compatible with Λ and let $A := Q^*$. Then there is a unique Q-invariant density p for Λ lying in $C(\Lambda)$. This is given through $p = p_f$, see (2.5), where f is given by the infinite convolution product*

$$f = \underset{N=0}{\overset{\infty}{\text{\Large$*$}}} \frac{1_{A^N \Omega_Q}}{\mathrm{vol}(A^N \Omega_Q)}. \quad (2.13)$$

Note that this convolution of characteristic functions defines a C^∞ function with compact support contained in Ω.

If $Q, Q^2, \ldots$, are all compatible with Λ, then it is instructive to look at the corresponding invariant densities $f = f_{(1)}, f_{(2)}, \ldots$. The sets $\{\Omega_{Q^n}\}$ are increasing and $\overline{\bigcup_n \Omega_{Q^n}} = \Omega$. The functions of the sequence $\{\hat{f}_{(k)}\}$ become increasingly concentrated around 0 and it is natural to expect $\lim_{s\to\infty} f_{(s)} = 1_\Omega/\mathrm{vol}(\Omega)$. This is illustrated in the case of our Example by the sequence of graphs of Fig. 1.1.

2.3 Example

We continue with Example 1.5 from above and determine the invariant density (or rather its Fourier transform) on $[-1, 1]$ corresponding to the

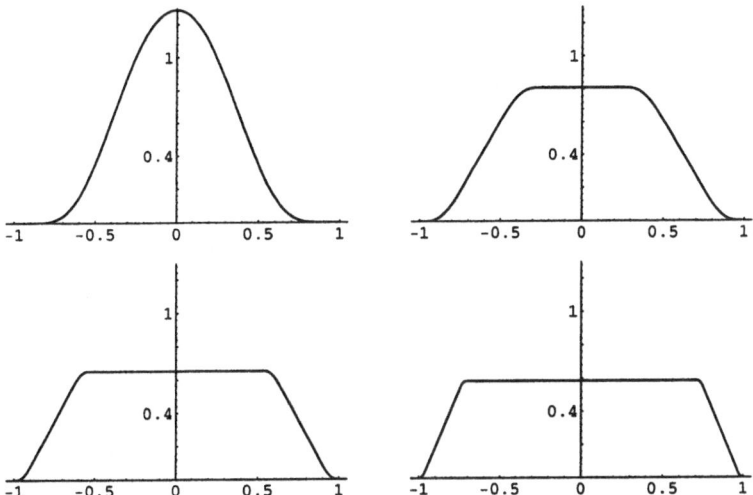

FIGURE 1.1. Invariant densities for the model set of Example 1.5, for inflation factors τ, τ^2, τ^3 and τ^4.

inflation factor τ. If f is this density, then Eq. (2.10) becomes

$$\hat{f}(k) = \frac{\tau^2}{2} \hat{1}_{[-1/\tau^2, 1/\tau^2]}(k) \hat{f}(-k/\tau). \qquad (2.14)$$

Routine calculation gives

$$\frac{1}{2a} \hat{1}_{[-a,a]}(k) = \frac{1}{2a} \int_{-a}^{a} e^{-2\pi i k x} \, dx = \frac{\sin(2\pi a k)}{2\pi a k}. \qquad (2.15)$$

So, we obtain

$$\hat{f}(k) = \frac{\sin((2\pi k)/(\tau^2))}{((2\pi k)/(\tau^2))} \hat{f}(-k/\tau) = \cdots = \prod_{N=2}^{\infty} \frac{\sin((2\pi k)/(\tau^N))}{((2\pi k)/(\tau^N))}. \qquad (2.16)$$

The calculations for other inflation factors are similar; Fig. 1.1 shows some of the resulting invariant densities.

2.4 Eigenvalues and eigenfunctions

According to [10], if the eigenvalues of A^t are $\{\alpha_1, \ldots, \alpha_n\}$, $0 < |\alpha_i| < 1$, then the spectrum of $\mathcal{A}$ is

$$\mathrm{spec}(\mathcal{A}) = \{\alpha^a \mid a \in \mathbb{Z}_{\geq 0}^n\}, \qquad (2.17)$$

where $\alpha^a := \alpha_1^{a_1} \cdots \cdots \alpha_n^{a_n}$. Furthermore, the multiplicities are those suggested by the notation

$$\mathrm{mult}(\lambda) = \#\{a \mid \alpha^a = \lambda\}. \qquad (2.18)$$

In particular, if $\alpha_1 = \cdots = \alpha_n = \alpha$, then $\lambda = \alpha^{|a|}$, $|a| = \alpha_1 + \cdots + \alpha_n$, and

$$\text{mult}(\lambda) = \binom{|a| + n - 1}{|a|},$$

a formula that is well-known from the n-dimensional harmonic oscillator.

Remark. In Refs. [9, 10] A is assumed to be diagonalizable and all the eigenvalues are assumed real. The diagonalizability of Q guarantees that of $Q^* = A$ [2]. By allowing complex-valued functions to enter the picture, it is not necessary to assume that the eigenvalues of A are real. Nor do we wish to impose such an assumption since we do not have this type of control over the spectrum of A.

It is quite easy to find eigenfunctions representing these eigenvalues in terms of the invariant density f. To do so, we choose an eigenbasis $\{v_1, \ldots, v_n\}$ for A^t in $\mathbb{C}^n$, using the fact that A^t is diagonalizable. Any $k \in \mathbb{C}^n$ can be written as $k = \kappa_1 v_1 + \cdots + \kappa_n v_n$ and the $\kappa_j = \kappa_j(k)$ are the corresponding coordinate functions (which we allow to be $\mathbb{C}$-valued). Now, fix $b \in \mathbb{Z}_{\geq 0}^n$, define $\kappa^b = \kappa_1^{b_1} \cdot \ldots \cdot \kappa_n^{b_n}$, and let $\widehat{u}(k) := (\kappa(k))^b \widehat{f}(k)$. Then we obtain from (2.10)

$$\widehat{\mathcal{A}u}(k) = \widehat{X_{\Omega_Q}}(k)\widehat{u}(A^t k) = \widehat{X_{\Omega_Q}}(k)(\kappa(A^t k))^b \widehat{f}(A^t k),$$

$$= \alpha^b (\kappa(k))^b \widehat{f}(k) = \alpha^b \widehat{u}(k). \tag{2.19}$$

Returning from the Fourier domain, this enables us to write down the eigenfunctions of $\mathcal{A}$. If $\{v_1^\dagger, \ldots, v_n^\dagger\}$ denotes the dual basis (i.e., $v_i^\dagger \cdot v_j = \delta_{ij}$), we define the directional derivative $D_j := v_j^\dagger \cdot \nabla$ and obtain

Proposition 2.2. *The partial derivatives*

$$D^b f, \quad b \in \mathbb{Z}_{\geq 0}^n, \tag{2.20}$$

are eigenfunctions of the refinement operator $\mathcal{A}$, *with eigenvalue* α^b.

Some of these derivatives, calculated for our guiding Example, are shown in Fig. 1.2.

3   Further Remarks

3.1   Diffraction and a product formula

The product formula (2.12) has an interpretation on the physical side of our picture. To see this, we define functions g_s and h_s by

$$g_s(k) = \frac{1}{\#T_s} \sum_{v \in T_s} e^{-2\pi i k \cdot v},$$

$$h_s(k) = \frac{1}{\#\Lambda_s} \sum_{w \in \Lambda_s} p(w) e^{-2\pi i k \cdot w}, \tag{3.1}$$

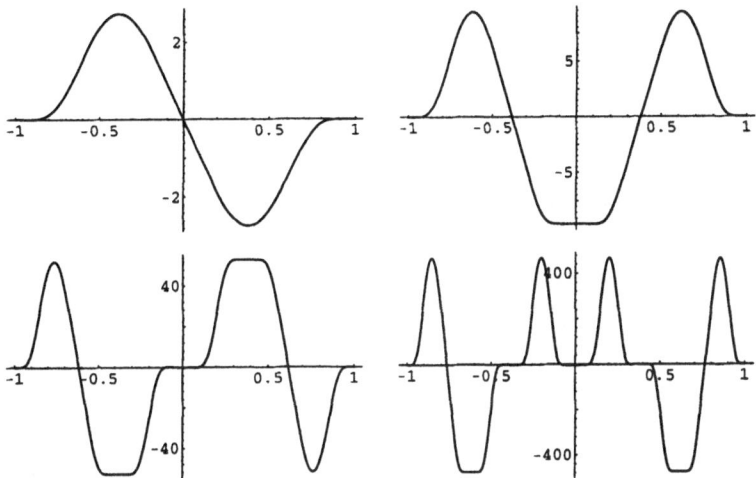

FIGURE 1.2. The first four derivatives of the invariant density of Example 1.5 for inflation factor τ.

and, furthermore, functions g and h by

$$g(k) = \lim_{s \to \infty} g_s(k),$$
$$h(k) = \lim_{s \to \infty} h_s(k). \qquad (3.2)$$

The existence of g as a well-defined function on $\mathbb{R}^m$ is a known consequence of the fact that T is a model set. We now have [2]—

Proposition 3.1. $h(k) = \prod_{N=0}^{\infty} g\big((Q^t)^N k\big).$

The significance of the functions g and h appears in the context of diffraction. Suppose that $w: L \to \mathbb{R}_{\geq 0}$ is some bounded nonnegative function that vanishes off Λ. Define the tempered distribution

$$\mu_w = \sum_{x \in \Lambda} w(x)\delta_x, \qquad (3.3)$$

where δ_x is the Dirac measure at x. The limit, as $s \to \infty$, of the averaged auto-correlation of this measure, which exists for model sets, is the *auto-correlation measure* of Λ (also called its *Patterson function*, though it is a distribution)

$$\gamma_w = \lim_{s \to \infty} \frac{1}{\#\Lambda_s} \sum_{x,y \in \Lambda_s} w(x)w(y)\delta_{x-y}. \qquad (3.4)$$

Its Fourier transform is a positive measure $\widehat{\gamma}_w$ which is the *diffraction pattern* of Λ. The point part of this measure is the *Bragg spectrum* of Λ. In the case that w is the indicator function (i.e., constant value 1) on a model set

Λ then $\widehat{\gamma}_w$ is a pure point measure and we say that Λ has a *pure point spectrum*. In any case, the Bragg spectrum can be calculated from the simpler function

$$g(\cdot\,; w)\colon\; g(k; w) = \lim_{s\to\infty} \frac{1}{\#\Lambda_s} \sum_{x\in\Lambda_s} w(x)e^{-2\pi i x\cdot k}, \tag{3.5}$$

provided that this limit exits everywhere. In fact [6]

$$\bigl|g(k; w)\bigr|^2 = \widehat{\gamma}_w\bigl(\{k\}\bigr), \quad \text{for all } k \in \mathbb{R}^m. \tag{3.6}$$

Our functions g and h correspond to the cases when $w = \mathbf{1}_\Lambda$ and w is the invariant density p, respectively. In particular, h allows us an exact description of the intensities of the Bragg spectrum of (Λ, p). Notice from the product formula that its support necessarily lies inside the support of the Bragg spectrum of Λ itself. It goes without saying that both g and h are highly discontinuous functions which, for a model set, are nonzero only on a dense point set of zero Lebesgue measure. This set is contained in the so-called *Fourier module* of the model set, i.e., in the set $\pi_1(\widetilde{L}^\circ)$, where $\widetilde{L}^\circ = \bigl\{y \in \mathbb{R}^{m+n} \mid x \cdot y \in \mathbb{Z} \text{ for all } x \in \widetilde{L}\bigr\}$ is the dual lattice of $\widetilde{L}$.

3.2 Hutchinson measures

Let f be the invariant density of $\mathcal{C}(\Omega)$ corresponding to the compatible similarity Q on $\Lambda = \Lambda(\Omega)$. There is a corresponding measure $\mu = \mu_f$, with support contained in Ω, defined by

$$\mu_f(Y) = \int_{\mathbb{R}^n} \mathbf{1}_Y(x)\mu_f(dx) = \int_{\mathbb{R}^n} \mathbf{1}_Y(x)f(x)\,dx. \tag{3.7}$$

We have $\mu_f(\Omega) = 1$. The measure μ_f is invariant in the sense that, if we define $t_v^* \cdot \mu_f$ by $t_v^* \cdot \mu_f(Y) = \mu_f\bigl((t_v^*)^{-1}(Y)\bigr)$, then

$$\mu_f = \lim_{s\to\infty} \frac{1}{\#T_s} \sum_{v\in T_s} t_v^* \cdot \mu_f. \tag{3.8}$$

Now fix some $s > 0$ and consider the finite set of contractions t_v^* on Ω, where $v \in T_s$. According to [7] there is a unique nonnegative Borel measure μ_s on Ω for which $\mu_s(\Omega) = 1$ and which is *invariant* in the sense that

$$\mu_s = \frac{1}{\#T_s} \sum_{v\in T_s} t_v^* \cdot \mu_s. \tag{3.9}$$

Furthermore, this measure is the unique fixed point of the process on the set of regular Borel measures on Ω with mass 1 that averages a measure by the right-hand side of (3.9). We call this a *Hutchinson measure*. Using the Levy continuity theorem [1] one can show—

Proposition 3.2. *The sequence of Hutchinson measures fulfills—*

(i) $\widehat{\mu}_s(k) = \prod_{N=0}^{\infty} g_s\big((A^t)^N k\big)$, *for any* $s > 0$;

(ii) $\{\widehat{\mu}_s\} \to \widehat{\mu}_f$, *where the convergence is uniform on compact sets*;

(iii) $\{\mu_s\} \to \mu_f$, *in the sense of weak convergence, i.e.,* $\{\mu_s(\varphi)\} \to \mu_f(\varphi)$ *for all* $\varphi \in C(\Omega)$.

In particular, for any function $\varphi \in C(\Omega)$, $\mu_s(\varphi)$, for large s, is a good estimate of $\mu_f(\varphi)$. Also, starting from *any* probability measure on Ω, the Hutchinson iteration of (3.9) will converge to μ_s and this procedure, for large s, will also give a good approximation to μ_f.

3.3 The topology of L and Λ

The space of functions $C(\Lambda)$ on which the averaging operator $\mathcal{A}$ acts seems strange. It is defined in terms of the topology of Ω in $\mathbb{R}^n$ and this is very different from the discrete topology that we see on Λ induced by the topology of its ambient space $\mathbb{R}^m$. The appropriate topology for Λ (and L) is defined intrinsically as follows [17]. For each compact set K of $\mathbb{R}^m$ define

$$N_K = N_K(\Lambda) := \{v \in L | v + (\Lambda \cap K) = \Lambda \cap (v + K)\}. \tag{3.10}$$

Thus N_K is the set of vectors v for which translation by v is a bijection of the K-patch of Λ and onto the $(v + K)$-patch of Λ. Note that the mapping $K \mapsto N_K$ is inclusion reversing.

Proposition 3.3 ([16, 17]). *Suppose that* $L^* \cap \partial\Omega = \emptyset$. *Then the collection of sets* $\{N_K | K \subset \mathbb{R}^m, K \text{ compact}\}$ *is a basis of neighborhoods of* 0 *for a topological group structure on* L. *Furthermore,* $\mathbb{R}^n$ *with its standard topology is the completion of* L *under the mapping* $(\ )^* : L \to \mathbb{R}^n$. *With this topology on* L, *the space* $C(\Lambda)$ *is precisely the space of continuous functions on* L *whose support lies in* Λ.

The intuition behind continuity of a function ϕ (defined on L) with respect to this topology is this: if translation by v is a bijection of two "large" patches of Λ, then $\phi(x + v) - \phi(x)$ is (uniformly) small.

4 Outlook

The existence of *positive* invariant densities naturally suggests probabilistic interpretations. In Ref. [3] we study the spectral properties of certain stochastic sets whose sites are selected from that of a model set on a probabilistic basis according to the density p. Effectively, this gives a distribution of points which, after ∗-mapping them into internal space, looks like our window shaped by the invariant density f. This may provide an alternative explanation of the recently made observations of such profiles in real data [11]. Furthermore, as such sets do have finite entropy density, they might be useful for further models of entropic stabilization of quasicrystals.

In this article, we have focused on one-component model sets. It is important for multi-component or multi-colored systems to be able to realize the similarity-averaging process and the existence of invariant densities in a matrix generalization of what we have done here. This means that we have a finite family of model sets based on cosets of a common $\mathbb{Z}$-module and matrices of similarity maps between them. In fact this set-up results in matrix continuous refinement operators and ultimately again in the existence of invariant densities. What is particularly interesting in this case is the appearance of a Markov matrix of weights for the contributions from the various windows relative to each other. An exposition of this will appear in Ref. [4].

Acknowledgments: It is our pleasure to thank R.Q. Jia, W. Allegretto, and A. Hof for valuable discussions. We thank the Volkswagen Stiftung for support through the RiP-program at Oberwolfach, where this paper was written. M.B. is supported by the German Science Foundation (DFG) through a Heisenberg-fellowship. R.V.M. thanks the Tata Institute of Fundamental Research, Bombay, and the Natural Sciences and Engineering Research Council of Canada for their support in this work.

5 REFERENCES

1. H. Bauer, *Probability theory*, de Gruyter, Berlin, 1996.

2. M. Baake and R.V. Moody, *Self-similar measures for quasicrystals*, Directions in Mathematical Quasicrystals (M. Baake and R.V. Moody (eds.), CRM Monograph Series, vol. 13, Amer. Math. Soc., Providence, RI, 2000.

3. M. Baake and R.V. Moody, *Diffractive point sets with entropy*, J. Phys. A **31** (1998), No. 45, 9023–9039.

4. M. Baake and R.V. Moody, *Multicomponent model sets and invariant densities*, Aperiodic 1997 (M. de Boisseau, J.-L. Verger-Gaugry, and R. Currat, eds.), World Scientific, Singapore, 1998, pp. 9–20.

5. N. Cotfas and J.L. Verger-Gaugry, *A mathematical construction of n-dimensional quasicrystals starting from G-clusters*, J. Phys. A **30** (1997), 4283–4291.

6. A. Hof, *Uniform distribution and the projection method*, Quasicrystals and Discrete Geometry (J. Patera, eds.) (Toronto, 1995), Fields Institute Monograph, Amer. Math. Soc., Providence, RI, 1998, pp. 201–206.

7. J.E. Hutchinson, *Fractals and selfsimilarity*, Indiana Univ. Math. J. **30** (1981), 713–747.

8. C. Janot, *Quasicrystals—A primer*, 2nd ed., Clarendon Press, Oxford, 1994.

9. R.Q. Jia, S.L. Lee, and A. Sharma, *Spectral properties of continuous refinement operators*, Proc. Amer. Math. Soc. **126** (1998), No. 3, 729–737.

10. Qingtang Jiang and S.L. Lee, *Spectral properties of matrix continuous refinement operators*, Adv. Comput. Math. **7** (1997), No. 3, 383–399.

11. D. Joseph, S. Ritsch, and C. Beeli, *Distinguishing quasiperiodic from random order in high-resolution* TEM *images*, Phys. Rev. B **55** (1997), 8175–8183.

12. L. Kuipers and H. Niederreiter, *Uniform distribution of sequences*, John Wiley and Sons, New York, 1974.

13. Y. Meyer, *Algebraic numbers and harmonic analysis*, North Holland, Amsterdam, 1972.

14. R.V. Moody, *Meyer sets and their duals*, The Mathematics of Long-Range Aperiodic Order (R. V. Moody, ed.) (Waterloo, 1995), NATO Adv. Sci. Inst. Ser. C Math. Phys. Sci., 489, Kluwer Acad. Publ., Dordrecht, 1997, pp. 403–441.

15. P.A.B. Pleasants, *Quasicrystals with arbitrary symmetry group*, Proc. 5th International Conference on Quasicrystals (C. Janot and R. Mosseri, eds.), World Scientific, Singapore, 1995, pp. 22–30.

16. M. Schlottmann, *Geometrische Eigenschaften quasiperiodischer Strukturen*, Dissertation, University Tübingen, 1993.

17. M. Schlottmann, *Cut-and-project sets in locally compact Abelian groups*, Quasicrystals and Discrete Geometry (J. Patera, ed.) (Toronto, 1995), Fields Institute Monograph, Vol.. 10, Amer. Math. Soc., Providence, RI, 1998, pp. 247–264.

18. H. Weyl, *Uber die Gleichverteilung von Zahlen mod. Eins*, Math. Ann. **77** (1916), 313–52.

2

Symmetry Operations in the Brain: Music and Reasoning

Mark Bodner and Gordon L. Shaw

ABSTRACT The importance of symmetry in the dynamics of physical and biological systems is well recognized. In the present paper, we discuss our research, which addresses the relevance of symmetry concepts to higher brain function. Starting from a highly structured model of cortical activity, trion model-specific types of symmetric spatial and temporal patterns are predicted to exist—patterns that play a fundamental role in how the mammalian brain thinks and reasons. We have recently shown that such families of symmetric spatial and temporal patterns are indeed present in neurophysiological data and that they are related to higher brain function. Experiments are described that support the predictions of the model and establish connections between patterns of cortical activity during reasoning and computation, and those that are present in music. Striking behavioral results show that music enhances spatial-temporal reasoning. This second point indicates that music might serve as a window to understanding higher brain function, as well as potentially having far-reaching ramifications with respect to education. We suggest that finding analytic solutions involving symmetries in models of higher brain function opens new lines of important research for mathematicians.

1 Introduction

Predictions [16, 17] from our structured neural model [23, 24, 36, 37, 39] of the cortex led us to the hypothesis that music could causally enhance spatial-temporal reasoning. We have shown [29, 30]—

(a) College students scored significantly higher on a spatial-temporal reasoning task after listening to a Mozart sonata, but not after listening to silence or to minimalist music.

(b) Preschool children who received private keyboard lessons for 6 months improved dramatically on a spatial-temporal reasoning task while appropriate control groups did not improve significantly [31].

Enhancement (a) lasted roughly 10 minutes and established the causal effect, while enhancement (b) lasted long enough to have major educational implications. Here we review the model, in particular, the "built-in" ability of the cortex to recognize symmetry relations [23] among the inherent spatial-temporal firing patterns, which we suggest is a crucial feature of the cortical relationship between music and spatial-temporal reasoning. Then we discuss recent studies that provide evidence that families of symmetric spatial and temporal patterns are present in neurophysiological data and that they are related to higher brain function [3]. This experimental evidence, we feel, provides strong impetus for taking seriously the idea that symmetry plays a fundamental role in how we think and reason. Furthermore, it is hoped that these results will stimulate interest in the mathematical community to carry out research in this area since analytical analysis of these symmetries is needed. Finally, we summarize the striking behavioral results [29–31], and make further predictions relevant to education.

Historical note

It is a great pleasure to acknowledge the major long-term influence that George Patera has had in all the research reported here throughout the development of the trion model and its relation to higher brain function. Discussions with George on the role of symmetries in the brain have been invaluable. The development of the MILA model (memory in Lie algebras) with George [26] represents a mathematical framework to study structured finite systems with adaptive temporal development. The weights of zero-level representations of affine Kac-Moody algebras form the underlying structure of the MILA model and permit some analytic analysis. We suggest that further development of Patera's pioneering work would now be of considerable interest in our goal of using music as a window in understanding higher brain function.

Symmetries

Symmetries have long been recognized as a vital component of physical and biological systems. It is apparent that as neuroanatomical and neurophysiological techniques have improved in the past decade, more and more structure has been found in the cortex. We expect this trend to continue. We proposed [16, 17, 23, 24, 36, 37, 39] that symmetry operations performed by the brain are a crucial feature of higher brain function and result from this spatial and temporal structure of the cortex. This modular structure with symmetry among the connections introduces symmetries among the "inherent" spatial-temporal firing patterns in the cortex. The symmetries of these inherent firing patterns can then be "exploited" to perform higher level computations or symmetry operations. Learning introduces small breaking of the symmetries in the connectivities, which enables a symmetry in the patterns to be recognized in the Monte Carlo evolution of the patterns [3, 22, 23, 26, 29–31]. Using the trion model of the cortex, we presented specific, simple examples of this in the recognition of rotational invariance and in the recognition of a time-reversed pattern (see Fig. 2.6 below). This then led to our main theme of pattern development, below.

1.1 Music and abstract reasoning

A profound dilemma of historical [1] origin is the similarity among such higher brain functions as music, mathematics, and chess. There are many correlational [11] and anecdotal reports [4, 8] of such relationships. Leng and Shaw in their model of higher brain function [17] proposed a causal link between music and spatial-temporal reasoning. The model was developed from the trion model [23, 24, 36, 37, 39], a highly structured mathematical realization of the Mountcastle [25] organization principle with the column as the basic neuronal network in the cortex. According to the model, newborns possess a structured cortex that yields an inherent repertoire of spatial-temporal firing patterns at the columnar level that can be excited and strengthened by small changes in connectivity via a Hebb [12, 20, 21] learning rule. These firing patterns evolve over time in a probabilistic manner from one to another in natural sequences related by specific symmetries, (see Fig. 2.3 below). *These inherent patterns form the common neural language of the cortex.* The results were striking when evolutions of the trion model firing patterns were mapped onto various pitches and instruments producing recognizable styles of music [16, 18]. (A cassette tape of trion music is available upon request.) This gave us the impetus [16, 17] to relate the neuronal processes involved in music and abstract spatial-temporal reasoning.

The key component of spatial-temporal reasoning may be the "built-in" ability of the columnar networks to recognize the symmetry relations [23] among cortical firing patterns in a sequential manner. *We refer to this sequential process of a temporal sequence of pattern recognition processes as pattern development* [17]. Pattern development mental processes may last some tens of seconds to minutes as compared to pattern recognition processes, such as face recognition, which might be accomplished in a fraction of a second. Music clearly involves this pattern development concept, as does spatial-temporal reasoning: the ability to create, maintain, transform, and relate complex mental images even in the absence of external sensory input or feedback [4, 6, 7, 17].

Although cognitive abilities such as music and spatial-temporal reasoning crucially depend on specific localized regions of the cortex, all higher cognitive abilities draw upon a wide range of cortical areas [28, 33]. Recent studies have demonstrated that sophisticated cognitive abilities are present in children as young as 5 months [41, 43]. Similarly, musical abilities are evident in infants [15, 42]. Music, then may serve as a "pre-language" [17] (with centers [27] distinct from language centers in the cortex), available at an early age, which can access inherent cortical spatial-temporal firing patterns and enhance the cortex's ability to accomplish pattern development.

The ideas of Leng and Shaw [17] then led to the behavioral experiments of Rauscher and Shaw to test the prediction that music training at an early age, when the child's cortex is very plastic, would enhance the ability to use pattern development in spatial-temporal reasoning. It became clear that the experiments they started with preschool children in September 1992 would take years at considerable financial cost [31]. Thus Rauscher and Shaw came up with the idea for the "Mozart effect" experiments [29, 30], which could be done relatively quickly. These experiments were the first to demonstrate a causal link between music enhancing spatial-temporal reasoning. The Mozart effect refers to the significant increase in spatial-temporal reasoning after listening to a Mozart Sonata (K.448), but not after listening to silence or to minimalist music. Mozart was chosen since he was composing at the age of four, and thus we expect that he was exploiting the inherent repertoire of spatial-temporal firing patterns in the cortex. However, this enhancement lasted roughly 10 minutes. The results of the study with a substantial number of preschool children showed that the group that received private keyboard lessons for 6 months improved dramatically on a spatial-temporal reasoning task, while appropriate control groups did not improve significantly [31]. Clearly these findings have major scientific implications. Although *much more work needs to be done*, as described in the concluding section, we already know that the enhancements found for the preschool children last at least a day. *Thus, even at this early stage of the research, the preschool results have enormous educational implications.*

2   Trion Model

Mountcastle [25] proposed that the well-established [10] cortical column is the basic network in the cortex and is composed of small irreducible processing units called minicolumns. A very simple pinwheel representation of the minicolumns in visual cortex had been suggested. The optical recording results by Bonhoeffer and Grinvald [5] not only show a strong similarity to these representations but find both helicities in the representation of the orientation minicolumns. We display this in a very highly idealized, structured and generalized scheme in Fig. 2.1. The column has the capability of being excited into complex spatial-temporal firing patterns. The assumption is that higher mammalian processes involve the creation and transformation of such complex spatial-temporal firing patterns (in contrast to a "code" which involves sets of neurons firing with high frequency). Evidence is accumulating in support of the viability of a spatial-temporal code being the "internal language" of the cortex.

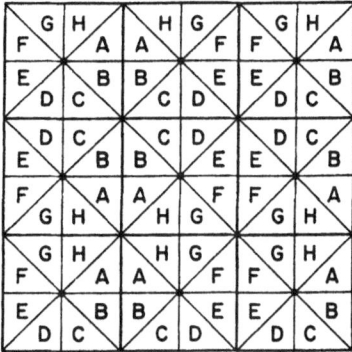

FIGURE 2.1. Highly schematic representation of the Mountcastle [25] principle of cortical organization. Each square represents a cortical column (horizontal dimension roughly 700μ, comprising the six vertical layers of dimension roughly 2000μ) while each triangle is a *distinguishable* minicolumn that encodes the relevant parameter in the stimuli (such as line orientation in the visual cortex) shown here by capital letters. We note the optical recording results by Bonhoeffer and Grinvald [5] in secondary visual cortex show a strong similarity to the cartoon idealized cortex shown here.

The trion model of the cortical column [23, 24, 36, 37, 39] is a mathematical realization of Mountcastle's organizational principle. It was developed starting from Little's [19] neural network analogy to the Ising spin system, and modified in a direction inspired by the ANNNI (axial next nearest neighbor Ising) model results of Fisher and Selke [9]. A trion, Fig. 2.2, represents an idealized *distinguishable* minicolumn or roughly 100 neurons, and has three levels of firing activity—above average, average, and below average.

A column with a small number of trions having structured connections yields a large repertoire of quasistable, periodic spatial-temporal firing patterns that can be excited, called magic patterns (or MPs). These inherent patterns are called magic patterns or MPs because of their ability to be learned or enhanced via a Hebb learning rule to a large cycling probability. The repertoire of (periodic) MPs is found by evolving all possible initial states (of the first two time steps) by following the most probable or deterministic path. The symmetries relating to MPs in a repertoire are discussed below. In a full probabilistic (or Monte Carlo) evolution, the MPs evolve in natural sequences from one to another. The probability of each MP remaining in that pattern can be enhanced by even a small change in connection strengths using a Hebbian learning rule. We consider these symmetry operations in the dynamics of the cortical column to be the basic elements of higher brain function.

The interactions among the trions are taken to be localized, competing (between excitation and inhibition) and highly structured, and the firing state of the network (cortical column) of the distinguishable trions at time $n\tau$ is updated in a probabilistic way related to the states of the two previous discrete time steps $(n-1)\tau$ and $(n-2)\tau$. We expect these time steps τ to be roughly 25–50 ms. The probability $P(S_i)$ of the ith trion having a firing level or state S_i at time $n\tau$ is given by

$$P(S_i) = \frac{g(S_i)\exp[BM_iS_i]}{\Sigma_S g(S)\exp[BM_iS]},$$
$$M_i = \Sigma_j[V_{ij}S'_j + W_{ij}S''_j] - V_i, \tag{2.1}$$

where S'_j and S''_j are the states of jth trion at the two earlier times $(n-1)\tau$ and $(n-2)\tau$, respectively. V_{ij} and W_{ij} are the interactions between trion i and j at time $n\tau$ from times $(n-1)\tau$ and $(n-2)\tau$, respectively; V_i is an effective firing threshold. The three possible firing states (of each trion) denoted by $+$, 0, $-$ for $S = 1, 0, -1$ represent, respectively, a large "burst" of firing, an average burst, and below average firing (see Fig. 2.2). The term $g(S)$ with $g(0) \gg g(+/-)$ takes into account the number of equivalent firing configurations of the *distinguishable* trion's internal *indistinguishable* neuronal constituents [35]. For example, in a trion representing a group of 90 neurons, firing levels of $+$, 0, $-$ could correspond to 90–61, 60–31, and 30–0 neurons firing, respectively. There are many more equivalent combinatorial ways of generating the 60–31 level from the *neurons*. This feature, $g(0) \gg g(+) = g(-)$, gives stability to the trion model firing patterns.

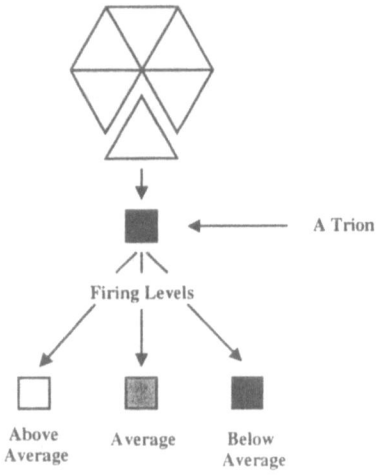

FIGURE 2.2. We identify a minicolumn with the idealized trion and the basic network of trions is the cortical column. As shown, the trion has three levels of firing activity.

The fluctuation parameter B is inversely proportional to the noise and results [38] from the statistical nature of neurotransmitter release from the synapses [14]. Studies of the trion model for learning and memory and higher brain function have been reported. Basically, the success of these studies is due to the fact that the localized (between excitation and inhibition) interactions with high symmetry yield a huge repertoire of MPs, any of which can be readily learned or enhanced [36, 39] with only small changes in the interaction strengths using the Hebb learning algorithm [12, 20, 21]:

$$\Delta V_{ij} = \varepsilon \sum_n^{\text{pattern}} S_i(n\tau) S_j \big((n-1)\tau\big),$$

$$\Delta W_{ij} = \varepsilon \sum_n^{\text{pattern}} S_i(n\tau) S_j \big((n-2)\tau\big), \qquad \varepsilon > 0, \tag{2.2}$$

which allows for both increases and decreases of interaction strengths. (Simply extending this learning rule to a third time step using the correlation

$$\Sigma S_i(n\tau) S_j \big((n-3)\tau\big)$$

significantly enhances the effects of learning with a smaller change in the total connectivity [16, 22].) Let us define the cycling probability $P_C(MP)$ that firing pattern for the columnar trion network remains in the MP for one cycle of the repeating MP. The $P_C(MP)$ is calculated by multiplying the probabilities $P(S_i)$, Eq. (2.1), of each trion i's being in the state S at time $n\tau$, given by that MP for its whole cycle length:

$$P_C(MP) = \Pi_n \Pi_i P\big(S_i(n\tau)\big). \tag{2.3}$$

Then as a result of learning an MP using the Hebb [12] algorithm (2.2), the cycling probability $P_C(MP)$, (2.3) is increased. Further, after learning, many more initial states will go to the learned MP (and some related MPs). Note that these MPs evolve in natural sequences from one to another in a probabilistic Monte Carlo calculation.

The simulations of a trion columnar network are simply performed:

(1) We specify the parameters of the trion network: The number of trions N, the degeneracy factors $g(S_i)$, the connectivities V_{ij} and W_{ij}, the firing thresholds V_i, and the fluctuation parameter B.

(2) A choice for the firing states for the initial two times steps is made. Since each of the N trions in each time step has three possible firing levels S, there are 3^{2N} possible initial choices.

(3) Given the firing states for each trion at the two earlier times $(n-1)\tau$ and $(n-2)\tau$, the probability $P(S_i)$ for ith trion being in state S_i at time $n\tau$ is calculated from Eq. (2.1).

Having made the choice of parameters, the repertoire of MPs or inherent, quasistable, periodic firing patterns is found as follows: For a given initial firing state, follow the procedure of always choosing the S for each trion which has the largest probability, determined by Eq. (2.1), or most probable path. Then the time evolution rapidly goes into a repeating spatial-temporal pattern or MP. Define the operator Γ which temporally evolves an MP according to its most probable path for its cycle length N_c. Then an MP is an eigenfunction of Γ with eigenvalue 1:

$$\Gamma MP = MP. \tag{2.4}$$

An explicit representation of Γ can be given. Going through all possible initial states gives all the MPs (the repertoire of MPs) as well as the number of initial states recalling each MP and the average time to recall an MP. An MP has the property of being readily learned or enhanced using the Hebb learning rule in Eq. (2.2) with only a relatively small change in the connections V and W. Arbitrary spatial-temporal patterns cannot be readily learned. Only an MP can be learned in a selective manner [39].

Thus far we have only considered the dynamics of a single cortical column. New properties occur when one considers a network of columns. Specifically, increases in $P_C(MP)$ in all of the constituent members, significantly greater than that observed for any single isolated column. This enhancement does not occur however for any "random" set of MPs which the columns may exhibit, but only for particular combinations of MPs. This emergent behavior has potential significance with respect to a number of issues. The fact that only specific combinations of MPs "fit together" has consequences for any cortical "language" which might consist of such patterns. An understanding of which patterns will enhance the individual $P_c(MP)$ for a particular network architecture may lead to important insights with respect to how knowledge is represented in our brains. Along this line, a particular example is given by short-term memory. Because adaptive behavior and the cognitive operations that support it commonly require the temporary retention of memories, cerebral mechanisms probably exist that ensure the temporary activation of memory networks (working memory). One mechanism that has been proposed is that activation is maintained in the short term by the reverberation of impulses in closed neural networks. A given temporal pattern exhibited by a network may not always occur with the same phase relative to the activation of the network. For example, if a pattern is an attractor state (e.g., limit cycle) of the network, the time of onset of a particular pattern may vary with respect to the onset of activation. If the pattern is cyclic, there could be variability in the phase of the cycle at which it begins. This type of behavior is clearly consistent with the results of the coupled column networks. The enhancement of cycling probability also is a possible explanation for the stability of short-term memory. Furthermore, the different patterns exhibited by the various constituent columns could be interpreted as representing different

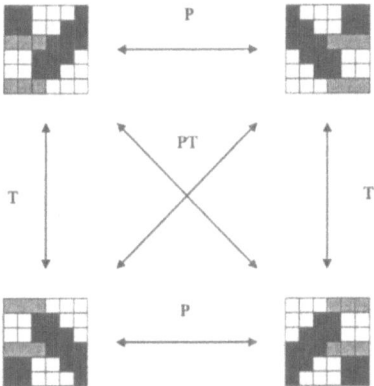

FIGURE 2.3. An explicit example [23] of the symmetry relations among MPs in a symmetry group in the repertoire for the connectivity (2.6). Each square represents a trion with three levels of firing activity as in Fig. 2.2. Each horizontal row represents a ring of interconnected (as in Fig. 2.2) trions (so that the sixth square wraps around to the first) and time evolves downward. Each of the four MPs here includes its spatial rotations (see Fig. 2.5).

stimulus features of the environment. This provides an elegant mechanism by which the brain can integrate different stimulus features into a single coherent representation (the so-called binding problem). It is therefore of great physiological and mathematical interest to determine which sets of MPs can be combined to enhance stability [32]. Analytic study of the symmetries involved would be of great interest.

Symmetries of the MPs in a repertoire

Consider a symmetry operator α acting on (2.4): $\alpha\Gamma MP = \alpha MP$. Then if α commutes with Γ, αMP is also an MP :

$$\alpha\Gamma MP = \alpha\Gamma\alpha^{-1}\alpha MP = \Gamma\alpha MP = \alpha MP \quad \text{for} \quad \alpha\Gamma\alpha^{-1} = \Gamma. \quad (2.5)$$

Thus, we expect for our structured connectivity in the trion model that there will be a number of symmetries α that will be useful to use to characterize a repertoire of MPs. To be explicit, let us examine a specific example of the repertoire of MPs for structured connections in an $N = 6$ network. Consider the connectivities and other parameters in Eq. (2.1) to be as follows:

$$V_{ii} = 2, \quad V_{i,i+1} = V_{i,i-1} = 1, \quad W_{ij} = -V_{ij}, \quad (2.6)$$

thresholds $V_i = 0$, all other V_{ij} equal to 0, $g(0)/g(+/-) = 500$. Then following the calculations in Section 2 above, each of the $3^{12} = 531,441$

possible choices for the initial states is followed to find the 155 MPs shown in Fig. 2.4 of Ref. [23].

These 155 MPs can be placed into 34 sets where the MPs in a set are related to each other by a spatial rotation among trions, R, among the (*distinguishable*) trions. It is evident from (2.6) and our ring boundary conditions that $\alpha = R$ commutes with Γ. Other symmetries among these MPs can be used to categorize groups of MPs. We see that a parity reflection, P; a time reversal operation, T; and the combination PT will relate different MPs, as shown in Fig. 2.3. That T commutes with Γ is not obvious. The 34 sets of MPs are placed in 20 groups with MPs in each group related by the symmetry operations α equal to P, T, and PT in addition to rotation R [23].

An additional symmetry operator changes firing level "spin" S to $-S$. In analogy with physical systems, let us define this as C, the "charge" conjugation operator. An example of a repertoire in which distinct MPs are related by C is seen in Ref. [37, Table 5].

There are 1804 MPs in the repertoire given by the connections [36],

$$V_{i,i+1} = V_{i,i-1} = 1, \quad W_{i,i+2} = W_{i,i-2} = -1, \tag{2.7}$$

thresholds $V_i = 0$, all other V_{ij} equal to 0, $g(0)/g(+/-) = 500$, $N = 6$ trions. Using these symmetry operations R, P, T, and PT, in addition to an additional one, R_T rotating in space and time, present for these special set of connections and number of trions, these 1804 MPs can be placed in 73 symmetry groups. This repertoire proved to be especially interesting when mapped onto music [16–18] and onto robotic motion [34]. An example of the MPs in two of these symmetry groups is given in Fig. 2.4. In one of these groups, the MPs with respect to spatial and temporal rotations have been arranged to make the symmetry relationships among the MPs more transparent. We leave it as an exercise for the reader to see these relationships among the complex spatial-temporal patterns in Fig. 2.4. This helps illustrate the power of these networks that may readily recognize these relationships.

We suspect that there may be additional general symmetries to be discovered, especially when several columnar networks are coupled together. We suggest that these groups or categories or MPs defined by symmetry operations are not only useful in understanding aspects of pattern recognition such as rotational invariance, but will prove invaluable in understanding the nature of the *sequences* of transitions of the MPs among themselves. A relevant example of an MP evolving into other MPs in Monte Carlo evolutions is shown in Fig. 2.5. This then forms the basis of inherent sequences of MPs.

We have called the above symmetries "global" in our columnar MPs, in contrast to "local" ones in which the temporal patterns for two specific trions might be interchanged (with a possible phase shift in time). For example, for the repertoire of Ref. [23, Table 4], $MP\,2$ is related to $MP\,1$ by

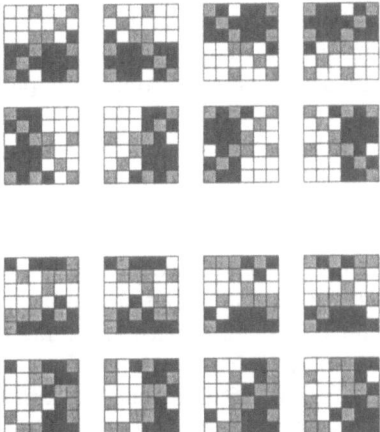

FIGURE 2.4. Example [23] of the MPs in two symmetry groups in the repertoire (see Fig. 2.3) for the connectivity (2.7). Each symmetry group consists of eight MPs (each including its spatial rotations) related by combinations of P, T, and a symmetry operation specific to this repertoire corresponding to a spatial-temporal rotation of 90^0 about the center of the MP. In the first symmetry group, the MPs with respect to spatial (and temporal) rotations have been arranged to make the symmetry relationships among the MPs more transparent. We leave it as an exercise for the reader to see these relationships among these MPs. This exercise will help illustrate the power of these trion networks that may readily recognize these symmetry relationships.

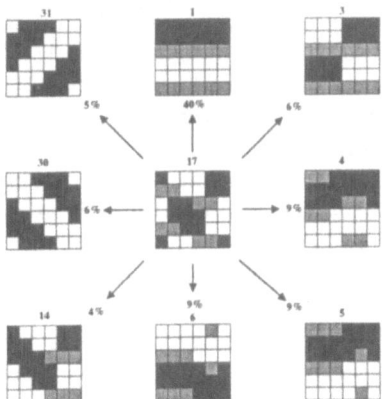

FIGURE 2.5. Monte Carlo calculations [23] for MP 17(0) {the number in $(\cdots)$ refers to the spatial rotation, see Fig. 2.5 in Ref. [23] for additional notation} in the repertoire for (2.6). The numbers at the end of the arrows give the percentage of 1000 Monte Carlo calculations that go to the eight other MPs (including all spatial rotations). Shown here are those MPs that are accessed with percentages greater than 3%. These relations can be substantially modified through learning.

shifting the patterns for trions 1 and 2 by 3 time steps (or applying C to just these two trions). We note then that all the MPs in this repertoire consist "simply" of combinations of just three temporal patterns for individual trions:

$$a = (+, +, 0, -, -, 0),$$
$$b = (+, +, +, -, -, -), \tag{2.8}$$
$$c = (0, 0, 0, 0, 0, 0).$$

It would be of strong interest to analytically determine the repertoire starting from this "alphabet" (2.8). (Note that this alphabet (2.8) holds for the repertoire from the connectivity (2.6) with any number of trions > 3.) We see from Table 6 of [37] that for connectivity $V_{ii} = 1$, $V_{ii+1} = V_{ii-1} = 1$, $W_{ij} = -V_{ij}$, the repertoire has 246 MPs and the alphabet consists of (2.8) plus $d = (-, +, -, +, -, +)$, $f = (+, -, 0, -, +, 0)$, $g = (+, 0, +, -, 0, -)$ and $h = (+, 0, 0, -, 0, 0)$. We are thinking of these trion temporal firing patterns (dependent on the connectivity) as "letters," the columnar MPs as "words." We have extended this concept "phrases" in coupled columnar trion networks [32] as the next step in developing a neural language of higher brain function.

Recognition of rotational invariance and time reversal

Here we discuss the recognition of spatially rotated and time-reversed objects in the trion model. This recognition of spatial rotational invariance is built into the highly structured trion model owing to its natural symmetry relations. Consider a simplified example in which a specific visual object, VO, is represented in the cortex by one of the MPs. Rotated VOs are represented by rotations of the MP of the standing VO. The VO seen in a normal standing position is learned with the Hebb rule and the symmetry among the connections is broken by a small amount. When a rotated VO is seen, the rotated VO MP evolves in a Monte Carlo calculation into the MP for the standing VO, thereby identifying it as a VO; the number of time steps to evolve is linearly related to the amount of rotation, in agreement with experiment [40]. A similar scenario is considered for recognition of time-reversed MPs. Here we do not give an explicit physical representation for the abstract MPs, although the mappings onto music [16, 18] and robotic motion [34] are relevant.

In addition to being able to identify rotations the network can also identify a time-reversal operation. An example of both [23] is shown in Fig. 2.6.

We suggest that these "built-in" recognitions of MPs related by symmetries constitute the basic operations in the cortex for higher brain function, and recent analysis of data obtained in neurophysiological experiments supports these idea in both human and nonhuman primates. The first of these experiments involved the recording of the activity of individual neurons in the cortex of monkeys during the performance of short-term memory tasks.

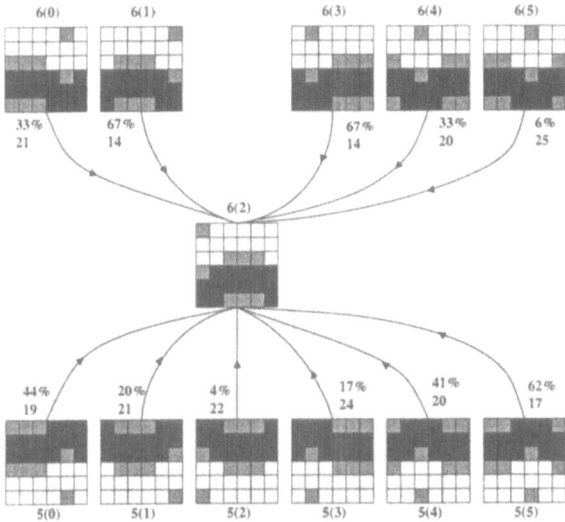

FIGURE 2.6. Example [23] of both rotation and time-reversal recognition in Monte Carlo calculations (see Fig. 2.5) after learning MP 6(2). MPs 5 are related to MPs 6 by T. Both the percentage of runs evolving to MP 6(2) and the average number of time steps taken are shown.

The time series consisting of the intervals between consecutive neuronal firings (spike train) were examined using a method of analysis that we developed to determine the presence of spatial and temporal symmetries [3]. The procedure consists of dividing time into slices of equal duration (bins) starting at the beginning of the time series. The number of times a neuron fires within each of the bins is then determined. To make contact with the trion model (see Fig. 2.2), the distribution of the numbers of neuronal firings occurring in the bins is divided into three levels corresponding to greater than average firing ($+1$), average firing (0), and less than average firing (-1). The spike train is thus mapped to a sequence of $+1's$ (white squares), 0's (gray squares), and -1's (black squares). An algorithm we

FIGURE 2.7. A particular family of symmetric temporal patterns that have been found [3] in the cortex of a monkey during the short-term retention of a stimulus. The entire pattern consists of these templates, along with each template's complete set of cyclically permuted versions in time. For example, if $D(0)$ is the non-permuted template shown above, then $D(1)$ is the template permuted one time step such that the last rectangle in the sequence becomes the first rectangle, the first rectangle becomes the second, etc.

developed was then applied to search these mapping for statistically significant sequences which were related to each other by symmetry operations.
As an example a particular sequence of 12 squares (1.8 seconds duration)
is shown in Fig. 2.7.

A symmetry family of pattern sequences is then defined as

$$F_S = (D, C, T, CT),$$

where D is the "direct" pattern sequence and all of its temporal cyclic
permutations, C is the sequence obtained from D by changing all $+1$'s to
-1's, T is obtained by temporally inverting D, and CT is the sequence
obtained by carrying out both the C and T symmetry operations on D.

The results of the analysis on the spike trains revealed that symmetry
families occurred primarily during the period when a monkey was hold-
ing information about a stimulus in short-term memory. Of particular importance is the fact that the presence of the entire family was often statistically significant without any of the individual family members, being
significant [3]. This supports the hypothesis that the symmetric relationships between patterns is exploited in cortical functions, rather than being
an epiphenomenon of network architecture. Furthermore, a given family
member pattern general repeated for several continuous cycles is consistent with the predictions of the trion model. A second experiment [2] was
conducted in which EEG recordings were obtained from a human subject
performing a mental rehearsal of a musical score. The analysis was modified
in this case so that the mapping was carried out on the average potential
occurring in each time bin rather than the number of neuronal firings.
Furthermore, since the EEG records were obtained from 19 different elec-
trodes distributed over the scalp, spatial as well as temporal patterns could
be detected. Thus we expanded the definition of a symmetry family to

$$F_S = (D, C, T, CT, P),$$

where P is the mirror reflection (parity symmetry) of D. Once again the
results of the analysis revealed the presence of these symmetric families of
patterns.

Trion music

We proposed that the trion model patterns represent a candidate for the
common internal language of the brain, that is, how one part of the brain
communicates with the other parts. We then mapped [16, 18] the Monte
Carlo evolutions of the model, or translated this internal language of the
brain, into music. The results were striking when the Monte Carlo evolutions were mapped onto pitches and instruments to produce music.

There are many, many possible choices of mappings! We presented in [18]
three examples to illustrate the mapping procedures. As we demonstrated,

they generate good approximations to different recognizable (western and eastern) human styles or music.

We then made the following generalizations [16, 17]: Music composition, performance and listening all involve the evolution of this inherent repertoire of spatial-temporal patterns (MPs in the trion model). It is the creative genius of the composer such as Mozart or the brilliance of the conductor such as Szell that produces a magnificent performance that involves brain function at its highest level. *We suggested what is involved here is the perfect use of these inherent spatial-temporal patterns common to* **mammalian cortex**. *As listeners, we need only appreciate the result of having these inherent patterns excited in our brains.* We can all appreciate music even though very few of us can compose it.

The assumption that this inherent repertoire of patterns is essentially present at birth is perhaps a necessary condition to understand the appreciation of music (and particular pieces of music) by infants [15]. In addition, there are inherent structures in the brain that are devoted to music just as there is such a structure for language. Further, this structure is accessible for use almost from birth without substantial learning. In this sense, we might consider music as a sort of "pre-language." Again the feature that millions of people over centuries will be captivated by certain composers and specific pieces of music speaks to the common universality of the repertoires of inherent spatial-temporal patterns.

These discussions gave the insight for us [16, 17] to relate the neuronal processes involved in music and abstract spatial-temporal reasoning, and thus to propose that music can enhance spatial-temporal reasoning.

3   Music Enhances Spatial-Temporal Reasoning

As discussed previously, Leng and Shaw had proposed [16, 17] that music training at an early age could enhance spatial-temporal reasoning in children. In particular, the neuronal processes of pattern development lasting some tens of seconds would employ in sequential manner the built-in basic ability of the cortex to recognize MPs related by symmetries. Music clearly involves this pattern development concept as does spatial-temporal reasoning. Although cognitive abilities such as music and spatial-temporal reasoning crucially depend on specific, localized regions of the cortex, all higher cognitive abilities draw upon a wide range of cortical areas [28, 33]. Musical abilities are evident in infants [15, 42]. Music, then, may serve a "pre-language" [17] (with centers [27] distinct from language centers in the cortex), available at an early age, which can access inherent cortical spatial-temporal firing patterns and enhance the cortex's ability to accomplish pattern development.

In addition to this potential of long-term enhancement of spatial-temporal reasoning in young children through music training, Rauscher and Shaw proposed that even listening to certain types of music might prime the neuronal pathways for pattern development, specifically, spatial-temporal reasoning, leading to the "Mozart effect" short-term enhancement experiments [29, 30].

Mozart effect

We performed an experiment to determine if short-term causal enhancements of spatial-temporal reasoning could be produced by merely listening to music [29]. Thirty-six UCI undergraduates listened to (the first) ten minutes of Mozart's Sonata for Two Pianos, K.448, and scored 8 to 9 points higher on the spatial IQ subtest of the Stanford-Binet Intelligence Scale than after they listened to taped relaxation instructions or silence. *This facilitation lasted only ten to fifteen minutes.*

In a follow-up study [30], seventy-nine UCI students participated for five consecutive days. We issued all students 16 Paper Folding and Cutting (PF&C) items on the first day of the experiment (Fig. 2.8), and then divided them into three groups with equivalent abilities (giving them 16

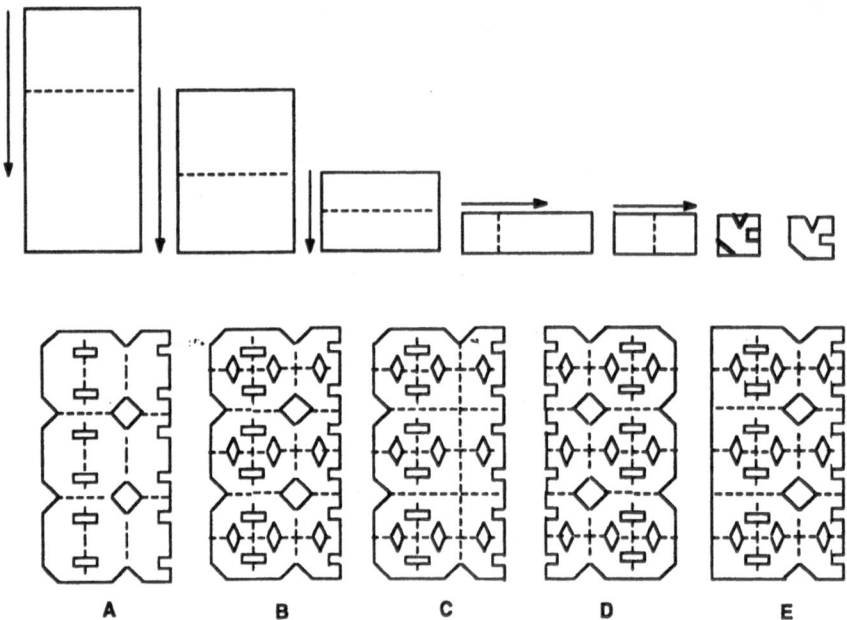

FIGURE 2.8. Example of the PF&C task used in our behavioral experiment. It depicts a picture of a paper before it was folded and cut (top left figure). The dotted lines and straight arrows represent the location and direction of folds. The solid lines represent cuts. Subjects were to choose which of the five choices below show how the paper would look unfolded.

different PF&C items on each following day). Of the three sub-tests in the spatial reasoning portion of Stanford-Binet's Intelligence Scale, we chose the PF&C task because it best fit our concept of spatial-temporal pattern development [17]. On subsequent days, we showed that listening to the same Mozart sonata K.448 enhanced performance on the PF&C tasks over listening to (i) silence; (ii) a minimalist work by Philip Glass; (iii) an audio-taped story; and (iv) a dance (trance) piece. Also, we showed that listening to the Mozart sonata did not enhance short-term memory. These experiments established the causal enhancement of spatial-temporal reasoning by listening to specific music.

The proposed mechanisms for this short-term enhancement of spatial-temporal reasoning by listening to music include the following:

(i) Listening to music helps "organize" the cortical firing patterns so that they do not wash out for other pattern development functions, in particular, the right hemisphere processes of spatial-temporal task performance.

(ii) Music acts as an "exercise" for exciting and priming the common repertoire and sequential flow of the cortical firing patterns responsible for higher brain functions. (We note that the potential mechanism (ii) would imply that care in experimental design must be taken by researchers who want to extend our findings since the task can prime itself.)

(iii) The cortical symmetry operations among the inherent patterns are enhanced and facilitated by music.

As previously discussed, we chose Mozart since he was composing at the age of four. Thus we expect that Mozart was exploiting the inherent repertoire of spatial-temporal firing patterns in the cortex. While one might explore the many possibilities with a large number of styles and composers, it would be perhaps more interesting to investigate the underlying neurophysiological bases using behavioral studies in conjunction with EEG and single neuron studies. We have begun such collaborative studies at the University of Vienna [33] and the University of California, Los Angeles, and the first results look very promising.

Music training enhances reasoning in preschool children

Predictions from our structured neuronal model of the brain [16, 17] led us to test the hypothesis that music training for young children enhances spatial-temporal reasoning. Seventy-eight preschool children participated in the recently published study [31]: The Keyboard group of children were given private keyboard music lessons for six months, and there were three control groups of children, including a group getting computer lessons. Four standard age-calibrated spatial reasoning tests were given at the beginning and at the end of the study: one test assessed spatial-temporal reasoning and three tests assessed spatial-recognition reasoning. A *highly significant improvement of large magnitude* was found for the Keyboard group in the spatial-temporal reasoning test (Object Assembly in which the child arranges pieces of a puzzle to create, e.g., a familiar animal by forming a mental image of the animal and rotating pieces to match the mental image; performance is helped by putting the pieces together in particular orders, defining the spatial-temporal nature of the task). No significant improvement was found on tests of spatial-recognition reasoning (such as matching, classifying, and recognizing similarities among objects). The control groups did not improve significantly on any of the tests. We suggest that spatial-temporal reasoning is crucial for such adult endeavors as higher mathematics, engineering, and chess. This experiment was not able to determine the temporal duration of the enhancement of spatial-temporal reasoning, but it was at least some days. *Thus these results have enormous educational implications.*

Discussion

Starting from the trion model, specific types of symmetric spatial and temporal patterns are predicted to exist that play a fundamental role in how the mammalian brain thinks and reasons. We have recently shown that such families of symmetric spatial and temporal patterns are indeed present in neurophysiological data and that they are related to higher brain function [2, 3].

Predictions [16, 17] from the trion model [23, 24, 36, 37, 39] led us to the hypothesis that music could causally enhance spatial-temporal reasoning. We have shown [29, 30] that college students scored significantly higher on a spatial-temporal reasoning task after listening to the Mozart Sonata K.448, as compared to silence and to other listening conditions. This enhancement lasted roughly ten minutes. Thus although this "Mozart effect" established the causal effect and has major scientific consequences, it does not as yet have educational implications. We have shown [31] that preschool children who received private keyboard lessons for six months improved dramatically on a spatial-temporal reasoning task while appropriate control groups did not improve significantly. This enhancement lasts at least one day, which is long enough to have major educational implications. The Mozart effect establishes the causal short-term enhancement of spatial-temporal reasoning; the pre-school music training study establishes an enhancement of at least some days (an increase by more than a factor of 100), making this of immediate educational use; clearly the goal would be to achieve a permanent enhancement in spatial-temporal reasoning.

It has been clearly documented [13] that young students have difficulty understanding the concepts of proportion (heavily used in math and science) and that no successful program has been developed to teach these concepts in the school system. We predict that an enhanced ability to evolve temporal sequences of spatial patterns as a result of music training will lead to an enhanced conceptual mastering of proportional reasoning. This is a strong proposal that is being investigated in our current research.

In this review, we have presented work performed over the last 12 years that we feel represents the essential first steps in establishing the principle of symmetry as being an essential property of higher brain function. Symmetry allows the recognition of patterns and temporal relationships and evolutions among patterns as in chess, mathematics, and music. In addition to the importance of symmetry in understanding higher brain function, the application of these principles we feel is precisely what is missing in present digital computers and would have important consequences in the development of the next generation of computational machines that can think and reason. The scientific, educational, and technological consequences of understanding and exploiting the role of symmetry in higher brain function are enormous.

Acknowledgments: This paper, as noted in the references cited, represents the work of a large number of colleagues and students over many years. However, G.S. wants to specifically acknowledge the absolutely vital role of two researchers on the main theme of this paper: the influence of music on how we think and reason: Xiao Leng on the cortical theory and Fran Rauscher on the behavioral experiments. He would especially like to thank George Patera for encouraging him over the past years to pursue the *seemingly* soft science of music as a window into higher brain function.

Our present research is supported by grants from the Ralph and Leona Gerard Foundation, Marjorie Rawlins, Edward and Vivian Thorp, the 42nd St. Fund, and the National Academy of Recording Arts and Sciences.

4   References

1. G.J. Allman, *Greek geometry from Thales to Euclid*, Arno, New York, 1976.

2. B. Bodner, J. Chen, G.L. Shaw, and K.H. Tachicki, *Symmetry in higher brain function: Theory and experiment*, to be presented at the Society of Neuroscience Annual Meeting, 1997.

3. M. Bodner, Y.-D. Zhou, G.L. Shaw, and J.M. Fuster, *Symmetric temporal patterns in cortical spike trains during performance of a short-term memory task*, Neurological Res. **19** (1997), 509.

4. W.S. Boettcher, S.S. Hahn, and G.L. Shaw, *Mathematics and music: A search for insight into higher brain function*, Leonardo Music J. **4**

(1994), 53.

5. T. Bonhoeffer and A. Grinvald, *Iso-orientation domains in cat visual cortex arranged in pinwheel-like patterns*, Nature **353** (1991), 429.

6. L. Brothers, G.L. Shaw, and E.L. Wright, *Durations of extended mental rehearsals are remarkably reproducible in higher level human performance*, Neurological Res. **15** (1993), 413.

7. W.G. Chase and H.A. Simon, *The mind's eye in chess*, Visual Information Processing (W.G. Chase, ed.), Academic Press, New York. 1973.

8. L.D. Cranberg and M.L. Albert, *The chess mind*, The Exceptional Brain (L.K. Obler and D. Fein, eds.), Guilford, New York, 1988.

9. M.E. Fisher and W. Selke, *Infinitely many commensurate phases in a simple Ising model*, Phys. Rev. Lett. **44** (180), 1502–1505.

10. P.S. Goldman-Rakic, *Introduction: The frontal lobes: Uncharted provinces of the brain*, Trends Neurosci. **7** (1984), 425.

11. M. Hassler, N. Birbaumer, and A. Feil, *Musical talent and visual-spatial abilities: A longitudinal study*, Psychology of Music **13** (1985), 99.

12. D.O. Hebb, *Organization of behavior*, Wiley, New York, 1949.

13. K.R. Karplus, S. Pulos, and E.K. Stage, *Early adolescent's proportional reasoning on 'rate' problems*, Educational Stud. Math. **14** (1983), 219.

14. B. Katz, *The release of neural transmitter substances*, Thomas, Springfield, 1969.

15. C.L. Krumhansl and P.W. Jusczyk, *Infants' perception of phrase structure in music*, Psychological Sci. **1** (1990), 70.

16. X. Leng, *Investigation of higher brain functions in music composition using models of the cortex based on physical system analogies*, Ph.D. thesis, University of California, Irvine, 1990.

17. X. Leng and G.L. Shaw, *Toward a neural theory of higher brain function using music as a window*, Concepts Neurosci. **2** (1991), 229.

18. X. Leng, G.L. Shaw, and E.L. Wright, *Coding of musical structure and the trion model of cortex*, Music Perception **8** (1990), 49.

19. W.A. Little, *Existence of persistent states in the brain*, Math. Biosci. **19** (1974), 101.

20. W.A. Little and G.L. Shaw, *A statistical theory of short and long-term memory*, Behav. Biol. **14** 1975, 115.

21. W.A. Little and G.L. Shaw, *Analytic study of the storage capacity of a neural network*, Math. Biosci. **39** (1978), 281.

22. J.V. McGrann, *Further theoretical investigations of the trion model of cortical organization*, Ph.D. thesis, University of California, Irvine, 1992.

23. J.V. McGrann, G.L. Shaw, K.V. Shenoy, X. Leng, and R.B. Mathews, *Computation by symmetry operations in a structured model of the brain*, Phys. Rev. E **49** (1994), 5830.

24. J.V. McGrann, G.L. Shaw, D.J. Silverman, and J.C. Pearson, *Higher temperature phases of a structured model of cortical organization*, Phys. Rev. A **43** (1991), 5678.

25. V.B. Mountcastle, *An organizing principle for cerebral function: The unit module and the distributed system*, The Mindful Brain (G.M. Edelman and V.B. Mountcastle, eds.), MIT, Cambridge, 1978.

26. J. Patera, G.L. Shaw, R. Slansky, and X. Leng, *Model of adaptive temporal development of structured finite systems*, Phys. Rev. A (3) **40** (1989), 1073–1087.

27. I. Peretz, R. Kolinsky, M. Tramo, R. Labrecque, C. Hublet, G. Demeurisse, and S. Belleville, *Functional dissociations following bilateral lesions of auditory cortex*, Brain **117** (1994), 1283.

28. H. Petsche, P. Richter, A. von Stein, S. Etlinger, and O. Filz, *EEG coherence and musical thinking*, Music Perception **11** (1993), 117.

29. F.H. Rauscher, G.L. Shaw, and K.N. Ky, *Music and spatial task performance*, Nature **365** (1993), 611.

30. F.H. Rauscher, G.L. Shaw, and K.N. Ky, *Listening to Mozart enhances spatial-temporal reasoning: Towards a neurophysiological basis*, Neurosci. Lett. **185** (1995), 44.

31. F.H. Rauscher, G.L. Shaw, L.J. Levine, E.L. Wright, W.R. Dennis, and R.L. Newcomb, *Music training causes long-term enhancement of preschool children's spatial-temporal reasoning*, Neurological Res. **19** (1997), 2.

32. M. Sardesai, C. Figge, M. Bodner, J.A. Quillfeldt, A. Ostling, and G.L. Shaw, *Reliable short-term memory in the trion model: Toward a cortical language and grammar*, Biol. Cybernetic (in press).

33. J. Sarnthein, A. von Stein, P. Rappelsberger, H. Petsche, F.R. Rauscher, and G.L. Shaw, *Persistent patterns of brain activity: An EEG coherence study of the positive effect of music on spatial-temporal reasoning*, Neurological Res. **19** (1997), 107.

34. S.H. Shanbhag, *Robotic motion and the trion model of the cortex*, unpublished report, 1991.

35. G.L. Shaw and K.J. Roney, *Analytic solution of a neural network theory based on an Ising spin system analogy*, Phys. Lett. A **74** (1979), 146–150.

36. G.L. Shaw, D.J. Silverman, and J.C. Pearson, *Model of cortical organization embodying a basis for a theory of information processing and memory recall*, Proc. Nat. Acad. Sci. U.S.A **82** (1985), 2364.

37. G.L. Shaw, D.J. Silverman, and J.C. Pearson, *Associative recall properties of the trion model of cortical organization*, Biol. Cybern. **53** (1986), 259.

38. G.L. Shaw and R. Vasudevan, *Persistent states of neural networks and the random nature of synaptic transmission*, Math. Biosci. **21** (1974), 207.

39. K.V. Shenoy, J. Kaufman, J.V. McGrann, and G.L. Shaw, *Learning by selection in the trion model of cortical organization*, Cerebral Cortex **3** (1993), 239.

40. R.N. Shepard and J. Metzler, *Mental rotation of three-dimensional objects*, Science **171** (1971), 701.

41. E.S. Spelke, *Principles of object perception*, Cognitive Sci. **14** (1990), 29.

42. S.A. Trehub, *Infants' perception of music patterns*, Perception and Psychophysics **41** (1987), 635.

43. K. Wynn, *Addition and subtraction by human infants*, Nature **358** (1992), 749.

3

Lie Modules of Bounded Multiplicities

D.J. Britten and F.W. Lemire

1 Introduction

Let $\mathcal{L}$ denote a finite-dimensional, simple Lie algebra over the complex numbers $\mathcal{C}$ and fix a Cartan subalgebra $\mathcal{H}$. An $\mathcal{L}$-module $\mathcal{V}$ is said to be $\mathcal{H}$-*diagonalizable* if $\mathcal{V} = \oplus \sum_{\mu \in \mathcal{H}^*} \mathcal{V}_\mu$ where $\mathcal{V}_\mu = \{v \in \mathcal{V} \mid hv = \mu(h)v$ for all $h \in \mathcal{H}\}$. Further $\mathcal{V}$ is said to have *bounded multiplicities* provided there exists a constant B such that $\dim \mathcal{V}_\mu \leq B$ for all $\mu \in \mathcal{H}^*$. The minimum such bound is said to be the *degree* of the module.

In this paper we survey recent results concerning the classification and explicit realization of all simple $\mathcal{L}$-modules having bounded multiplicities relative to $\mathcal{H}$. This set clearly includes all finite-dimensional, simple $\mathcal{L}$-modules and so we focus our attention on the infinite-dimensional cases. It turns out that not every simple Lie algebra admits an infinite-dimensional module of finite degree. In fact we have—

Theorem 1.1 ([1] Prop. 1.4). *Let $\mathcal{V}$ be an infinite-dimensional, simple $\mathcal{H}$-diagonalizable $\mathcal{L}$-module of finite degree; then*

(i) *the Gel'fand Kirillov dimension of $\mathcal{V}$ is equal to the rank of $\mathcal{L}$, and,*

(ii) *$\mathcal{L}$ must be of type A_n or C_n.*

With this result in mind the paper is organized as follows. In Section 2, by way of motivating the problem considered in this paper, we review Fernando's work [7] on the classification of all simple $\mathcal{L}$-modules having finite-dimensional (not necessarily bounded!) weight spaces relative to $\mathcal{H}$. Fernando has shown that this classification can be reduced to the classification of all simple, *torsion-free* modules, that is, $\mathcal{L}$-modules M such that for all $x \in \mathcal{L} \backslash \mathcal{H}$ and all $0 \neq m \in M$ the space $\mathrm{Span}_\mathcal{C}\{x^n \cdot m \mid n = 0, 1, \dots\}$ is infinite-dimensional. In Section 3 we determine, up to equivalence, all simple, completely pointed modules, that is, modules of degree one, for the Lie algebras of type A_n and C_n. In Section 4 we consider tensor product modules formed by tensoring a completely pointed module and a finite-dimensional module. This produces $\mathcal{H}$-diagonalizable modules having bounded multiplicities. We conjecture that, up to equivalence, every simple A_n and C_n module of finite degree can be realized in this manner and indicate the progress that has been made in this direction.

2 Simple $\mathcal{L}$ Modules with Finite-Dimensional Weight Spaces

Let $\mathcal{M}(\mathcal{L}, \mathcal{H})$ denote the category of all $\mathcal{H}$-diagonalizable $\mathcal{L}$-modules having finite-dimensional (not necessarily bounded) weight spaces. This category is a natural, nontrivial generalization of category $\mathcal{O}$. Fernando [7] has shown that the classification of all simple modules in $\mathcal{M}(\mathcal{L}, \mathcal{H})$ can be reduced to the classification of all simple torsion-free modules for semi-simple subalgebras $\mathcal{L}'$ in $\mathcal{L}$.

This reduction can be briefly summarized as follows. For any simple module $\mathcal{V} \in \mathcal{M}(\mathcal{L}, \mathcal{H})$ one can uniquely associate a parabolic subalgebra $\mathcal{P}$ of $\mathcal{L}$ containing $\mathcal{H}$. Let $\mathbf{u}$ be the nilradical of $\mathcal{P}$ and denote by $\mathbf{l}$ the reductive, ad$(\mathcal{H})$-stable Levi complement of $\mathbf{u}$ in $\mathcal{P}$. Let X denote the $\mathbf{u}$-invariants in $\mathcal{V}$, that is, $X = \mathcal{V}^{\mathbf{u}} = \{v \in \mathcal{V} \mid \mathbf{u} \cdot v = 0\}$. Then the Levi complement $\mathbf{l}$ admits a decomposition $\mathbf{l} = \mathbf{r} \oplus \mathbf{t}$, where $\mathbf{r}$ is an ad$(\mathcal{H})$-stable reductive ideal of $\mathbf{l}$ that acts locally finitely on X and $\mathbf{t}$ is an ad$(\mathcal{H})$-stable semisimple ideal of $\mathbf{l}$ such that X is a torsion-free $\mathbf{t}$-module. This decomposition of $\mathbf{l}$ yields a corresponding decomposition of the $\mathbf{l}$-module X into a tensor product $X_{\text{fin}} \otimes X_{\text{free}}$ where X_{fin} is a simple finite-dimensional $\mathbf{r}$-module and X_{free} is a simple torsion-free $\mathbf{t}$-module in $\mathcal{M}(\mathbf{t}, \mathbf{t} \cap \mathcal{H})$. The simple module $\mathcal{V}$ is completely determined by the parabolic subalgebra $\mathcal{P}$ of $\mathcal{L}$, the decomposition $\mathbf{l} = \mathbf{r} \oplus \mathbf{t}$ of $\mathbf{l}$ and the modules X_{fin} and X_{free}. Since simple finite-dimensional Lie modules are well known, the classification of simple modules in $\mathcal{M}(\mathcal{L}, \mathcal{H})$ effectively reduces to the classification of simple, torsion-free modules.

We note that if M is a simple torsion-free module in $\mathcal{M}(\mathcal{L}, \mathcal{H})$ then every element $x \in \mathcal{L}$ acts bijectively on M and hence all nonzero weight spaces of M have the same dimension. It follows then that the problem of classifying all simple modules of finite degree studied in this paper is a natural generalization of the classification of all simple torsion-free modules.

3 Completely Pointed Modules

In this section we restrict attention to simple, completely pointed modules, that is, modules having all one dimensional weight spaces. In Ref. [1] we provide a unified construction of all completely pointed modules for the algebras of type A_n and C_n. To this end we first construct a certain family of simple modules for the Weyl algebra.

Let $\mathcal{A}_n$ denote the Weyl algebra over $\mathcal{C}$ presented by the generators $p_i, q_i, i = 1, \ldots, n$ and the relations

$$[q_i, q_j] = [p_i, p_j] = 0 \quad 1 \leq i \neq j \leq n,$$
$$[p_i, q_j] = \delta_{ij} \qquad 1 \leq i,\ j \leq n.$$

For any fixed n-tuple $\vec{a} = (a_1, \ldots, a_n) \in \mathcal{C}^n$, we define a formal complex linear space by setting

$$W(\vec{a}) = \mathrm{Span}_{\mathcal{C}}\{x(\vec{b}) \mid b_i - a_i \in \mathcal{Z}, \text{ if } a_i \in \mathcal{Z} \text{ then } b_i < 0 \iff a_i < 0\},$$

We provide $W(\vec{a})$ with an $\mathcal{A}_n$-module structure as follows:

$$q_i \cdot x(\vec{b}) = \begin{cases} (b_i + 1)x(\vec{b} + \varepsilon_i) & \text{if } b_i \in \mathcal{Z}_{<0}, \\ x(\vec{b} + \varepsilon_i) & \text{otherwise,} \end{cases}$$

$$p_i \cdot x(\vec{b}) = \begin{cases} x(\vec{b} - \varepsilon_i) & \text{if } b_i \in \mathcal{Z}_{<0}, \\ b_i x(\vec{b} - \varepsilon_i) & \text{otherwise.} \end{cases}$$

For each n-tuple $\vec{a}$ the module $W(\vec{a})$ is easily seen to be a simple $\mathcal{A}_n$-module. To relate this to our original goal of constructing simple, completely pointed modules for $\mathrm{sl}(n)$ and $\mathrm{sp}(2n)$ we note that there are natural embeddings of the general linear Lie algebra $\mathrm{gl}(n)$ and the symplectic Lie algebra $\mathrm{sp}(2n)$ into the Weyl algebra $\mathcal{A}_n$. In fact, by viewing $\mathrm{gl}(n)$ as the Lie algebra of $n \times n$ complex matrices with its standard basis of matrix units $e_{i,j}$, we obtain a Lie algebra embedding: $e_{i,j} \mapsto q_i p_j \in \mathcal{A}_n$. A simple Lie algebra $\mathcal{L}$ of type A_{n-1} may be identified with $\mathrm{sl}(n)$, the traceless $n \times n$ complex matrices, and restriction of the above map to $\mathrm{sl}(n)$ determines a Lie algebra embedding of $\mathrm{sl}(n)$ into $\mathcal{A}_n$. It is also well known that the complex linear span of the degree two monomials $\{q_i p_j, p_i p_j, q_i q_j \mid 1 \leq i, j \leq n\}$ in $\mathcal{A}_n$ under the commutator product $[w, x] = wx - xw$ determines a Lie subalgebra that is isomorphic to $\mathrm{sp}(2n)$. As a result of these embeddings, the modules $W(\vec{a})$, constructed above, can be viewed as modules for $\mathrm{sl}(n)$ and $\mathrm{sp}(2n)$. Clearly $W(\vec{a})$ is not simple as either an $\mathrm{sl}(n)$ or $\mathrm{sp}(2n)$-module; however, in each case we can select certain distinguished submodules that are simple and completely pointed.

For the symplectic algebra $\mathrm{sp}(2n)$ we start by fixing a basis of simple roots $\pi = \{\alpha_1, \ldots, \alpha_n\}$ corresponding to the Chevalley basis of the Cartan subalgebra $\mathcal{H}$ given by $h_{\alpha_i} = q_i p_i - q_{i+1} p_{i+1}$ for $i = 1, \ldots, n-1$ and $h_{\alpha_n} = q_n p_n + \frac{1}{2}$. Let $\{\omega_1, \ldots, \omega_n\}$ denote the corresponding fundamental weights. Define

$$M(\vec{a}) = \mathrm{Span}_{\mathcal{C}}\left\{x(\vec{b}) \in W(\vec{a}) \,\bigg|\, \sum_{i=1}^{n}(b_i - a_i) \in 2\mathcal{Z}\right\};$$

then we have—

Theorem 3.1 ([1] Props. 2.12, 3.6, 5.20). $M(\vec{a})$ *is a simple, completely pointed, infinite-dimensional* $\mathrm{sp}(2n)$ *module relative to the Cartan subalgebra* $\mathcal{H}$. *Moreover, if* V *is a simple, completely pointed, infinite-dimensional* C_n-*module, then* V *is equivalent to one of the following:*

(i) $M(-1,\ldots,-1)$—*the simple* C_n-*module of highest-weight* $-\frac{1}{2}\omega_n$.

(ii) $M(-1,\ldots,-1,-2)$—*the simple* C_n-*module of highest-weight* $\omega_{n-1} - 3/2\omega_n$.

(iii) $M(-1,\ldots,-1,a_k,\ldots,a_n)$ *with* $a_i \in C \setminus \mathcal{Z}$ *for* $i = k,\ldots,n$ *where the real part of* a_i *may be assumed to be in the interval* $[0,1)$.

An interesting observation is that all of these modules have the same central character, namely, $\mathcal{X}_{-1/2\omega_n}$, that is the central character of the C_n-module having highest-weight $-\frac{1}{2}\omega_n$. We also draw attention to the fact that every simple, completely pointed, torsion-free C_n-module is equivalent to a module $M(a_1,\ldots,a_n)$ where the parameters $a_i \in C \setminus \mathcal{Z}$.

For the algebra $\mathrm{sl}(n)$ we fix a basis of simple roots $\pi = \{\alpha_1,\ldots,\alpha_{n-1}\}$ corresponding to the Chevalley basis of the Cartan subalgebra $\mathcal{H}$ given by $h_{\alpha_i} = q_i p_i - q_{i+1}p_{i+1}$ for $i = 1,\ldots,n-1$. Let $\{\omega_1,\ldots,\omega_{n-1}\}$ denote the corresponding fundamental weights. Define

$$N(\vec{a}) = \mathrm{Span}_C\left\{x(\vec{b}) \in W(\vec{a}) \mid \sum_{i=1}^{n}(b_i - a_i) = 0\right\},$$

then we have:

Theorem 3.2 ([1] Props. 2.12, 3.4, 5.8). $N(\vec{a})$ *is a simple, completely pointed, infinite-dimensional* $\mathrm{sl}(n)$-*module relative to the Cartan subalgebra* $\mathcal{H}$. *Moreover, if* V *is a simple, completely pointed, infinite-dimensional* $\mathrm{sl}(n)$-*module, then* V *is equivalent to* $N(a_1,\ldots,a_n)$ *where* $a_i = -1$ *for* $i = 1,\ldots,k-1$, $a_i \in C \setminus \mathcal{Z}$ *for* $i = k,\ldots,\ell$ *and* $a_i = 0$ *for* $i = \ell+1,\ldots,n$.

In this list the completely pointed, highest-weight modules correspond to

$$N(c,0,\ldots,0), \quad N(-1,\ldots,-1,-(c+1)) \quad \text{and}$$
$$N(-1,\ldots,-1,c,-(c+1),0,\ldots,0)$$

which are the C_n-modules with highest-weights $c\omega_1$, $c\omega_{n-1}$ and $c\omega_i - (c+1)\omega_{i+1}$ respectively. Also the completely pointed, torsion-free C_n-modules correspond to the modules $N(a_1,\ldots,a_n)$ where the parameters a_i are all noninteger complex numbers.

4 Completely Pointed Modules Tensored with Finite-Dimensional Modules

In order to construct modules of finite degree greater than 1 we consider the tensor product of an infinite-dimensional, completely pointed module with a finite-dimensional simple module. Such a tensor product module has degree equal to the dimension of the finite-dimensional module but will not in general be simple. We now consider the decompositions of these tensor product modules. Since the situation is quite different for the algebras of type C_n and A_n, we consider each case separately.

4.1 C_n-modules of finite degree

We start by considering the problem of determining all simple, *highest-weight*, infinite-dimensional C_n-modules having finite degree. We denote by $L(\lambda)$ the simple module having highest-weight λ. From our previous results there exist only two simple, completely pointed, infinite-dimensional highest-weight C_n-modules, namely, $L\left(-\frac{1}{2}\omega_n\right)$ and $L\left(\omega_{n-1}-\frac{3}{2}\omega_n\right)$. In [3] we show that the tensor products of these modules with finite-dimensional, simple modules are completely reducible. In fact we have—

Theorem 4.1 ([3] Theorem 5.5). *Assume that $\lambda = \lambda_1\omega_1 + \cdots + \lambda_n\omega_n$ is a dominant integral weight of C_n and let $L(\lambda)$ denote the associated simple highest-weight C_n-module. Then*

$$L\left(-\frac{1}{2}\omega_n\right) \otimes L(\lambda) = \oplus \sum_{\mu \in \mathcal{J}_\lambda} L\left(-\frac{1}{2}\omega_n + \mu\right).$$

The indexing set $\mathcal{J}_\lambda$ is given in terms of the epsilon basis $\{\varepsilon_i\}$ of $\mathcal{H}^*$ by

$$\mathcal{J}_\lambda = \left\{\mu \mid \lambda - \mu = \sum_{j=1}^{n} d_j\varepsilon_j\right\},$$

where $(d_j \in \mathcal{Z}_+$ with $0 \le d_j \le \lambda_j$ for $j = 1, \ldots, n-1$, $0 \le d_n \le 2\lambda_n + 1$ and $\sum_{i=1}^{n} d_i \in 2\mathcal{Z}$.
A similar decomposition can be written for $L\left(\omega_{n-1} - 3/2\omega_n\right) \otimes L(\lambda)$.

It follows from this theorem that for any dominant integral weight λ and each $\mu \in \mathcal{J}_\lambda$ the simple module with highest-weight $-\frac{1}{2}\omega_n + \mu$ has finite degree. It is also important to observe that for any dominant integral weight λ the central characters $\mathcal{X}_{-1/2\omega_n+\mu}$ for $\mu \in \mathcal{J}_\lambda$ are distinct.
In Ref. [4] we have directly determined all simple highest-weight modules having finite degree:

Theorem 4.2 ([4] Theorem 2.15). *Let $\nu = \nu_1\omega_1 + \cdots + \nu_n\omega_n$ be an arbitrary weight. Then the simple highest-weight C_n-module $L(\nu_1\omega_1 + \cdots +$*

$\nu_n\omega_n)$ is infinite-dimensional and has bounded multiplicities iff the coefficient ν_i are nonnegative integers for $i = 1, \ldots, n-1$ and ν_n is one half of an odd integer such that $\nu_{n-1} + 2\nu_n + 3 > 0$.

Combining the previous two theorems we conclude that every infinite-dimensional highest-weight C_n-module having bounded multiplicities occurs as a summand of the tensor product $L(-\frac{1}{2}\omega_n) \otimes L(\lambda)$ for an appropriate choice of the dominant integral weight λ. Therefore, in some sense, the module $L(-\frac{1}{2}\omega_n)$ "generates" all infinite-dimensional highest-weight C_n-modules having bounded multiplicities. A similar statement is true for the completely pointed module $L(\omega_{n-1} - 3/2\omega_n)$.

We now turn to the general case. Using decompositions of the tensor product for the highest-weight cases together with the fact that the central characters of all completely pointed simple C_n-modules are equal we are now able to prove the following:

Theorem 4.3 ([4] Theorem 3.10). *Assume that $\lambda = \lambda_1\omega_1 + \cdots + \lambda_n\omega_n$ is a dominant integral weight of C_n. Then for any n-tuple of noninteger complex numbers $\vec{a} = (a_1, \ldots, a_n)$ the tensor product modules $M(\vec{a}) \otimes L(\lambda)$ are completely reducible with*

$$M(a_1, \ldots, a_n) \otimes L(\lambda) = \oplus \sum_{\mu \in \mathcal{J}_\lambda} M\left(-\frac{1}{2}\omega_n + \mu\right), \qquad (4.1)$$

where for each $\mu \in \mathcal{J}_\lambda$, $M(-\frac{1}{2}\omega_n + \mu)$ denotes a simple, torsion-free C_n-module having central character $\mathcal{X}_{-1/2\omega_n+\mu}$ and degree equal to the degree of the highest-weight module $L(-\frac{1}{2}\omega_n + \mu)$.

We note that using virtually the same proof as in the previous theorem we can show that the tensor product module $M(-1, \ldots, -1, a_k, \ldots, a_n) \otimes L(\lambda)$ is also completely reducible where the summands are again simple C_n-modules (necessarily having finite degree) with central characters $\{\mathcal{X}_{-1/2\omega_n+\mu} \mid \mu \in \mathcal{J}_\lambda\}$.

The results on highest-weight modules lead us to the conjecture that every simple C_n-module having bounded multiplicities is equivalent to a submodule of the tensor product of a completely pointed module and a finite-dimensional module. In particular this includes the conjecture that every simple, torsion-free C_n-module of finite degree can be realized as a submodule of the tensor product of a completely pointed, torsion-free module and a finite-dimensional module. The latter conjecture has been verified by Chen [6] for the case of simple, torsion-free C_2-modules of degree 2.

4.2   sl(n)-modules of finite degree

As in the symplectic case we approach the study of simple, infinite-dimensional sl(n)-modules of finite degree by considering tensor product modules of completely pointed modules with finite-dimensional modules. As a guide for the general case we first observe that for each nonnegative integer k the simple $sl(n)$-module with highest-weight $k\omega_1$ is finite-dimensional and completely pointed. For notational purposes it is convenient to associate with each n-tuple $\vec{\pi} = (\pi_1, \dots, \pi_n)$ of integers a weight $\lambda_{\vec{\pi}} = \sum_{i=1}^{n-1}(\pi_i - \pi_{i+1})\omega_i$. For any $(n-1)$-tuple of integers $\vec{p} = (p_1, \dots, p_{n-1})$ we write $\vec{p} \prec \vec{\pi}$ iff $\pi_1 \geq p_1 \geq \pi_2 \geq \cdots \geq p_{n-1} \geq \pi_n$.

Now if we fix an n-tuple of integers $\vec{\pi}$ with $\pi_1 \geq \pi_2 \geq \cdots \geq \pi_n = 0$, then $\lambda_{\vec{\pi}}$ is a dominant integral weight. If we further assume that $k > n\pi_1$ and $\sum_{i=1}^{n} \pi_i = N$, then, by applying Pieri's formula, we have that the tensor product module $T(k; \vec{\pi}) = L\big((k-N)\omega_1\big) \otimes L(\lambda_{\vec{\pi}})$ decomposes as

$$T(k; \vec{\pi}) = \oplus \sum_{\vec{p} \prec \vec{\pi}} L(\lambda_{\hat{p}}),$$

where for each $\vec{p} \prec \vec{\pi}$ we define $\hat{p}$ to be the n-tuple $(k - \sum_{i=1}^{n-1} p_i, p_1, \dots p_{n-1})$. We observe that the central characters $\{\mathcal{X}_{\lambda_{\hat{p}}} = \mathcal{X}_{\vec{p}}^k \mid \vec{p} \prec \vec{\pi}\}$ are all distinct.

Now fix an n-tuple $\vec{a} = (a_1, \dots, a_n)$ of noninteger complex numbers. We recall that the simple, completely pointed module $N(\vec{a} - \vec{\pi})$ is a torsion-free module having central character equal to the central character of $L\big((\sum a_i - N)\omega_1\big)$ and has a realization that is closely related to the realization of the modules $L\big((k-N)\omega_1\big)$. Let $\mathcal{X}_{\vec{p}}^{\vec{a}}$ denote the central character $\mathcal{X}_{\vec{p}}^k$ where k is replaced by $\sum_{i=1}^{n} a_i$. Then in Ref. [5] we show that—

Theorem 4.4 ([5] Prop. 3.4). *For any n-tuple $\vec{a} = (a_1, \dots, a_n)$ of complex noninteger scalars such that the central characters $\{\mathcal{X}_{\vec{p}}^{\vec{a}} \mid \vec{p} \prec \pi\}$ are distinct, the tensor product module $T(\vec{a}; \vec{\pi})$ is completely reducible. In fact,*

$$T(\vec{a}; \vec{\pi}) = \oplus \sum_{\vec{p} \prec \vec{\pi}} T(\vec{a}; \vec{\pi})_{(\vec{p};\vec{a})},$$

where, for each $\vec{p} \prec \vec{\pi}$, $T(k; \vec{\pi})_{(\vec{p};\vec{a})}$ denotes the submodule of $T(k; \vec{\pi})$ determined by the central character $\mathcal{X}_{\vec{p}}^{\vec{a}}$.

We observe that if there exists $\vec{p}, \vec{q} \prec \vec{\pi}$ with $\vec{p} \neq \vec{q}$ and $\mathcal{X}_{\vec{p}}^{\vec{a}} = \mathcal{X}_{\vec{q}}^{\vec{a}}$ the tensor product module $T(\vec{a}; \vec{\pi})$ is not, in general, completely reducible. An example of this situation is given in Ref. [5, Section 4].

In Ref. [2] we proved that every simple, torsion-free $sl(n)$-module of finite degree occurs as a submodule of an appropriately chosen tensor product module $T(\vec{a}; \vec{\pi})$. This result leads us to conjecture that a similar result is true for $sl(n)$. In fact it appears plausible that every simple, infinite-dimensional $sl(n)$-module of finite degree can be realized as a distinguished submodule of the tensor product of a simple, infinite-dimensional, completely pointed module and an appropriately chosen finite-dimensional module.

Acknowledgments: The work of the authors was supported in part by the Natural Science and Engineering Research Council of Canada under grants A8471 and A7742 respectively

5   REFERENCES

1. G.M. Benkart, D.J. Britten, and F.W. Lemire, *Modules with bounded weight multiplicities for simple Lie algebras*, Math. Z. **225** (1997), no. 2, 333–353.

2. D.J. Britten, V. Futorny, and F.W. Lemire, *Simple A_2 modules with a finite-dimensional weight space*, Comm. Algebra **23** (1995), 467–510.

3. D.J. Britten, J. Hooper, and F.W. Lemire, *Simple C_n-modules with multiplicities 1 and applications*, Canad. J. Phys. **72** (1994), 326–335.

4. D.J. Britten and F. W. Lemire, *On modules of bounded multiplicities for the symplectic algebra*, Trans. Amer. Math. Soc. **351** (1999), no. 8, 3413–3431.

5. D.J. Britten and F.W. Lemire, *A Pieri-like formula for torsion-free modules*, preprint.

6. L. Chen, *Simple torsion-free C_2-modules having finite-dimensional weight spaces*, Master's thesis, University of Windsor, 1995.

7. S.L. Fernando, *Lie algebra modules with finite-dimensional weight spaces* I, Trans. Amer. Math. Soc. **322** (1990), 757–781.

4

Moving Frames and Coframes

Mark Fels and Peter J. Olver

Dedicated to Jiři Patera and Pavel Winternitz
on the occasion of their 60th birthdays

First introduced by Gaston Darboux, and then brought to maturity by Élie Cartan, [4, 5], the theory of moving frames ("repères mobiles") is widely acknowledged to be a powerful tool for studying the geometric properties of submanifolds under the action of a transformation group. While the basic ideas of moving frames for classical group actions are now ubiquitous in differential geometry, the theory and practice of the moving frame method for more general transformation group actions has remained relatively undeveloped. The famous critical assessment by Weyl in his review [27] of Cartan's seminal book [5] retains its perspicuity to this day:

> I did not quite understand how he [Cartan] does this in general, though in the examples he gives the procedure is clear. . . . Nevertheless, I must admit I found the book, like most of Cartan's papers, hard reading.

Implementations of the method of moving frames for certain groups having direct geometrical significance—including the Euclidean, affine, and projective groups—can be found in both Cartan's original treatise, [5], as well as many standard texts in differential geometry, e.g., [13, 24, 28]. The method continues to attract the attention of modern day researchers and has been successfully extended to a few additional examples, including, for instance, holomorphic curves in projective spaces and Grassmannians. The papers of Griffiths [12], Green [11], Chern [7], and the lecture notes of Jensen [14] are particularly noteworthy attempts to place Cartan's intuitive constructions on a firm theoretical and differential geometric foundation.

Recently [9, 10], the authors introduced two new methods that enable us to implement algorithmically both the practical and theoretical construction of moving frames for general transformation groups. The first algorithm, which we call the method of "moving coframes," not only reproduces all of the classical moving frame constructions, often in a simpler and more direct fashion, but can be readily applied to a wide variety of new situations, including infinite-dimensional pseudo-groups, intransitive group actions, restricted reparametrization problems, and joint group actions, to name a few. The second "regularized" method is applicable to general finite-dimensional transformation group actions, and provides a completely rigorous justification of the general theory. The regularized method bypasses many of the complications inherent in traditional approaches by completely avoiding the usual process of normalization during the general computation. Once a moving frame and coframe, along with the complete system of invariants, are constructed in the regularized framework, one can easily restrict these invariants to particular classes of submanifolds, producing (in nonsingular cases) the standard moving frame. Perhaps Griffiths is the closest in spirit to our guiding philosophy; we fully agree with his statement, [12, p. 777], that

> The effective use of frames ... goes far beyond the notion that 'frames are essentially the same as studying connections in the principal bundle of the tangent bundle.'

Indeed, by de-emphasizing the group theoretical basis for normalization, which, in the past, has hindered the theoretical foundations from covering all the situations to which the practical algorithm could be applied, our formulation of the framework goes beyond what Griffiths envisioned, and successfully realizes Cartan's original vision [4, 5]. Significant applications include a new and more general proof of the fundamental theorem on classification of differential invariants, a general classification theorem for syzygies of differential invariants, as well as new explicit commutation formulae for the associated invariant differential operators. We demonstrate a simple but striking generalization of a "replacement theorem" attributed to T.Y. Thomas [25]. Refined versions of known general theorems on the equivalence, symmetry, and rigidity of submanifolds are further direct consequences of our approach.

In this paper, we shall review the results of our investigations, referring the reader to [9, 10] for proofs, further details, as well as numerous examples and applications. We begin by presenting the basics of the regularized theory and its applications to differential invariants, illustrated by an example arising in classical invariant theory. The moving coframe method is then briefly discussed and applied to two examples—first, the classical case of equi-affine geometry of curves in the plane, and second, an infinite-dimensional pseudo-group originally studied by Lie.

Throughout this paper G will denote an r-dimensional Lie group acting smoothly on an m-dimensional manifold M. Let $G_S = \{g \in G \mid g \cdot S = S\}$ denote the *isotropy subgroup* of a subset $S \subset M$, and $G_S^* = \bigcap_{x \in S} G_x$ its *global isotropy subgroup*, which consists of those group elements that fix all points in S. The group G acts *freely* if $G_z = \{e\}$ for all $z \in M$, *effectively* if $G_M^* = \{e\}$, and *effectively on subsets* if $G_U^* = e$ for every open $U \subset M$. We further incorporate the adjective "locally" in these concepts by replacing $\{e\}$ by a general discrete subgroup of G. If G does not act effectively, one can, without any loss of generality, replace G by the effectively acting subgroup G/G_M^*, which acts in essentially the same manner as G does, cf. [21]. Clearly, if G acts effectively on subsets, then G acts effectively. Analytic continuation demonstrates that the converse is true in the analytic category, although not for general smooth actions. To avoid pathology, we shall always assume that G acts locally effectively on subsets. A group acts *semi-regularly* if all its orbits have the same dimension. The action is *regular* if, in addition, each point $x \in M$ has arbitrarily small neighborhoods whose intersection with each orbit is connected.

Let $J^n = J^n(M, p)$ denote the nth-order (extended) jet bundle consisting of equivalence classes of p-dimensional submanifolds $S \subset M$ under the equivalence relation of nth-order contact, [20, Chapter 3]. We let $j_n S \subset J^n$ denote the n-jet of the submanifold S. We introduce local coordinates $z = (x, u)$ on M, considering the first p components $x = (x^1, \ldots, x^p)$ as independent variables, and the latter $q = m - p$ components $u = (u^1, \ldots, u^q)$ as dependent variables. The induced local coordinates on J^n are denoted by $z^{(n)} = (x, u^{(n)})$, with components u_J^α, where $J = (j_1, \ldots, j_k)$, $1 \le j_\nu \le p$, representing the partial derivatives of the dependent variables with respect to the independent variables.

Since G preserves the order of contact between submanifolds, there is an induced action of G on the jet bundle J^n known as its nth *prolongation*, and denoted by $G^{(n)}$. We choose a basis $\{\mathbf{v}_1, \ldots, \mathbf{v}_r\}$ for the Lie algebra $\mathfrak{g}$ of the infinitesimal generators on M, and let $\{\mathrm{pr}^{(n)}\mathbf{v}_1, \ldots, \mathrm{pr}^{(n)}\mathbf{v}_r\}$ denote the corresponding basis for the Lie algebra $\mathfrak{g}^{(n)}$ of the prolonged group action $G^{(n)}$. The prolonged generators are obtained by truncating, at order n, the infinitely prolonged vector fields

$$\mathrm{pr}\,\mathbf{v}_\kappa = \sum_{i=1}^{p} \xi_\kappa^i(x, u) \frac{\partial}{\partial x^i} + \sum_{\alpha=1}^{q} \sum_{k = \#J \ge 0} \varphi_{J,\kappa}^\alpha(x, u^{(k)}) \frac{\partial}{\partial u_J^\alpha}, \qquad (1.1)$$

where

$$\varphi_{J,\kappa}^\alpha = D_J Q_\kappa^\alpha + \sum_{i=1}^{p} \xi_\kappa^i u_{J,i}^\alpha,$$

$$Q_\kappa^\alpha(x, u^{(1)}) = \varphi_\kappa^\alpha(x, u) - \sum_{i=1}^{p} \xi_\kappa^i(x, u) u_i^\alpha.$$

Here $Q_\kappa = (Q_\kappa^1, \ldots, Q_\kappa^q)$ is the usual characteristic of $\mathbf{v}_\kappa$, and $D_J = D_{j_1} \ldots D_{j_k}$ denotes total differentiation of order $k = \#J$.

Let $s_n = \max\{\dim \mathfrak{g}^{(n)}|_{z^{(n)}}\}$ denote the maximal orbit dimension of the prolonged action $G^{(n)}$ on J^n. The *stable orbit dimension* is $s = \max s_n$. The *stabilization order* of G is the minimal n such that $s_n = s$. The *regular subset* $\mathcal{V}^n \subset J^n$ is the open subset consisting of all prolonged group orbits of dimension equal to the stable orbit dimension, while the *singular subset* is $\mathcal{S}^n = J^n \setminus \mathcal{V}^n$. Note that, by this definition, $\mathcal{V}^n = \{\emptyset\}$ and $\mathcal{S}^n = J^n$ if n is less than the stabilization order of $G^{(n)}$. Ovsiannikov's stabilization theorem, [21, 23], completely characterizes the stable orbit dimension. A correct version can be stated as follows; see [22] for details.

Theorem 1.1. *A Lie group G acts locally effectively on subsets of M if and only if its stable orbit dimension equals its dimension, $s = r = \dim G$, which means that G acts locally freely on the regular subset $\mathcal{V}^n \subset J^n$.*

A submanifold $S \subset M$ is called *totally singular* if all its jets never intersect the regular subset. Such submanifolds can be geometrically characterized as follows.

Theorem 1.2. *A submanifold $S \subset M$ is totally singular, meaning that $\mathrm{j}_n S \subset \mathcal{S}^n$ for all $n = 0, 1, \ldots$, if and only if its isotropy subgroup G_S does not act locally freely on S.*

Our approach to the theory of moving frames is based on the following simple but remarkably powerful device. In general, any complicated transformation group action can be "regularized" by lifting it to a suitable bundle sitting over the original manifold. Let $\mathcal{B}^n = G \times J^n$ denote the trivial (left) principal G–bundle over the jet space. The nth-order *regularization* of the prolonged action $G^{(n)}$ is the action of G on $\mathcal{B}^n$ given by

$$g \cdot (h, z^{(n)}) = (g \cdot h, g^{(n)} \cdot z^{(n)}). \tag{1.2}$$

The key, elementary result is that regularizing any group action immediately eliminates *all* singularities and irregularities, e.g., lower-dimensional orbits, nonembedded orbits, etc.

Theorem 1.3. *For any $n \geq 0$, the regularized action (1.2) defines a regular, free action of G on the bundle $\mathcal{B}^n = G \times J^n$.*

Recall that a *differential invariant* is a function $I \colon J^n \to \mathbb{R}$ that is invariant under the action of $G^{(n)}$. Similarly, a *lifted differential invariant* is defined as a function $L \colon \mathcal{B}^n \to \mathbb{R}$ that is invariant under the regularized action (1.2). Remarkably, all the lifted differential invariants are trivial to construct; they are the components of the order n evaluation map $w^{(n)} \colon \mathcal{B}^n \to J^n$ that is given by $w^{(n)}(g, z^{(n)}) = (g^{(n)})^{-1} \cdot z^{(n)}$.

Proposition 1.1. *Every lifted differential invariant can be locally written as a function of the fundamental lifted differential invariants $w^{(n)}(g, z^{(n)})$.*

In particular, an ordinary differential invariant $I \colon J^n \to \mathbb{R}$ also defines a lifted differential invariant $L = I \circ \pi_n \colon \mathcal{B}^n \to \mathbb{R}$. Conversely, any lifted invariant $L(g, x, u^{(n)})$ that does not depend on the g coordinates automatically defines an ordinary differential invariant.

Theorem 1.4. *Let $I(z^{(n)})$ be an ordinary differential invariant. Then we can write $I(z^{(n)}) = I(w^{(n)})$ as the same function of the lifted differential invariants.*

In Riemannian geometry, Theorem 1.4 reduces to the striking Thomas replacement theorem, [25, p. 109]. See [2] for recent applications of Thomas' result.

The introduction of local coordinates $z = (x, u)$ on M also partitions the fundamental zeroth-order lifted invariants $w = (w^1, \ldots, w^m) = g^{-1} \cdot z$ into two components, $w = (y, v)$, where $y = (y^1, \ldots, y^p)$ will be considered as "lifted independent variables," and $v = (v^1, \ldots, v^q)$ as "lifted dependent variables." The lifted differential invariants can be found via a process of invariant differentiation, which we now describe.

The identification of independent variables on M induces a splitting of the differential forms on J^n into horizontal and contact components [cf. 1, 21]. Given a differential function $F(x, u^{(n)})$, let

$$d_H F = \sum_{i=1}^{p} (D_i F) \, dx^i, \tag{1.3}$$

denote the horizontal component of its exterior derivative, known as the *total differential* of F. Formula (1.3) extends without change to lifted functions $F(g, x, u^{(n)})$. Let

$$\eta^i = d_H y^i = \sum_{j=1}^{p} (D_j y^i) \, dx^j, \quad i = 1, \ldots, p,$$

denote the horizontal differentials of the lifted independent variables. We then rewrite (1.3) in invariant form

$$d_H F = \sum_{j=1}^{p} (\mathcal{E}_j F) \eta^j \text{ where } \mathcal{E}_j F = \frac{\mathbf{D}(y^1, \ldots, y^{j-1}, F, y^{j+1}, \ldots y^p)}{\mathbf{D}(y^1, \ldots, y^p)}.$$

We can identify the *lifted invariant differential operator* $\mathcal{E}_j = D_{y^j}$ with total differentiation with respect to the lifted invariant y^j. In column vector notation, $\mathcal{E} = (\mathbf{D}y)^{-T} \cdot \mathbf{D}$, where $\mathbf{D}y$ is the total Jacobian matrix of y and $\mathbf{D} = (D_1, \ldots, D_p)^T$ is the "total gradient operator." A very important point is that, unlike the usual invariant differential operators, the lifted invariant differential operators *always* mutually commute, $[\mathcal{E}_j, \mathcal{E}_k] = 0$.

Proposition 1.2. *The components* $w^{(n)} = (y, v^{(n)})$ *of the fundamental lifted invariants are found by successively applying the invariant differential operators* $\mathcal{E}_j = D_{y^j}$ *associated with the first p lifted invariants* $y = (y_1, \ldots, y^p)$ *to the remaining zeroth-order invariants* $v = (v^1, \ldots, v^q)$, *so that* $v_J^\alpha = \mathcal{E}_J v^\alpha$, *where* $\mathcal{E}_J = \mathcal{E}_{j_1} \cdots \cdots \mathcal{E}_{j_k}$.

The primary use of a moving frame is that it enables one to pass from lifted invariant objects, which are trivial, to their ordinary invariant counterparts back on the original manifold and its jet spaces. This allows us systematically to analyze the invariants via the particularities of the moving frame. We first discuss the theory of completely determined moving frames, meaning ones that do not depend on any group parameters.

Definition 1.1. An nth-order (left) *moving frame* is a map $\rho^{(n)}\colon J^n \to G$ that is (locally) G-equivariant with respect to the prolonged action $G^{(n)}$ on J^n, and the left action $h \mapsto g \cdot h$ of G on itself.

We can identify a moving frame with an equivariant section $\sigma^{(n)}\colon J^n \to \mathcal{B}^n = G \times J^n$ given by $\sigma^{(n)}(x, u^{(n)}) = \left(\rho^{(n)}(x, u^{(n)}), x, u^{(n)}\right)$. Note that any nth-order moving frame also defines a moving frame on all higher-order jet bundles by composition with the standard projections $\pi_n^k\colon J^k \to J^n$, $k > n$.

Theorem 1.5. *If G acts effectively on subsets, then an nth-order moving frame exists in a neighborhood of a point $z^{(n)} \in J^n$ if and only if $z^{(n)} \in \mathcal{V}^n$ is a regular jet.*

In particular, the minimal order at which any moving frame exists is the stabilization order of the group. In practical implementations, Cartan's normalization procedure for constructing moving frames amounts to choosing a (local) cross section $\mathcal{K}^n \subset \mathcal{V}^n$ to the regular prolonged group orbits. Let $\mathcal{O}^n$ denote the $G^{(n)}$ orbit passing through the regular jet $z^{(n)} \in \mathcal{O}^n \subset \mathcal{V}^n$, and suppose that $\mathcal{O}^n$ intersects the cross section at the unique point $k^{(n)} \in \mathcal{O}^n \cap \mathcal{K}^n$; we can view $k^{(n)}$ as the "canonical form" of the jet $z^{(n)}$. Finally, let $g = \rho^{(n)}(z^{(n)})$ denote the group element that maps $k^{(n)}$ to $z^{(n)} = g^{(n)} \cdot k^{(n)}$. The resulting map $\rho^{(n)}\colon J^n \to G$ from the jet space to the group is the moving frame defined by the chosen cross section.

Assuming $G^{(n)}$ acts locally freely, the simplest local cross sections are obtained by setting $r = \dim G$ of the jet coordinates $z^{(n)} = (x, u^{(n)})$ to be constant. Let $z_1, \ldots, z_r$ denote the chosen coordinates, so that each z_ν is either one of the x^i's or one of the u_J^α's. Let $w_1, \ldots, w_r$ be the corresponding lifted invariants, so that w_ν is the corresponding y^i or v_J^α. The normalization constants $c_1, \ldots, c_r$ are chosen so that the *normalization equations*

$$w_1(g, x, u^{(n)}) = c_1, \quad \cdots \quad w_r(g, x, u^{(n)}) = c_r, \tag{1.4}$$

can be (locally) uniquely solved for $g = \rho^{(n)}(x, u^{(n)})$ in terms of the jet coordinates; the resulting map defines the moving frame associated with the chosen cross section.

Definition 1.2. The *fundamental nth-order normalized differential invariants* associated with a moving frame $\rho^{(n)}$ of order n (or less) are given by

$$I^{(n)}(z^{(n)}) = w^{(n)} \circ \sigma^{(n)}(z^{(n)}) = \rho^{(n)}(z^{(n)})^{-1} \cdot z^{(n)}.$$

Note that $I^{(n)}(z^{(n)}) = k^{(n)} \in \mathcal{K}^n$ can be identified with the canonical form of the jet $z^{(n)}$. In terms of the invariant local coordinates $w^{(n)} =$

$(y, v^{(n)})$ on $\mathcal{B}^n$, the fundamental normalized differential invariants are

$$J^i(x, u^{(n)}) = y^i\big(\rho^{(n)}(x, u^{(n)}), x, u\big),$$
$$i = 1, \ldots, p,$$
$$I_K^\alpha(x, u^{(k)}) = v_K^\alpha\big(\rho^{(n)}(x, u^{(n)}), x, u^{(k)}\big),$$
$$\alpha = 1, \ldots, q, \ \ k = \#K \geq 0. \tag{1.5}$$

In particular, the normalized differential invariants corresponding to the components w_i being normalized via (1.4) will be constant. We shall call these the *phantom differential invariants*. The other components of $w^{(n)}$ will define a complete system of functionally independent differential invariants defined on the domain of definition of the moving frame map.

Theorem 1.6. *Let n be greater than or equal to the order of the moving frame. Every nth-order differential invariant can be locally written as a function of the normalized nth-order differential invariants $I^{(n)}$. The function is unique provided it does not depend on the phantom invariants.*

Given an arbitrary differential function $F: J^n \to \mathbb{R}$, then $L = F \circ w^{(n)}: \mathcal{B}^n \to \mathbb{R}$ defines a lifted differential invariant, and hence $J = L \circ \sigma^{(n)} = F \circ I^n$ defines a differential invariant, called the *invariantization* of F with respect to the given moving frame. Thus, a moving frame provides a natural way to construct a differential invariant from any differential function! Theorem 1.4 just says that if F itself is a differential invariant, then $F \circ w^{(n)}$ is independent of the group parameters, and hence $J = F$. In other words, invariantization defines a projection, depending on the moving frame, from the space of differential functions to the space of differential invariants.

The higher-order differential invariants can also be obtained by invariant differentiation. The *normalized contact-invariant coframe* is the pullback of the lifted contact-invariant coframe: $\boldsymbol{\omega} = (\sigma^{(n)})^* \boldsymbol{\eta} = (\mathbf{D}y \circ \sigma^{(n)})\, d\boldsymbol{x}$. The associated invariant differential operators $\mathcal{D} = (\mathbf{D}y \circ \sigma^{(n)})^{-T} \cdot \mathbf{D}$ are obtained by normalizing the lifted invariant differential operators $\mathcal{E}$. The invariant differential operators $\mathcal{D}_j$ do not necessarily commute; explicit commutation formulae are presented below.

The invariant differential operators will map differential invariants to higher-order differential invariants. However, unlike their lifted counterparts, they do not directly produce the normalized differential invariants; in other words, $\mathcal{D}_K I^\alpha$ is *not*, in general, equal to I_K^α. The moving frame method will effectively resolve the computational difficulties in the usual (unlifted) theory. The fundamental *recurrence formulae* for the differential invariants (1.5) are

$$\mathcal{D}_j J^i = \delta_j^i + M_j^i, \quad \mathcal{D}_j I_K^\alpha = I_{K,j}^\alpha + M_{K,j}^\alpha. \tag{1.6}$$

The "correction terms" M_j^i, $M_{K,j}^\alpha$ can be effectively computed using the following algorithm. For any n greater than or equal to the order of the moving frame, let $q^{(n)} = p + q\binom{p+n}{n} = \dim J^n$. Let $V = V^{(n)}$ denote the $r \times q^{(n)}$ matrix whose entries are the coefficients ξ^i, φ_J^α of the nth-order prolonged infinitesimal generators (1.1). Let $W = V \circ I^{(n)}$ be its invariantized version, obtained by replacing the jet coordinates $z^{(n)}$ by the associated differential invariants $I^{(n)}$. We perform a Gauss–Jordan row reduction on the matrix W so as to reduce the $r \times r$ minor whose columns correspond to the chosen normalization variables $z_1, \ldots, z_r$ to be the identity matrix; let P be the resulting $r \times q^{(n)}$ matrix of invariants. Let $S = (S_i^\kappa)$ denote the $p \times r$ matrix whose entries are the total derivatives $S_i^\kappa = D_i z_\kappa$ of the normalization coordinates. Let $T = S \circ I^n$ be its invariantization. Then the correction terms in Eq. (1.6) are the entries of the $p \times q^{(n)}$ matrix product $M = -T \cdot P$.

A *syzygy* is a functional dependency $H(\ldots \mathcal{D}_J I_\nu \ldots) \equiv 0$ among the fundamental differentiated invariants. The normalization procedure not only gives us a generating system of fundamental differential invariants, but also classifies all syzygies among the normalized differential invariants.

Theorem 1.7. *A generating system of differential invariants consists of*

(a) *all nonphantom differential invariants J^i and I^α coming from the unnormalized zeroth-order lifted invariants y^i, v^α; and*

(b) *all nonphantom differential invariants of the form $I_{J,i}^\alpha$ where I_J^α is a phantom differential invariant.*

In other words, every other differential invariant can, locally, be written as a function of the generating invariants and their invariant derivatives, $\mathcal{D}_K J^i$, $\mathcal{D}_K I_{J,i}^\alpha$. All syzygies among the differentiated invariants are differential consequences of the following three fundamental types:

(i) $\mathcal{D}_j J^i = \delta_j^i + M_j^i$, *when J^i is nonphantom;*

(ii) $\mathcal{D}_J I_K^\alpha = c + M_{K,J}^\alpha$, *when I_K^α is a generating differential invariant, while $I_{J,K}^\alpha = c$ is a phantom differential invariant; and*

(iii) $\mathcal{D}_J I_{LK}^\alpha - \mathcal{D}_K I_{LJ}^\alpha = M_{LK,J}^\alpha - M_{LJ,K}^\alpha$, *where I_{LK}^α and I_{LJ}^α are generating differential invariants the multi-indices $K \cap J = \emptyset$ are disjoint and nonzero, while L is an arbitrary multi-index.*

A similar algorithm produces the commutation formulae

$$[\mathcal{D}_i, \mathcal{D}_j] = \sum_{i=1}^{p} A_{ij}^k \mathcal{D}_k, \quad i, j = 1, \ldots, p, \tag{1.7}$$

among the normalized invariant differential operators. Let X^k denote the $r \times p$ matrix whose entries are the total derivatives $X_{\kappa i}^k = D_i \xi_\kappa^k$ of the kth independent variable coefficients. Let $Y^k = X^k \circ I^{(1)}$ denote its invariantization. Let B^k the result of performing the same Gauss–Jordan reduction on Y^k as was done on W; in other words, if $P = E \cdot W$, then $B^k = E \cdot Y^k$. Then the coefficient matrix $A^k = (A^k_{ij})$ in Eq. (1.7) is the skew-symmetric part of the matrix product $C^k = T \cdot B^k$, i.e., $A^k = C^k - (C^k)^T$.

Example 1.1. Let $M = \mathbb{R}^3$, with coordinates x^1, x^2, u. Let $G = \mathrm{GL}(2)$, and consider the action $(x^1, x^2, u) \mapsto (\alpha x^1 + \beta x^2, \gamma x^1 + \delta x^2, \lambda u)$, where $\lambda = \alpha\delta - \beta\gamma$. This action plays a key role in the classical invariant theory of binary forms, when u is a homogeneous polynomial [21]. The order zero invariants are obtained by inverting the group transformations:

$$y^1 = \lambda^{-1}(\delta x^1 - \beta x^2), \quad y^2 = \lambda^{-1}(-\gamma x^1 + \alpha x^2), \quad v = \lambda^{-1}u.$$

The lifted contact-invariant coframe and associated invariant differential operators are

$$\eta^1 = d_H y^1 = \lambda^{-1}(\delta\, dx^1 - \beta\, dx^2), \quad \mathcal{E}_1 = \alpha D_1 + \gamma D_2,$$
$$\eta^2 = d_H y^2 = \lambda^{-1}(-\gamma\, dx^1 + \alpha\, dx^2), \quad \mathcal{E}_2 = \beta D_1 + \delta D_2.$$

The higher-order lifted differential invariants are then $v_{jk} = (\mathcal{E}_1)^j (\mathcal{E}_2)^k v$; in particular,

$$v_1 = \frac{\alpha u_1 + \gamma u_2}{\lambda}, \quad v_2 = \frac{\beta u_1 + \delta u_2}{\lambda}, \quad v_{11} = \frac{\alpha^2 u_{11} + 2\alpha\gamma u_{12} + \gamma^2 u_{22}}{\lambda},$$

$$v_{12} = \frac{\alpha\beta\delta u_{11} + (\alpha\delta + \beta\gamma)u_{12} + \gamma\delta u_{22}}{\lambda}, \quad v_{22} = \frac{\beta^2 u_{11} + 2\beta\delta u_{12} + \delta^2 u_{22}}{\lambda}.$$

Let us choose the cross section $\mathcal{K}^1 = \{x^1 = 1, x^2 = 0, u_1 = 1, u_2 = 0\}$. The normalization equations

$$y^1 = 1, \quad y^2 = 0, \quad v_1 = 1, \quad v_2 = 0,$$

are then solved for the group parameters, leading to a first-order moving frame

$$\begin{pmatrix} \alpha & \beta \\ \gamma & \delta \end{pmatrix} = \begin{pmatrix} x^1 & -u_2 \\ x^2 & u_1 \end{pmatrix},$$

which is well-defined on surfaces $u = u(x, y)$ provided $x^1 u_1 + x^2 u_2 \neq 0$. The resulting normalized differential invariants are

$$J^1 = 1, \quad J^2 = 0, \quad I = \frac{u}{x^1 u_1 + x^2 u_2}, \quad I_1 = 1, \quad I_2 = 0,$$

$$I_{11} = \frac{(x^1)^2 u_{11} + 2x^1 x^2 u_{12} + (x^2)^2 u_{22}}{x^1 u_1 + x^2 u_2},$$

$$I_{12} = \frac{-x^1 u_2 u_{11} + (x^1 u_1 - x^2 u_2)u_{12} + x^2 u_1 u_{22}}{x^1 u_1 + x^2 u_2},$$

$$I_{22} = \frac{(u_2)^2 u_{11} - 2u_1 u_2 u_{12} + (u_1)^2 u_{22}}{x^1 u_1 + x^2 u_2}.$$

The Replacement Theorem 1.4 shows that each of these can be rewritten as the *identical* function of the lifted invariants; e.g., $I = (x^1 u_1 + x^2 u_2)^{-1} u = (y^1 v_1 + y^2 v_2)^{-1} v$. According to Theorem 1.7, we can take $I, I_{11}, I_{12}, I_{22}$ as our generating system, meaning that all higher-order differential invariants can be obtained by successively applying the invariant differential operators to them. The normalized coframe is

$$\omega^1 = \frac{u_1\, dx^1 + u_2\, dx^2}{x^1 u_1 + x^2 u_2} = \frac{d_H u}{x^1 u_1 + x^2 u_2}, \qquad \omega^2 = \frac{-x^2\, dx^1 + x^1\, dx^2}{x^1 u_1 + x^2 u_2}.$$

The associated invariant differential operators are well known in classical invariant theory: $\mathcal{D}_1 = x^1 D_1 + x^2 D_2$ is the scaling process and $\mathcal{D}_2 = -u_2 D_1 + u_1 D_2$ is the Jacobian process.

Let us now illustrate the algorithm for determining recurrence formulae, syzygies, and commutation formulae. The prolonged infinitesimal generator coefficient matrix and its invariantized counterpart are, up to second-order,

$$V = \begin{pmatrix} x^1 & 0 & u & 0 & u_2 & u_{11} & 0 & u_{22} \\ x^2 & 0 & 0 & 0 & -u_1 & 0 & -u_{11} & -2u_{12} \\ 0 & x^1 & 0 & -u_2 & 0 & -2u_{12} & -u_{22} & 0 \\ 0 & x^2 & u & u_1 & 0 & u_{11} & 0 & -u_{22} \end{pmatrix},$$

$$W = \begin{pmatrix} 1 & 0 & I & 0 & 0 & -I_{11} & 0 & I_{22} \\ 0 & 0 & 0 & 0 & -1 & 0 & -I_{11} & -2I_{12} \\ 0 & 1 & 0 & 0 & 0 & -2I_{12} & -I_{22} & 0 \\ 0 & 0 & I & 1 & 0 & I_{11} & 0 & -I_{22} \end{pmatrix}.$$

Since we are normalizing x^1, x^2, u_1, u_2, we also need the matrices

$$S = \begin{pmatrix} 1 & 0 & u_{11} & u_{12} \\ 0 & 1 & u_{12} & u_{22} \end{pmatrix}, \qquad T = \begin{pmatrix} 1 & 0 & I_{11} & I_{12} \\ 0 & 1 & I_{12} & I_{22} \end{pmatrix}.$$

We use Gauss–Jordan reduction on the invariantized coefficient matrix W making the chosen normalization columns—in the present case columns 1, 2, 4 and 5—into an identity matrix, and then premultiply the resulting matrix by T. The entries of the resulting matrix product

$$\begin{pmatrix} 1 & 0 & I_{11} & I_{12} \\ 0 & 1 & I_{12} & I_{22} \end{pmatrix} \begin{pmatrix} 1 & 0 & I & 0 & 0 & -I_{11} & 0 & I_{22} \\ 0 & 1 & 0 & 0 & 0 & -2I_{12} & -I_{22} & 0 \\ 0 & 0 & I & 1 & 0 & I_{11} & 0 & -I_{22} \\ 0 & 0 & 0 & 0 & 1 & 0 & I_{11} & 2I_{12} \end{pmatrix}$$

$$= \begin{pmatrix} 1 & 0 & I(1+I_{11}) & I_{11} & I_{12} & (I_{11}-1)I_{11} & I_{11}I_{12} & 2I_{12}^2-(I_{11}-1)I_{22} \\ 0 & 1 & II_{12} & I_{12} & I_{22} & (I_{11}-2)I_{12} & (I_{11}-1)I_{22} & I_{12}I_{22} \end{pmatrix},$$

are minus the required correction terms:

$$\mathcal{D}_1 J^1 = \delta_1^1 - 1 = 0, \qquad\qquad \mathcal{D}_2 J^1 = \delta_2^1 - 0 = 0,$$
$$\mathcal{D}_1 J^2 = \delta_1^2 - 0 = 0, \qquad\qquad \mathcal{D}_2 J^2 = \delta_2^2 - 1 = 0,$$
$$\mathcal{D}_1 I = I_1 - I(1 + I_{11}) = 1 - I(1 + I_{11}), \quad \mathcal{D}_2 I = I_2 - II_{12} = -II_{12},$$
$$\mathcal{D}_1 I_1 = I_{11} - I_{11} = 0, \qquad\qquad \mathcal{D}_2 I_1 = I_{12} - I_{12} = 0,$$
$$\mathcal{D}_1 I_2 = I_{12} - I_{12} = 0, \qquad\qquad \mathcal{D}_2 I_2 = I_{22} - I_{22} = 0,$$
$$\mathcal{D}_1 I_{11} = I_{111} + (1 - I_{11})I_{11}, \qquad \mathcal{D}_2 I_{11} = I_{112} + (2 - I_{11})I_{12},$$
$$\mathcal{D}_1 I_{12} = I_{112} - I_{11}I_{12}, \qquad\qquad \mathcal{D}_2 I_{12} = I_{122} + (1 - I_{11})I_{22},$$
$$\mathcal{D}_1 I_{22} = I_{122} + (I_{11} - 1)I_{22} - 2I_{12}^2, \qquad \mathcal{D}_2 I_{22} = I_{222} - I_{12}I_{22}.$$

The formulae for $\mathcal{D}_1 I$ and $\mathcal{D}_2 I$ provide the syzygies of the second type, and show that we can use I to generate I_{11} and I_{12}. (There are no syzygies of the first type since we normalized all the lifted independent variables.) There are three fundamental syzygies of the third type:

$$\mathcal{D}_1 I_{12} - \mathcal{D}_2 I_{11} = -2I_{12},$$
$$\mathcal{D}_1 I_{22} - \mathcal{D}_2 I_{12} = 2(I_{11} - 1)I_{22} - 2I_{12}^2,$$
$$(\mathcal{D}_1)^2 I_{22} - (\mathcal{D}_2)^2 I_{11} = 2I_{22}\mathcal{D}_1 I_{11} + (5I_{12} - 2)\mathcal{D}_1 I_{12} + (3I_{11} - 5)\mathcal{D}_1 I_{22}$$
$$- (2I_{11} - 5)(I_{11} - 1)I_{12} + 4(I_{11} - 1)I_{12}^2.$$

Finally, the commutation formulae can be determined directly:

$$[\mathcal{D}_1, \mathcal{D}_2] = -I_{12}\mathcal{D}_1 + (I_{11} - 1)\mathcal{D}_2. \qquad (1.8)$$

The alternative method is to first construct the matrices

$$X^1 = Y^1 = \begin{pmatrix} 1 & 0 \\ 0 & 1 \\ 0 & 0 \\ 0 & 0 \end{pmatrix}, \quad B^1 = \begin{pmatrix} 1 & 0 \\ 0 & 0 \\ 0 & 0 \\ 0 & -1 \end{pmatrix},$$

$$X^2 = Y^2 = \begin{pmatrix} 0 & 0 \\ 0 & 0 \\ 1 & 0 \\ 0 & 1 \end{pmatrix}, \quad B^2 = \begin{pmatrix} 0 & 0 \\ 1 & 0 \\ 0 & 1 \\ 0 & 0 \end{pmatrix},$$

where X^i are obtained by differentiating the coefficients ξ^i of ∂_{x^i} in the infinitesimal generators; Y^i are their invariantizations, which are the same because the X^i happen to be constant, and B^i is obtained from Y^i by applying the same Gauss–Jordan row operations as were done to W above. Multiplying B^i by T to obtain C^i, and then skew symmetrizing to obtain

A^i yields

$$C^1 = T \cdot B^1 = \begin{pmatrix} 1 - I_{12} & \\ 0 & I_{22} \end{pmatrix},$$

$$A^1 = C^1 - (C^1)^T = \begin{pmatrix} 0 & -I_{12} \\ I_{12} & 0 \end{pmatrix},$$

$$C^2 = T \cdot B^2 = \begin{pmatrix} 0 & I_{11} \\ 1 & I_{12} \end{pmatrix},$$

$$A^2 = C^2 - (C^2)^T = \begin{pmatrix} 0 & I_{11} - 1 \\ 1 - I_{11} & 0 \end{pmatrix}.$$

The $(1,2)$ entry of A^i provides the coefficient of $\mathcal{D}_i$ in (1.8).

In applications to equivalence problems and geometry, we restrict the moving frame and associated invariants to a submanifold of the appropriate dimension.

Definition 1.3. A p-dimensional submanifold parametrized by $\iota\colon X \to S \subset M$ is called *regular* with respect to a moving frame $\rho^{(n)}\colon \mathrm{J}^n \to G$ if its n-jet $\mathrm{j}^n S$ lies in the domain of definition of $\rho^{(n)}$. In this case, the restricted *moving frame* on the submanifold is defined as the composition $\lambda^{(n)} = \rho^{(n)} \circ \mathrm{j}_n\iota\colon X \to G$.

Theorem 1.8. *A submanifold $S \subset M$ admits an nth-order moving frame if and only if S is regular of order n; i.e., $\mathrm{j}_n S \subset \mathcal{V}^n$. Thus, in the analytic category, a submanifold S admits a moving frame (of some sufficiently higher-order) if and only if its isotropy subgroup G_S acts freely on S itself.*

Let S be a regular submanifold for a moving frame $\rho^{(n)}$. For any $k \geq n$, the kth-order *differential invariant classifying manifold* $\mathcal{C}^{(k)}(S)$ associated with a submanifold $\iota\colon X \to M$ is the manifold parametrized by the normalized differential invariants of order k, namely, $\mathrm{J}^{(k)} = I^{(k)} \circ \mathrm{j}_k\iota$. For simplicity, let us assume that $\mathcal{C}^{(k)}(S)$ is an embedded submanifold of its classifying space $Z^{(k)} \simeq \mathrm{J}^k$ of dimension t_k for $k \geq n$. Note that t_k equals the number of functionally independent invariants obtained by restricting the normalized kth-order differential invariants to S. In the fully regular case, then, we have

$$t_n < t_{n+1} < t_{n+2} < \cdots < t_s = t_{s+1} = \cdots = t \leq p,$$

where t is the *differential invariant rank* and s is the *differential invariant order* of S. We can now state the fundamental equivalence and symmetry theorems.

Theorem 1.9. *Let $S, \bar{S} \subset M$ be regular p-dimensional submanifolds with respect to a moving frame map $\rho^{(n)}$. Then S and $\bar{S}$ are (locally) congruent, $\bar{S} = g \cdot S$, if and only if they have the same differential invariant order* *s and their classifying manifolds of order* $s+1$ *are identical:* $C^{(s+1)}(\bar{S}) = C^{(s+1)}(S)$.

Theorem 1.10. *Let* $S \subset M$ *be a regular p-dimensional submanifold of differential invariant rank* t *with respect to a moving frame* $\rho^{(n)}$. *Then its isotropy group* G_S *is an* $(r - t)$-*dimensional subgroup of* G *acting locally freely on* S.

A submanifold S is *order k congruent* to a submanifold $\bar{S}$ at $z \in S$ if there is a group transformation $g \in G$ such that S and $g \cdot \bar{S}$ have order k contact at z. Note that the group transformation $g = g(z)$ may vary from point to point. The *rigidity order* of S is the minimal k for which order k congruence implies congruence, so $\bar{S} = g \cdot S$ for fixed $g \in G$. It turns out that this also means that the only congruent submanifold $\bar{S} = g \cdot S$ that has kth-order contact with S at a point is S itself.

Theorem 1.11. *If* $S \subset M$ *is a regular submanifold of differential invariant order* s *with respect to a moving frame, then* S *has rigidity order at most* $s + 1$.

We now describe the method of moving coframes, which provides an alternative approach, based on invariant differential forms, that also extends to pseudo-group actions. For simplicity, let us assume that G acts transitively on M. Choose a base point $z_0 \in M$. A smooth map $\rho^{(0)} : M \to G$ is called a *compatible lift* with base point z_0 if it satisfies

$$\rho^{(0)}(z) \cdot z_0 = z. \tag{1.9}$$

We will call the general compatible lift $\rho^{(0)}(z, h)$ the *moving frame of order zero*. It is computed by solving the system of m equations (1.9) for m of the group parameters in terms of the coordinates z on M and the remaining $r - m = \dim G - \dim M$ group parameters, which we denote by h. Unlike the preceding moving frames, unless G acts locally freely on M, the order zero moving frame *will* depend on some of the group parameters. We can use $\rho^{(0)}$ to pullback the left-invariant Maurer–Cartan forms on G, leading to the *moving coframe* of order zero. We can determine lifted invariants by analyzing the linear dependencies among the horizontal components of the moving coframe forms. Group-dependent invariants can be normalized to convenient constant values by solving for some of the unnormalized parameters. We successively eliminate parameters by substituting the normalization formulae into the moving coframe and recomputing dependencies. After the parameters have all been normalized, the differential invariants will appear through any remaining dependencies among the final moving coframe elements. Let us illustrate the basic method by a classical example; see [9] for more details and applications.

Example 1.2. The equi-affine geometry of curves in the plane is governed by the special affine group $\mathrm{SA}(2) = \mathrm{SL}(2) \ltimes \mathbb{R}^2$, acting on $M = \mathbb{R}^2$ according to

$$g: \mathbf{x} \longmapsto A\mathbf{x} + \mathbf{a}, \quad \mathbf{x} \in M, \quad A \in \mathrm{SL}(2), \quad \mathbf{a} \in \mathbb{R}^2.$$

We adopt a vector notation for the matrix $A = (\boldsymbol{\alpha}\boldsymbol{\beta}) \in \mathrm{SL}(2)$, where $\det A = \boldsymbol{\alpha} \wedge \boldsymbol{\beta} = 1$. The Maurer–Cartan forms on SA(2) are

$$\mu_1 = \boldsymbol{\alpha} \wedge d\boldsymbol{\alpha}, \quad \mu_2 = \boldsymbol{\beta} \wedge d\boldsymbol{\alpha} = \boldsymbol{\alpha} \wedge d\boldsymbol{\beta}, \quad \mu_3 = \boldsymbol{\beta} \wedge d\boldsymbol{\beta},$$
$$\nu_1 = \boldsymbol{\alpha} \wedge d\mathbf{a}, \quad \nu_2 = \boldsymbol{\beta} \wedge d\mathbf{a}.$$

Choose the base point to be $\mathbf{x}_0 = 0$. Solving the compatible lift equations $\mathbf{x} = g \cdot \mathbf{x}_0 = \mathbf{a}$ yields the zeroth-order moving frame $\mathbf{a} = \mathbf{x}$. Substituting into the Maurer–Cartan forms, we find that, for a parametrized curve $\mathbf{x}(t)$, the forms ν_1, ν_2 restrict to the following two horizontal forms:

$$\nu_1 = (\boldsymbol{\alpha} \wedge \mathbf{x}_t) \, dt, \quad \nu_2 = (\boldsymbol{\beta} \wedge \mathbf{x}_t) \, dt.$$

Their ratio produces the lifted invariant $(\boldsymbol{\alpha} \wedge \mathbf{x}_t)/(\boldsymbol{\beta} \wedge \mathbf{x}_t)$, which is normalized to 0 by setting $\boldsymbol{\alpha} = \lambda \mathbf{x}_t$ for some scalar parameter λ. This implies that $\mu_1 = \lambda^2 (\mathbf{x}_t \wedge \mathbf{x}_{tt}) \, dt$. Assuming $\mathbf{x}_t \wedge \mathbf{x}_{tt} \neq 0$, the latter form can be normalized to equal $-\nu_2$ by setting

$$-\boldsymbol{\beta} \wedge \mathbf{x}_t = \lambda^2 (\mathbf{x}_t \wedge \mathbf{x}_{tt}), \quad \text{or} \quad \boldsymbol{\beta} = \lambda^2 \mathbf{x}_{tt} + \mu \mathbf{x}_t,$$

for some scalar μ. Unimodularity implies $\lambda = (\mathbf{x}_t \wedge \mathbf{x}_{tt})^{-1/3}$. Therefore

$$-\nu_2 = ds = \sqrt[3]{\mathbf{x}_t \wedge \mathbf{x}_{tt}} \, dt,$$

reproduces the equi-affine arc length element. Furthermore, $\mu_2 = \boldsymbol{\beta} \wedge d\boldsymbol{\alpha} = J \, ds$, where the lifted invariant

$$J = \mu(\mathbf{x}_t \wedge \mathbf{x}_{tt})^{1/3} + \frac{\mathbf{x}_t \wedge \mathbf{x}_{ttt}}{3(\mathbf{x}_t \wedge \mathbf{x}_{tt})^{4/3}},$$

is normalized to zero in the obvious manner. Therefore, the final moving frame is given by

$$\boldsymbol{\alpha} = \frac{d\mathbf{x}}{ds} = \frac{\mathbf{x}_t}{\sqrt[3]{\mathbf{x}_t \wedge \mathbf{x}_{tt}}}, \quad \boldsymbol{\beta} = \frac{d^2\mathbf{x}}{s^2} = \frac{\mathbf{x}_{tt}}{(\mathbf{x}_t \wedge \mathbf{x}_{tt})^{2/3}} - \frac{\mathbf{x}_t}{3(\mathbf{x}_t \wedge \mathbf{x}_{tt})^{5/3}},$$

$$\mathbf{a} = \mathbf{x}.$$

Finally, $\mu_3 = \kappa \, ds$, where

$$\kappa = \mathbf{x}_{ss} \wedge \mathbf{x}_{sss} = \frac{(\mathbf{x}_t \wedge \mathbf{x}_{tttt}) + 4(\mathbf{x}_{tt} \wedge \mathbf{x}_{ttt})}{3(\mathbf{x}_t \wedge \mathbf{x}_{tt})^{5/3}} - \frac{5(\mathbf{x}_t \wedge \mathbf{x}_{ttt})^2}{9(\mathbf{x}_t \wedge \mathbf{x}_{tt})^{8/3}},$$

defines the equi-affine curvature. All higher-order differential invariants are obtained by differentiating κ with respect to the equi-affine arc length ds. This reproduces the basic invariants of the equi-affine geometry of curves,

[13]; see also [3, 8], for applications in computer vision. The classical Frenet equations are a simple reformulation of the final moving frame formulae. We identify the linear part $A = (\mathbf{e}_1, \mathbf{e}_2) = (\mathbf{x}_s, \mathbf{x}_{ss})$ of the final moving frame with the equi-affine frame at a point $\mathbf{x}(t)$ on the curve, so that $\mathbf{e}_1 = \mathbf{x}_s$ is the unit affine tangent vector, whereas $\mathbf{e}_2 = \mathbf{x}_{ss}$ is the unit equi-affine normal. Combining this with the SL(2) Maurer–Cartan matrix $A^{-1}\, dA = \left(\begin{smallmatrix} 0 & 1 \\ \kappa & 0 \end{smallmatrix}\right)\, ds$ leads to the complete Frenet equations of planar equi-affine geometry:

$$\frac{d\mathbf{x}}{ds} = \mathbf{e}_1, \quad \frac{d\mathbf{e}_1}{ds} = \mathbf{e}_2, \quad \frac{d\mathbf{e}_2}{ds} = \kappa \mathbf{e}_1.$$

Note that the chosen normalizations are governed by the cross section

$$\mathcal{K}^3 = \{x = u = u_x = 0, u_{xx} = 1, u_{xxx} = 0\},$$

to the group orbits on J^3. In fact, it is not hard to apply the regularized method directly in this example. In general, the more complicated the group action, the more efficient the moving coframe approach becomes. Curves whose 2-jets pass through the singular locus $\mathbf{x}_t \wedge \mathbf{x}_{tt} = 0$ can be covered by higher-order moving frames, except for the straight lines, which are the totally singular curves in equi-affine geometry.

Finally, we demonstrate how the moving coframe method can be adapted to the case of infinite Lie pseudo-groups. By definition, a Lie pseudo-group consists of an infinite-dimensional family of invertible (local) transformations that form the general solution to an involutive system of partial differential equations, cf. [6, 17]. We can always characterize the pseudo-group transformations $\psi \colon M \to M$ as the projections of bundle maps $\Psi \colon \mathcal{B} \to \mathcal{B}$, defined on a principal fiber bundle $\mathcal{B} \to M$, that preserve a system of one-forms $\boldsymbol{\zeta} = \{\zeta_1, \ldots \zeta_k\}$, so that $\Psi^* \boldsymbol{\zeta} = \boldsymbol{\zeta}$. The forms $\boldsymbol{\zeta}$ will play the role of the moving coframe forms for the pseudo-group, and the fiber coordinates of the bundle $\mathcal{B}$ will play the role of the undetermined group parameters. Of course, in this case $\boldsymbol{\zeta}$ does not form a full coframe on $\mathcal{B}$. A compatible lift, or moving frame of order zero, is just an arbitrary section $\sigma^{(0)} \colon M \to \mathcal{B}$. With these provisos, the normalization and reduction procedure is implemented as in the finite-dimensional situation.

Example 1.3. Consider the intransitive pseudo-group consisting of (local) diffeomorphisms on $M = \mathbb{R}^3$ of the form

$$\overline{x} = f(x), \quad \overline{y} = y, \quad \overline{u} = \frac{u}{f'(x)}. \tag{1.10}$$

This pseudo-group was introduced by Lie [18, p. 373], in his study of second-order partial differential equations integrable by the method of Darboux, and was further investigated by Medolaghi [19], Vessiot [26], and Kumpera [16]. Following a general procedure presented in Ref. [15], a zeroth-order moving coframe consists of the one-forms

$$\zeta_1 = u\, dx, \quad \zeta_2 = \alpha\, dx + \frac{du}{u}, \quad \zeta_3 = dy,$$

which are defined on a rank one bundle $\mathcal{B} \to M$ with fiber coordinate α. Indeed, any transformation that satisfies $\Psi^* \zeta_i = \zeta_i$, $i = 1, 2, 3$, projects to a pseudo-group transformation (1.10). For surfaces $u = u(x, y)$, the linear dependency $\zeta_2 = -(u\alpha + u_x)\zeta_1 - (u_y/u)\,dy$, produces the normalization $\alpha = -u_x/u$, along with the basic first-order differential invariant $I = u_y/u$. The final invariant moving coframe is

$$\zeta_1 = u\,dx, \quad \zeta_2 = \frac{du - u_x\,dx}{u}, \quad \zeta_3 = dy.$$

The invariant total differential operators are thus $\mathcal{D}_1 = u^{-1}D_x$, $\mathcal{D}_2 = D_y$. A complete system of differential invariants consists of y, I, and the higher-order invariant derivatives $(\mathcal{D}_1)^j (\mathcal{D}_2)^k I$.

Acknowledgments: We would like to thank Ian Anderson for sharing his insights, inspiration, and critical comments while this work was in progress. Mark Fels' work is supported in part by an NSERC Postdoctoral Fellowship and Peter Olver's by NSF grant DMS 95–00981.

REFERENCES

1. I.M. Anderson, *Introduction to the variational bicomplex*, Contemp. Math., **132** (1992), 51–73.

2. I.M. Anderson and C.G. Torre, *Two component spinors and natural coordinates for the prolonged Einstein equation manifolds*, preprint, Utah State University, 1997.

3. E. Calabi, P.J. Olver, C. Shakiban, A. Tannenbaum, and S. Haker, *Differential and numerically invariant signature curves applied to object recognition*, Internat. J. Comput. Vision. (To appear)

4. É. Cartan, *La méthode du repère mobile, la théorie des groupes continus, et les espaces généralisés*, Exposés de Géométrie, Vol. 5, Hermann, Paris, 1935.

5. É. Cartan, *La théorie des groupes finis et continus et la géométrie différentielle traitées par la méthode du repère mobile*, Cahiers Scientifiques, Vol. 18, Gauthier–Villars, Paris, 1937.

6. É. Cartan, *Sur la structure des groupes infinis de transformations*, Oeuvres Complètes, part. II, Vol. 2, Gauthier–Villars, Paris, 1953, pp. 571–624.

7. S.-S. Chern, *Moving frames*, The Mathematical Heritage of Élie Cartan (Lyon, 1984), Astérisque, 1985, Numero Hors Serie, pp. 67–77.

8. O. Faugeras, *Cartan's moving frame method and its application to the geometry and evolution of curves in the euclidean, affine and projective planes*, Applications of Invariance in Computer Vision (J.L. Mundy, A. Zisserman, D. Forsyth, eds.), Lecture Notes in Comput. Sci., Vol. 825, 1994, Springer, New York, pp. 11–46.

9. M. Fels, P.J. Olver, *Moving coframes. I. A practical algorithm*, Acta Appl. Math. **15** (1998), No. 2, 161–213.

10. M. Fels, P.J. Olver, *Moving coframes. II. Regularization and theoretical foundations*, Acta Appl. Math. **55** (1999), No. 2, 127–208.

11. M.L. Green, *The moving frame, differential invariants and rigidity theorems for curves in homogeneous spaces*, Duke Math. J. **45** (1978), 735–779.

12. P.A. Griffiths, *On Cartan's method of Lie groups and moving frames as applied to uniqueness and existence questions in differential geometry*, Duke Math. J. **41** (1974), 775–814.

13. H.W. Guggenheimer, *Differential geometry*, McGraw–Hill, New York, 1963.

14. G.R. Jensen, *Higher order contact of submanifolds of homogeneous spaces*, Lecture Notes in Math., Vol. 610, Springer, New York, 1977.

15. N. Kamran, *Contributions to the study of the equivalence problem of Élie Cartan and its applications to partial and ordinary differential equations*, Acad. Roy. Belg. Cl. Sci. Mém. Collect. 8° (2) **45** (1989), No. 7, 1–122.

16. A. Kumpera, *Invariants différentiels d'un pseudogroupe de Lie*, J. Differential Geom. **10** (1975), 289–416.

17. S. Lie, *Die Grundlagen für die Theorie der unendlichen kontinuierlichen Transformationsgruppen*, Leipzig. Berich. **43** (1891), 316–393; Gesammelte Abhandlungen, Vol. 6, B.G. Teubner, Leipzig, 1927, pp. 300–364.

18. S. Lie, *Zur allgemeinen Theorie der partiellen Differentialgleichungen beliebeger Ordnung*, Leipz. Berich. **47** (1895), 53–128; Gesammelte Abhandlungen, Vol. 4, B.G. Teubner, Leipzig, 1929, pp. 320–384.

19. P. Medolaghi, *Classificazione delle equazioni alle derivate parziali del secondo ordine, che ammettono un gruppo infinito di trasformazioni puntuali*, Ann. Mat. Pura Appl. (4) **1** (1898), No. 3, 229–263.

20. P.J. Olver, *Applications of Lie groups to differential equations*, 2nd ed., Grad. Texts in Math., Vol. 107, Springer, New York, 1993.

21. P.J. Olver, *Equivalence, invariants, and symmetry*, Cambridge University Press, Cambridge, 1995.

22. P.J. Olver, *Singularities of prolonged group actions on jet bundles*, University of Minnesota, 1997.

23. L.V. Ovsiannikov, *Group analysis of differential equations*, Academic Press, New York, 1982.

24. S. Sternberg, *Lectures on differential geometry*, Prentice-Hall, Englewood Cliffs, N.J., 1964.

25. T.Y. Thomas, *The differential invariants of generalized spaces*, Chelsea Publ. Co., New York, 1991.

26. E. Vessiot, *Sur l'intégration des systèmes différentiels qui admettent des groupes continus de transformations*, Acta. Math. **28** (1904), 307–349.

27. H. Weyl, *Cartan on groups and differential geometry*, Bull. Amer. Math. Soc. **44** (1938), 598–601.

28. T.J. Willmore, *Riemannian geometry*, Oxford University Press, Oxford, 1993.

5

The Fibonacci-Deformed Harmonic Oscillator

Jean-Pierre Gazeau and Bernard Champagne

ABSTRACT Given an infinite countable, strictly increasing sequence of positive real numbers, it is possible to associate to it a triplet ⟨Identity I, Creation $a^\dagger$, Annihilation a⟩ of operators acting on some Hilbert space *and* an overcomplete set of coherent states labeled with complex numbers ("holomorphic map"). The triplet generates a quantum algebra that reduces to the q-oscillator algebra when the sequence is the set of q-deformed natural numbers.

We shall present here another type of quantum algebra. The latter is based on the quasiperiodic sequence of numbers (the half-infinite Fibonacci chain) that are the counterparts of the positive integers in "golden-mean" basis. As a byproduct, we shall give a survey on that Fibonacci zoology: new remarkable numbers and related special functions.

1 About Strictly Increasing Sequences of Positive Numbers

The story starts out with a strictly increasing sequence of positive numbers that is countably infinite:

$$\sigma = \{x_0 < x_1 < x_2 < \cdots < x_n < x_{n+1} < \cdots\}. \tag{1.1}$$

Such a set can, for instance, represent an infinite bounded-below set of Schrödinger spectral levels for a confining potential.

So we have a one-to-one correspondence

$$n \rightarrow x_n, \tag{1.2}$$

between $\mathbb{N}$ and σ, (possibly) together with a successor function, denoted by $n \rightarrow \varphi(n)$ and defined by

$$x_{n+1} = x_n + \varphi(n). \tag{1.3}$$

From now on, we put $x_0 \stackrel{\text{def}}{=} 0$ and we suppose that $\sigma \equiv \sigma(X_N)$ is the spectrum of a positive self-adjoint operator X_N acting on some separable

Hilbert space $\mathcal{H}$:

$$X_N|n\rangle \overset{\text{def}}{=} x_n|n\rangle, \tag{1.4}$$

with $\langle m \mid n\rangle \overset{\text{def}}{=} \delta_{m,n}$, and where we suppose that the set of states $\{|n\rangle, n \in \mathbb{N}\}$ is a basis of $\mathcal{H}$. Consistently with the notation X_N, we define the number operator N by

$$N|n\rangle \overset{\text{def}}{=} n|n\rangle, \quad \text{with} \quad n \in \mathbb{N}. \tag{1.5}$$

The above definition of X_N goes in hand with the existence of lowering and raising operators a and $a^\dagger$. They are given through their respective actions on the basis elements

$$a|n\rangle \overset{\text{def}}{=} \sqrt{x_n}|n-1\rangle, \tag{1.6a}$$

$$a^\dagger|n\rangle \overset{\text{def}}{=} \sqrt{x_{n+1}}|n+1\rangle. \tag{1.6b}$$

The fact that $a|0\rangle = 0$ results from our convention $x_0 \overset{\text{def}}{=} 0$. Hence we have

$$X_N = a^\dagger a. \tag{1.7}$$

Next, we define the (finite-difference) derivation of X_N as being the positive self-adjoint operator:

$$X'_N|n\rangle \overset{\text{def}}{=} \varphi(n)|n\rangle; \tag{1.8}$$

i.e., its spectrum is made of spectral widths of X_N resulting from decay transitions $n+1 \to n$. More generally,

$$X_N^{(k)}|n\rangle \overset{\text{def}}{=} \varphi^{(k-1)}(n)|n\rangle, \tag{1.9}$$

where $\varphi^{(k)}(n)$ is the kth finite-difference derivative of the successor function $\varphi(n)$. It is recursively defined by

$$\varphi^{(k+1)}(n) \equiv \varphi^{(k)}(n+1) - \varphi^{(k)}(n). \tag{1.10}$$

With these notations, the following commutation rules are easily derived:

$$[a, a^\dagger] = X'_N, \tag{1.11}$$

$$[a, X_N] = X'_N a, \tag{1.12}$$

$$[a^\dagger, X_N] = -a^\dagger X'_N. \tag{1.13}$$

Note that (1.12) and (1.13) hold true for any diagonal operator Δ:

$$[a, \Delta] = \Delta' a, \quad [a^\dagger, \Delta] = -a^\dagger \Delta'.$$

For instance, $[a, N] = N'a = a$, and $[a^\dagger, N] = -a^\dagger N'$.

2 Quantum Algebra Associated with the Spectrum $\sigma = \{x_n\}$

We see from (1.11)–(1.13) how the Lie structure of the algebra $\mathcal{A}$ freely generated by the triplet $\langle \mathbb{I}, a, a^\dagger \rangle$ rests on the properties of the successor function φ. If the latter is polynomial of degree p, its $(p+1)$th derivative cancels out and $\mathcal{A}$ can be finite-dimensional. The simplest case $\varphi(n) = \alpha = $ constant corresponds to the harmonic oscillator:

$$X'_N = \alpha \mathbb{I}. \tag{2.1}$$

Generically $\varphi^{(k)} \neq 0$ for any k. As a matter of fact, the standard q-deformation of the harmonic oscillator, which is based on the commutation rule

$$aa^\dagger - qa^\dagger a = \mathbb{I}, \tag{2.2}$$

corresponds to the sequence

$$x_0 = 0, \quad x_n = [n]_q \equiv 1 + q + \cdots + q^{n-1}. \tag{2.3}$$

Equivalently, the q-oscillator corresponds to the successor function

$$\varphi(n) = q^n, \tag{2.4}$$

together with the initial condition $x_0 = 0$. The algebraic structure of $\mathcal{A}$ is then remarkably repetitive, since

$$\varphi^{(k)}(n) = (q-1)^k \varphi(n), \tag{2.5}$$

and so

$$X_N^{(k+1)} = (q-1)^k X'_N. \tag{2.6}$$

Starting from (1.11)–(1.13), if we try to go through the set of operators appearing on the right-hand sides of successive commutation relations, we become very soon aware of the difficulty of organizing in an explicit way the elements of $\mathcal{A}$. For instance,

$$[a, X'_N a] = X''_N a^2, \tag{2.7}$$

$$[a, a^\dagger X'_N] = (X'_n)^2 + a^\dagger X''_N a. \tag{2.8}$$

However, one can decide to deal systematically with normal-ordered products of operators a, $a^\dagger$ and diagonal operators Δ (i.e., $\Delta|n\rangle = \delta_n|n\rangle$):

$$(a^\dagger)^l \Delta (a)^m, \qquad \text{where} \quad l, m \in \mathbb{N}. \tag{2.9}$$

For instance, the operators that appear in the right-hand side (r.h.s.) of Eq. (2.7) and Eq. (2.8) are of such a type, whereas, e.g., $aX''_N a$ is not. Instead of the latter, we shall consider the alternative expression:

$$aX''_N a = X''_N a^2 + X'''_N a^2, \tag{2.10}$$

where the terms in the r.h.s. are of the type (2.9). We can easily understand that a generic element of the infinite-dimensional Lie algebra $\mathcal{A}$ is a linear superposition of terms of the form (2.9), where Δ is monomial in X_N and its successive derivatives

$$\Delta = (X_N)^{p_0}(X'_N)^{p_1}\ldots(X_N^{(k)})^{p_k}. \tag{2.11}$$

Indeed, putting a product of operators a, $a^\dagger$, and diagonal ones into a normal-ordered form, e.g.,

$$a\Delta = (\Delta + \Delta')a,$$

introduces derivatives of diagonal operators in the final result. Now, the derivative of a diagonal operator that is like (2.11) is itself a sum of operators of the same type. This stems from a counterpart of the Leibnitz rule for finite difference calculus:

$$(FG)' = F'G' + FG' + F'G, \tag{2.12}$$

where F and G are diagonal operators. The general commutation relations between two operators of the type (2.9),

$$[(a^\dagger)^l F a^m, (a^\dagger)^{l'} G a^{m'}], \tag{2.13}$$

where F and G are diagonal, are computed in terms of normal-ordered operators by making use of the following building block rules:

$$a^l \Delta = \overset{l\rightarrow}{\Delta} a^l \tag{2.14}$$

$$\Delta (a^\dagger)^l = (a^\dagger)^l \overset{l\rightarrow}{\Delta}, \tag{2.15}$$

$$a^m (a^\dagger)^l = \begin{cases} (a^\dagger)^{l-m} \prod_{r=0}^{m-1} \overset{(l-r)\rightarrow}{X_N}, & \text{if } m \le l, \\ \prod_{r=0}^{l-1} \overset{(m-r)\rightarrow}{X_N} a^{m-l}, & \text{if } m > l. \end{cases} \tag{2.16}$$

Here, the notation "$l \rightarrow$" means that the eigenvalues δ_n of the diagonal operator Δ are shifted on the right by l:

$$\overset{l\rightarrow}{\Delta} |n\rangle = \delta_{n+l}|n\rangle. \tag{2.17}$$

The results (2.14)–(2.16) stem from the finite-difference formula:

$$\overset{l\rightarrow}{\Delta} = \sum_{r=0}^{l} \binom{l}{r} \Delta^{(r)}. \tag{2.18}$$

Another way to delimit the structural properties of $\mathcal{A}$ is to parallel the standard q-deformation (2.2) of the canonical commutation rules. Indeed, if we choose the scale $\varphi(0) \overset{\text{def}}{=} x_1 = 1$, (1.11) is equivalent to the following one:

$$[a, a^\dagger]_\rho \equiv aa^\dagger - \rho a^\dagger a = \mathbb{I}, \tag{2.19}$$

where ρ is a diagonal operator defined by

$$\rho|n\rangle \overset{\text{def}}{=} r_n|n\rangle. \tag{2.20}$$

The number r_0 is here arbitrary, and if $n > 0$,

$$r_n = \frac{x_{n+1} - 1}{x_n} = 1 - \frac{1 - \varphi(n)}{x_n}. \tag{2.21}$$

Note that ρ commutes with both $aa^\dagger$ and $a^\dagger a$. The q-deformation is easily recovered from Eq. (2.3) or Eq. (2.4): $r_n = q$ and we put $r_0 \overset{\text{def}}{=} 0$. We do not know at the present time what could lie at a deeper mathematical level behind this rather pedestrian presentation of $\mathcal{A}$. Of course, we should not expect too much in the generic case, except we strongly suspect that $\mathcal{A}$ is then simply the infinite Lie algebra $\mathfrak{gl}(\infty)$. On the other hand, a large amount of work has already been done on the q-deformation case and its related extensions. Note that all of the latter deal with continuous deformations of the Weyl–Heisenberg algebra. We shall consider below deformations that are *not* continuous (we should adopt a more realistic terminology, like shearing!) although the successor function φ can still be expressed in a rather simple way.

3   The β-Natural Spectrum

There exist sequences $\sigma = \{x_n\}_{n\in\mathbb{N}}$ of real numbers that are "almost" like the naturals in the sense that they are quasiperiodic with a finite number of values for the interval lengths

$$\{\varphi(n) = x_{n+1} - x_n\}_{n\in\mathbb{N}}.$$

This means that the range of the successor function is a finite set of strictly positive numbers.

Some of those sequences are scaling invariant, $\beta\sigma \subset \sigma$ and are biunivocally determined by the nonnatural number $\beta > 1$. The numbers x_n of the sequence are the naturals in "basis β."

First, what does it mean to write a (positive) real number x in "basis β" with β nonnatural? The procedure can, for instance, be traced back to Rényi [4]. There exist $j \equiv j(x) \in \mathbb{Z}$ and a sequence of positive integers $\{\xi_l \equiv \xi_l(x)\}_{l \leq j}$ assuming their values in the set

$$\xi_l \in \{0, 1, \ldots, [\beta]\} \overset{\text{def}}{=} \text{ integer part of } \beta\}, \tag{3.1}$$

such that x can be written

$$x = \sum_{l=-\infty}^{j} \xi_l \beta^l \equiv \xi_j \xi_{j-1} \ldots \xi_0 . \xi_{-1} \xi_{-2} \cdots . \tag{3.2}$$

The leading power j is the maximal integer such that

$$\beta^j \leq x < \beta^{j+1}. \tag{3.3}$$

The expansion coefficients are determined by using the (Rényi) algorithm,

$$\xi_l = \left[\beta T_\beta^{j-l}(x/\beta^{j+1}) \right], \tag{3.4}$$

where

$$T_\beta(x) \stackrel{1}{=} \beta x, \quad \text{for } x \in [0,1). \tag{3.5}$$

The set of allowed semi-infinite words (3.2) is lexicographically ruled as follows:

> "no such word and all its right-handed shifted is lexicographically larger or equal to the so-called Rényi expansion of 1:
>
> $$d(1,\beta) \equiv 0.t_1 t_2 \ldots t_l \ldots, \tag{3.6}$$
>
> where
>
> $$t_l = \left[\beta T_\beta^{l-1}(1) \right]. \text{"} \tag{3.7}$$

The set of positive β-integers is then the strictly increasing sequence of all allowed β-expansions (3.2) having *no* negative powers of β in the expansion:

$$\mathbb{N}_\beta = \{ x = \xi_j \beta^j + \xi_{j-1} \beta^{j-1} + \cdots + \xi_1 \beta + \xi_0 \equiv \xi_j \xi_{j-1} \ldots \xi_1 \xi_0 \}. \tag{3.8}$$

The self-similarity $\beta \mathbb{N}_\beta \subset \mathbb{N}_\beta$ is a simple consequence of this construction. The first term of $\mathbb{N}_\beta$ is $x_0 = 0$, of course, and we denote the $(n+1)$th term by x_n. We thus have a bijection from $\mathbb{N}$ onto $\mathbb{N}_\beta$:

$$x \to x_n = \xi_j \beta^j + \cdots + \xi_1 \beta + \xi_0, \tag{3.9}$$

where $j \equiv j(n)$ and $\xi_l \equiv \xi_l(n)$, $0 \leq l \leq j$. Note that if β were natural > 1, the coefficients (3.1) would be allowed to assume their values in the set $\{0, 1, \ldots, \beta - 1\}$ and (3.9) would just be the expansion of natural numbers in basis β, i.e., $\mathbb{N}_\beta = \mathbb{N}$. The counterpart of Eq. (3.6) is then

$$d(1,\beta) = 0.(\beta - 1)(\beta - 1)(\beta - 1) \cdots \equiv 0.(\beta - 1)^\omega, \tag{3.10}$$

where $(* * *)^\omega$ means that the word between parentheses is indefinitely repeated.

For a generic nonnatural β, even for a nonnatural rational, the sequence $\mathbb{N}_\beta$ is quite unpredictable. The range of the successor function $\varphi(n)$ is indeed given by the set

$$\text{Range}[\varphi] = \{T_\beta{}^i(1),\ i \geq 0\}, \tag{3.11}$$

and this dynamical system can give rise to any kind of infinite countable set filling (maybe densely) the interval $(0, 1]$. Now, it can be shown that if β is a Pisot–Vijayaraghavan algebraic integer [2], i.e., is a solution to

$$X^m = a_{m-1}X^{m-1} + \cdots + a_1 X + a_0, \quad \text{with } a_i \in \mathbb{Z}, \tag{3.12}$$

and with Galois conjugates (other roots of Eq. (3.12)) lying within the open unit disc, then the Rényi expansion (3.6) of 1 is finite,

$$d(1, \beta) = 0.t_1 t_2 \ldots t_p 00 \ldots,$$

or eventually periodic,

$$d(1, \beta) = 0.t_1 t_2 \ldots t_r (t_{r+1} t_{r+2} \ldots t_{r+p})^\omega,$$

and the range (3.11) of φ is finite with $(r + p)$ elements. This range is given by

$$\text{Range}[\varphi] = \{1, \beta - t_1, \beta^2 - t_1\beta - t_2, \ldots,$$
$$\beta^{p+r-1} - t_1\beta^{p+r-2} - \cdots - t_{p+r-1}\}. \tag{3.13}$$

The set $\mathbb{N}_\beta$ is then quasiperiodic with a finite set of lengths separating two subsequent elements of the sequence.

At this point, it should be stressed that the existence of an infinite uncountable set of numeration systems $\mathbb{N}_\beta$ makes it possible to pick one of them as a labeling point of an arbitrary sequence $\sigma = \{x_n\}$ of the type (1.1). More precisely, it is a matter of determining $\beta > 1$ such that

$$\sigma \subset c\beta^q \mathbb{N}_\beta, \tag{3.14}$$

for a certain scaling factor c and a certain power $q \in \mathbb{Z}$.

4   The Fibonacci Deformation of Weyl Algebra

The simplest (and smallest) of the Pisot numbers is the well-known golden mean $\tau \equiv (1 + \sqrt{5})/2 \doteq 1.618$. It is solution to $X^2 = X + 1$ and the conjugate root is $-1/\tau = (1 - \sqrt{5})/2 \doteq .618$. The Rényi expansion of 1 is given by

$$d(1, \tau) = 0.11. \tag{4.1}$$

So the set of τ-naturals reads

$$\mathbb{N}_\beta$$
$$= \{x = \xi_j \tau^j + \xi_{j-1}\tau^{j-1} + \cdots + \xi_1\tau + \xi_0, \xi_l \in \{0,1\}, \xi_l\xi_{l+1} = 0\}. \quad (4.2)$$

The first terms of this sequence, which is remarkably "close" to the sequence of the naturals, are given by

$$\mathbb{N}_\tau = \{x_0 = 0, x_1 = 1, x_2 = \tau \equiv 10, x_3 = \tau^2 \equiv 100, x_4 = \tau^2 + 1 \equiv 101,$$
$$x_5 = \tau^3 \equiv 1000, x_6 = \tau^3 + 1 \equiv 1001, x_7 = \tau^3 + \tau \equiv 1010, \dots\}. \quad (4.3)$$

The correspondence $\mathbb{N} \to \mathbb{N}_\tau$ introduced in Eq. (3.9) is here simply given by replacing in the so-called Fibonacci expansion of the natural n,

$$n = \xi_j F_j + \xi_{j-1}F_{j-1} + \cdots + \xi_0 F_0, \quad (4.4)$$

with $F_j = F_{j-1} + F_{j-2}$, $F_0 = 1$, $F_1 = 2$, the Fibonacci numbers F_j by τ^j:

$$n \to x_n = \xi_j\tau^j + \xi_{j-1}\tau^{j-1} + \cdots + \xi_1\tau + \xi_0.$$

The successor function φ assumes only two values, namely, 1 and $1/\tau$, and can be easily seen to be equal to

$$\varphi(n) \equiv x_{n+1} - x_n = 1/\tau^{\xi_0(n)} = \begin{cases} 1, & \text{if } x_n \text{ is divisible by } \tau, \\ 1/\tau, & \text{if not.} \end{cases} \quad (4.5)$$

It follows that the derivative X_N' of the operator X_N is "almost" the identity. The spectrum

$$\sigma(X_N') \equiv \sigma_1 = \{1, 1/\tau, 1, 1, 1/\tau, 1, 1/\tau, 1, \dots\}, \quad (4.6)$$

is called a half-infinite Fibonacci chain in the language of the theory of substitution sequences. It can be seen as the semi-infinite word

$$\Sigma_0 \equiv A_0 B_0 A_0 A_0 B_0 A_0 B_0 A_0 \dots, \quad \text{where} \quad A_0 \overset{\text{def}}{=} 1, \ B_0 \overset{\text{def}}{=} 1/\tau, \quad (4.7)$$

obtained from A_0 through infinitely repeated substitutions

$$\varsigma : \begin{cases} A_0 \overset{\varsigma}{\to} A_0 B_0, \\ B_0 \overset{\varsigma}{\to} A_0, \end{cases} \quad (4.8)$$

i.e.,

$$\Sigma_0 = \varsigma^\infty(A_0). \quad (4.9)$$

The computation of the successive derivatives of φ shows a quasiperiodic structure globally similar to the Fibonacci chain σ_i:

$$\varphi^{(k)}(n) \to \sigma(X_N^{(k+1)}) \equiv \sigma_{k+1} \to \Sigma_k \equiv A_k B_k A_k A_k B_k A_k \dots. \quad (4.10)$$

The cells A_k and B_k in Eq. (4.10) have respective lengths that are Fibonacci numbers

$$|A_k| = F_{l(k)}, \quad |B_k| = F_{l(k)-1}. \tag{4.11}$$

So the lengths increase monotonically in stages in such a way that

$$|B_k|/|A_k| \xrightarrow{k \to \infty} 1/\tau, \tag{4.12}$$

and the lengths of successive stages form a Fibonacci sequence! For instance,

$$|A_0| = |B_0| = 1; \ |A_1| = 2, |B_1| = 1; \ |A_2| = |A_3| = 3, |B_2| = |B_3| = 2,$$
$$|A_4| = |A_5| = |A_6| = 5, |B_4| = |B_5| = |B_6| = 3, \text{ etc. } \dots \tag{4.13}$$

However, the internal structure of those cells shows increasing disorder as it can be seen in the first examples $k \le 10$ (see Table 5.1). Note that for each cell, the sum of all numbers forming it is zero. Few things can be told at the moment regarding the quantum algebra structure lying behind this sequence of operators $X_N^{(k)}$. Note the form assumed by the diagonal operator $\rho = \{r_n\}$ of Eq. (2.19):

$$r_n = 1 + \frac{1 - \tau^{\xi_0(n)}}{\tau^{\xi_0(n)} x_n} = \begin{cases} 1, & \text{if } \xi_0(n) = 0, \\ 1 - \frac{1}{\tau^2 x_n}, & \text{if } \xi_0(n) = 1. \end{cases}$$

Note also that $r_n \xrightarrow{k \to \infty} 1$.

TABLE 5.1. The first σ_k.

$\sigma_0 = \{0, 1, \tau, \tau^2, \tau^2 + 1, \tau^3, \tau^3 + 1, \tau^3 + \tau, \tau^4, \tau^4 + 1, \tau^4 + \tau, \tau^4 + \tau^2,$
$\tau^4 + \tau^2 + 1, \tau^5, \tau^5 + 1, \tau^5 + \tau, \tau^5 + \tau^2, \tau^5 + \tau^2 + 1, \tau^5 + \tau^3,$
$\tau^5 + \tau^3 + 1, \tau^5 + \tau^3 + \tau, \tau^6, \tau^6 + 1, \tau^6 + \tau, \tau^6 + \tau^2, \tau^6 + \tau^2 + 1,$
$\tau^6 + \tau^3, \tau^6 + \tau^3 + 1, \tau^6 + \tau^3 + \tau, \tau^6 + \tau^4, \tau^6 + \tau^4 + 1,$
$\tau^6 + \tau^4 + \tau^2 + 1, \tau^6 + \tau^4 + \tau, \tau^6 + \tau^4 + \tau^2, \tau^7, \tau^7 + 1, \tau^7 + \tau,$
$\tau^7 + \tau^2, \tau^7 + \tau^2 + 1, \tau^7 + \tau^3, \tau^7 + \tau^3 + 1, \tau^7 + \tau^3 + \tau, \tau^7 + \tau^4,$
$\tau^7 + \tau^4 + 1, \tau^7 + \tau^4 + \tau, \tau^7 + \tau^4 + \tau^2, \tau^7 + \tau^4 + \tau^2 + 1, \tau^7 + \tau^5,$
$\tau^7 + \tau^5 + 1, \tau^7 + \tau^5 + \tau, \tau^7 + \tau^5 + \tau^2, \tau^7 + \tau^5 + \tau^2 + 1,$
$\tau^7 + \tau^5 + \tau^3, \tau^7 + \tau^5 + \tau^3 + 1, \tau^7 + \tau^5 + \tau^3 + \tau, \tau^8, \dots\}$

$\sigma_1 = \{\underbrace{1}_{A_0}, \underbrace{1/\tau}_{B_0}, 1, 1, 1/\tau, 1, 1/\tau, 1, 1, 1/\tau, 1, 1, 1/\tau, 1, 1/\tau, 1, 1, 1/\tau, 1, 1/\tau,$
$1, 1, 1/\tau, 1, 1, 1/\tau, 1, 1/\tau, 1, 1, 1/\tau, 1, 1, 1/\tau, 1, 1/\tau, 1, 1, 1,$
$1/\tau, 1, 1/\tau, 1, 1, 1/\tau, 1, 1, 1/\tau, 1, 1/\tau, 1, 1, 1/\tau, 1,$
$1/\tau, 1, \dots\}$

$1/1 = |B_0|/|A_0|$

$\sigma_2 = \{\underbrace{-1, 1}_{A_1}, \underbrace{0}_{B_1}, -1, 1, -1, 1, 0, -1, 1, 0, -1, 1, -1, 1, 0, -1, 1, -1,$
$1, 0, -1, 1, 0, -1, 1, -1, 1, 0, -1, 1, 0, -1, 1, -1, 1, 0,$
$-1, 1, -1, 1, 0, -1, 1, 0, -1, 1, -1,$

TABLE 5.1. (continued)

$$1, 0, -1, 1, -1, 1, \ldots \}/\tau^2$$

$1/2 = |B_1|/|A_1|$

$\sigma_3 = \{\underbrace{2, -1, -1}_{A_2}, \underbrace{2, -2, 2}_{B_2}, -1, -1, 2, -1, -1, 2, -2, 2, -1, -1, 2, -2, 2, -1,$
$-1, 2, -1, -1, 2, -2, 2, -1, -1, 2, -1, -1, 2,$
$-2, 2, -1, -1, 2, -2, 2, -1, -1, 2, -1, -1, 2,$
$-2, 2, -1, -1, 2, -2, 2, \ldots \}/\tau^2$

$2/3 = |B_2|/|A_2|$

$\sigma_4 = \{\underbrace{-3, 0, 3}_{A_3}, \underbrace{-4, 4}_{B_3}, -3, 0, 3, -3, 0, 3, -4, 4, -3, 0, 3, -4, 4, -3, 0, 3, -3,$
$0, 3, -4, 4, -3, 0, 3, -3, 0, 3, -4, 4, -3, 0, 3, -4.$
$4, -3, 0, 3, -3, 0, 3, -4, 4, -3, 0, 3,$
$-4, 4, \ldots \}/\tau^2$

$2/3 = |B_3|/|A_3|$

$\sigma_5 = \{\underbrace{3, 3, -7, 8, -7}_{A_4}, \underbrace{3, 3, -6}_{B_3}, 3, 3, -7, 8, -7, 3, 3, -7, 8, -7, 3, 3, -6, 3, 3,$
$-7, 8, -7, 3, 3, -6, 3, 3, -7, 8, -7, 3, 3,$
$-7, 8, -7, 3, 3, -6, 3, 3, -7, 8, -7, 3, 3,$
$-7, 8, \ldots, \}/\tau^2$

$3/5 = |B_4|/|A_4|$

$\sigma_6 = \{\underbrace{0, -10, 15, -15, 10}_{A_5}, \underbrace{0, -9, 9}_{B_5}, 0, -10, 15, -15, 10, 0, -10, 15, -15, 10,$
$0, -9, 9, 0, -10, 15, -15, 10, 0, -9, 9, 0,$
$-10, 15, -15, 10, 0, -10, 15, -15, 10,$
$0 - 9, 9, 0, -10, 15, -15, 10, 0,$
$-10, 15, \ldots, \}/\tau^2$

$3/5 = |B_5|/|A_5|$

$\sigma_7 = \{\underbrace{-10, 25, -30, 25, -10}_{A_6}, \underbrace{-9, 18, -9}_{B_6}, -10, 25, -30, 25, -10, -10, 25,$
$-30, 25, -10, -9, 18, -9, -10, 25, -30, 25, -10, -9, 18,$
$-9, -10, 25, -30, 25, -10, -10, 25, -30, 25, -10, -9, 18,$
$-9, -10, 25, -30, 25, -10, -10, 25, \ldots, \}/\tau^2$

$3/5 = |B_5|/|A_5|$

$\sigma_8 = \{\underbrace{35, -55, 55, -35, 1, 27, -27, -1}_{A_7}, \underbrace{35, -55, 55, -35, 0,}_{B_7}$
$35, -55, 55, -35, 1, 27, -27, -1, 35, -55, 55, -35, 1, 27, -27,$
$-1, 35, -55, 55, -35, 0, 35, -55, 55, -35, 1, 27, -27, -1, 35,$
$-55, 55, -35, 0, 35, \ldots, \}/\tau^2$

$5/8 = |B_7|/|A_7|$

$\sigma_9 = \{\underbrace{-90, 110, -90, 36, 26, -54, 26, 36}_{A_8}, \underbrace{-90, 110, -90, 35, 35,}_{B_8}$
$-90, 110, -90, 36, 26, -54, 26, 36, -90, 110, -90, 36, 26, -54,$
$26, 36, -90, 110, -90, 35, 35, -90, 110, -90, 36, 26, -54, 26, 36,$
$-90, 110, -90, 35, 35, \ldots, \}/\tau^2$

$5/8 = |B_8|/|A_8|$

TABLE 5.1. (continued)

$$\sigma_{10}= \{ \underbrace{200, -200, 126, -10, -80, 80, 10, -126, 200, -200, 125, 0, -125,}_{A_9}$$

$$\underbrace{\vphantom{200}}_{B_9}$$

$$200, -200, 126, -10, -80, 80, 10, -126, 200, -200, 126, -10,$$
$$-80, 80, 10, -126, 200, -200, 125, 0, -125, 200, -200, 126,$$
$$-10, -80, 80, 10, -126, 200, -200, 125, 0, \ldots , \}/\tau^2$$

$$5/8 = |B_9|/|A_9|$$

5   Coherent States and Some Special Functions

One can associate to any sequence $\{x_n\}$ of the type (1.1) another one that represents its "factorial." Paralleling the "q-language" we shall put

$$[x_n]! \equiv c_0 x_1 x_2 \ldots x_n, \quad [x_0]! \equiv c_0 > 0. \tag{5.1}$$

Following Odzijewicz, let us introduce a continuous family of coherent states $\{\zeta_z\}_{z \in \mathcal{D}_R}$ in $\mathcal{H}$ [3],

$$|\zeta_z\rangle \equiv \sum_{n=0}^{+\infty} \frac{z^n}{\sqrt{[x_n]!}} |n\rangle. \tag{5.2}$$

They are labeled with complex number z lying within the disc $\mathcal{D}_R = \{z \in \mathbb{C} \mid |z| < R\}$ with radius R given by

$$R = \limsup_{n \to +\infty} \sqrt[n]{[x_n]!}. \tag{5.3}$$

It follows from $\langle m \mid n \rangle = \delta_{m,n}$ and Eq. (5.2) that the value of the inner product of two coherent states is

$$\langle \zeta_{z'} \mid \zeta_z \rangle = \sum_{n=0}^{+\infty} \frac{(z\bar{z}')^n}{[x_n]!} \equiv \exp_F(z\bar{z}'). \tag{5.4}$$

Here, $\exp_F$ is the "exponential" function associated to the sequence $\{x_n\}$:

$$\exp_F(t) = \sum_{n=0}^{+\infty} \frac{t^n}{[x_n]!}. \tag{5.5}$$

Its convergence radius is precisely R. If nontrivial, i.e., if $R > 0$, the family $\{\zeta_z\}$ is clearly overcomplete in $\mathcal{H}$. In order to examine the (over-) completeness properties of these coherent states and hence the reproducing nature of the kernel (5.4), one considers the formal operator

$$A \equiv \int_{\mathcal{D}_R} |\zeta_z\rangle\langle\zeta_z| \, d\mu(z, \bar{z}), \tag{5.6}$$

where the measure μ is defined as

$$d\mu(z, \bar{z}) \equiv -\frac{1}{\exp_F(|z|^2)}\frac{dz \wedge d\bar{z}}{2\pi i}. \tag{5.7}$$

It can be shown that, as a weak integral, the right-hand side of Eq. (5.6) defines A as a bounded positive operator with $\|A\| \leq 1$, and that A^{-1} exists, although it may not be bounded. Since by a simple computation

$$\langle m|A|n \rangle = \frac{1}{[x_n]!} \int_0^{R^2} \frac{t^n}{\exp_F(t)}dt\delta_{m,n}, \tag{5.8}$$

A is diagonal in the basis $\{|n\rangle\}$ and can be written

$$A = \sum_{n=0}^{+\infty} \lambda_n |n\rangle\langle n|. \tag{5.9}$$

The set $\{\lambda_n\}_{n\in\mathbb{N}}$ is the spectrum of A, with

$$\lambda_n = \frac{1}{[x_n]!} \int_0^{R^2} \frac{t^n}{\exp_F(t)}dt \equiv \frac{\Gamma_F(n+1)}{[x_n]!}, \tag{5.10}$$

with self-explanatory notations for the Gamma function associated to the sequence $\{x_n\}$. While $\{\lambda_n\}$ is upper bounded, it is not assured that $\underline{\lim}_n \lambda_n$ is positive. If it is so, then A^{-1} is bounded and A is what is called a frame in $\mathcal{H}$ [1].

The ζ_z are called coherent states not only because of their overcompleteness and their "quasi"-reproducing properties exemplified by the existence of the operator (5.6), but also because they are eigenvectors of the operator a:

$$a|\zeta_z\rangle = z|\zeta_z\rangle. \tag{5.11}$$

The coherent states ζ_z also induce a mapping

$$W: \mathcal{H} \rightarrow L^2_{\text{Hol}}[\mathcal{D}_R, d\mu],$$

from the Hilbert space $\mathcal{H}$ into the Hilbert space of μ-square integrable functions which are holomorphic on $\mathcal{D}_R$. Explicitly,

$$(W \cdot \phi)(z) \stackrel{\text{def}}{=} \langle \zeta_{\bar{z}} | \phi \rangle, \tag{5.12}$$

and W is an isometry if $A = \mathbb{I}$. The images of the operator a and $a^\dagger$ are denoted, respectively, by

$$\partial_F \equiv WaW^{-1}, \quad Z \equiv Wa^\dagger W^{-1}. \tag{5.13}$$

They make sense of course on the range of W. Their respective actions are given by

$$(Z \cdot f)(z) = zf(z), \quad \partial_F \cdot z^n = x_n z^{n-1}. \tag{5.14}$$

Let us now specify this C.S. machinery to the case where the sequence $\{x_n\}$ is the set $\mathbb{N}_\tau$ of τ-naturals. For the sake of simplicity, we choose the constant c_0 in Eq. (5.1) equal to 1. We easily see that the radius R introduced in Eq. (5.3) is here infinite: $\mathcal{D}_R = \mathbb{C}$. It is first interesting to have some insight on the associated exponential

$$\exp_F(t) = \sum_{n=0}^{+\infty} \frac{t^n}{[n]!} = 1 + t + \frac{t^2}{\tau} + \frac{t^3}{\tau^3} + \frac{t^4}{\tau^3(\tau^2+1)} + \dots \tag{5.15}$$

and on its inverse $1/\exp_F(t)$. The first terms of the latter series are given by

$$\frac{1}{\exp_F(t)} = 1 - t + \frac{t^2}{\tau^2} - \frac{t^4}{\tau^3(\tau^2+1)} - \frac{t^5}{\tau^6(\tau^2+1)} + \dots \tag{5.16}$$

The graph of $\exp_F$ is shown in Fig. 5.1. In Fig. 5.2, we show the graph of the corresponding Gaussian $G_F(t) \equiv 1/\exp_F(t^2)$. We also have numerically computed the Gaussian integral

$$\int_{-\infty}^{+\infty} \frac{1}{\exp_F(t^2)} \, dt, \tag{5.17}$$

the value of which is denoted by $\sqrt{[\pi]}$ by referring to the standard case. We find

$$[\pi] \doteq 2.9155. \tag{5.18}$$

We next consider the spectrum $\{\lambda_n\}_{n\in\mathbb{N}}$ of the operator A defined by Eq. (5.6). The first values of λ_n computed by Eq. (5.10) are given by

$$\lambda_0 \doteq 0.8107, \ \lambda_1 \doteq 0.8771, \ \lambda_2 \doteq 0.8728,$$
$$\lambda_3 \doteq 0.8347, \ \lambda_4 \doteq 0.8494, \dots \tag{5.19}$$

The numerical computation of the limit of λ_n gives

$$\lim_{n\to\infty} \lambda_n \doteq 0.8539. \tag{5.20}$$

What is quite striking is that this limit is found to be equal to $\frac{1}{2}\sqrt{[\pi]}$; i.e.,

$$\lim_{n\to\infty} \frac{\Gamma_F(n+1)}{[x_n]!} = \lim_{n\to\infty} \frac{1}{x_1 x_2 \cdots x_n}, \int_0^{+\infty} \frac{t^n}{\exp_F(t)} \, dt$$
$$= \int_0^{+\infty} \frac{1}{\exp_F(t^2)} \, dt. \tag{5.21}$$

No explanation of this remarkable fact can be given at the moment. Finally, we see from the behavior of λ_n that A is a "frame," which means that its inverse is itself bounded. Its spectrum is discrete and is contained within the interval

$$\sigma(A) \subset [\lambda_0, \lambda_1]. \tag{5.22}$$

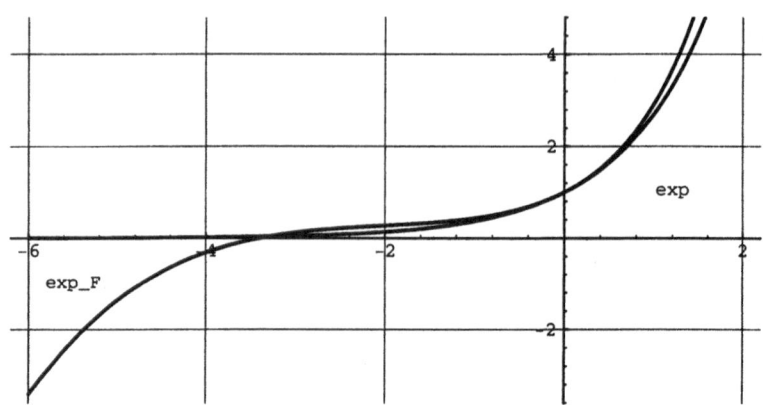

FIGURE 5.1. The function $\exp_F$ and the standard exponential exp.

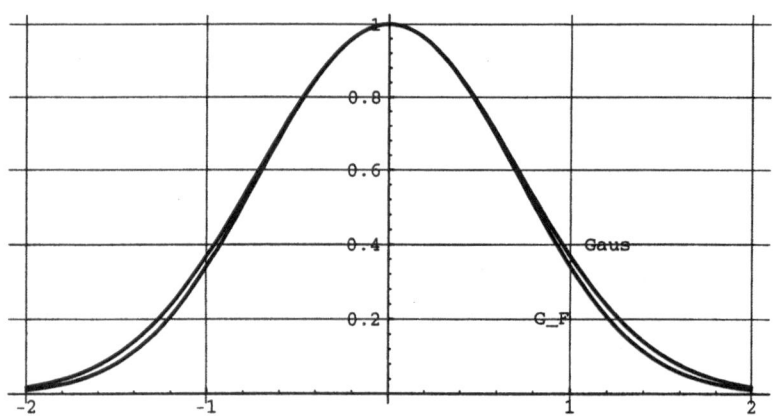

FIGURE 5.2. The Gaussian G_F and the standard Gaussian gauss.

Acknowledgments: The authors thank L. Frappat, J. Letourneux, J.F. van Diejen, L. Vinet, and T. Ali for stimulating discussions.

REFERENCES

1. S.T. Ali, J.-P. Antoine, J.-P. Gazeau, and U.A. Mueller, *Coherent states and their generalizations: A mathematical overview*, Rev. Math. Phys. **7** (1995), 1013–1104.

2. A. Bertrand, *Développements en base de Pisot et répartition modulo 1*, C. R. Acad. Sci., Paris, **285** (1977), 419–421.

3. A. Odzijewicz, *Quantum algebras and q-special functions related to coherent state quantization of the disk*, Comm. Math. Phys. **192** (1998), no. 1, 183–215.

4. A. Rényi, *Representations for real numbers and their ergodic properties*, Acta Math. Acad. Sci. Hungar. **8** (1957), 477–493, (1957).

6

Continuous and Discrete Linearizable Systems: The Riccati Saga

B. Grammaticos and A. Ramani

ABSTRACT We investigate the extensions of the discrete Riccati equation, as a linearizable system, to higher dimensions. We first study the continuous and discrete (second-order) Gambier equation, which is a coupling of two Riccati equations in cascade. In the N-dimensional case, three new integrable mappings are obtained: they are the linearizable discretizations of the well-known projective, matrix, and conformal Riccati systems.

1 Introduction

Riccati equations have always played an important role in the domain of nonlinear systems. Already the "standard" first-order, one-component differential equation

$$y' = \alpha y^2 + \beta y + \gamma \qquad (1.1)$$

is quite remarkable. It can be transformed into a linear, second-order equation through a simple Cole–Hopf transformation:

$$y = -\frac{w'}{\alpha w}, \qquad (1.2)$$

leading to

$$w'' + \left(\beta + \frac{\alpha'}{\alpha}\right)w' + \alpha\gamma w = 0. \qquad (1.3)$$

(This result is also a minor disappointment since it shows that, although the Riccati equation is integrable, it does not define any new transcendent.) As a consequence of the linearization (1.3) the Riccati equation possesses a (nonlinear) superposition property. Any four solutions of Eq. (1.1) can be related through the expression

$$\frac{(y_4 - y_1)(y_3 - y_2)}{(y_4 - y_2)(y_3 - y_1)} = k, \qquad (1.4)$$

where k is an integration constant. Moreover, the Riccati equation is the only first-order differential equation of the form $y' = P(y, z)$, with P polynomial in y, that possesses the Painlevé property. Indeed the dominant singularity of Eq. (1.1) is just

$$y \sim -\frac{1}{\alpha(z - z_0)} + \ldots \tag{1.5}$$

and y can be given as a Laurent series involving one free parameter (depending on the initial conditions), namely, the location z_0 of the singularity.

It was quite natural, as soon as the interest of the integrability community focused on discrete systems, to look for the discrete form of the Riccati equation. One very interesting way to derive the Riccati mapping is to ask for a discrete system that possesses the same nonlinear superposition property as the continuous one. Starting from Eq. (1.4), we assume that the dependent variable $y \equiv y_4$ and the *given* functions y_1, y_2, y_3 depend on the discrete variable n. Next, we up-shift (1.4) (i.e., we compute it at the point $n + 1$) and eliminate the constant k. The result is the homographic mapping

$$\overline{y} = \frac{ay + b}{cy + d}, \tag{1.6}$$

where $y \equiv y(n)$, $\overline{y} \equiv y(n+1)$ and the a, b, c, d can be expressed in terms of the y_1, y_2, y_3. Expression (1.6) is the well-known integrable discretization of the Riccati equation. It can be linearised through the discrete equivalent of the Cole–Hopf transformation

$$y = \frac{P}{Q}, \tag{1.7}$$

which leads to

$$P = \overline{Q} - \frac{d}{c}Q, \tag{1.8}$$

$$\overline{\overline{Q}} - \left(\frac{\overline{d}}{\overline{c}} + \frac{a}{c}\right)\overline{Q} + \left(\frac{ad - bc}{c^2}\right)Q = 0. \tag{1.9}$$

One interesting question is whether the Riccati mapping satisfies the discrete integrability criterion, i.e., whether it possesses the singularity confinement property. In Ref. [9] we have analyzed mappings of the form $\overline{y} = f(y, n)$, where f is a rational function of y. We have shown that all mappings of the form

$$\overline{y} = a_0 + \sum_k \frac{b_k}{(c_k y + d_k)^{m_k}}, \tag{1.10}$$

with integer m_k, do have confined singularities. However, the mapping (1.10) does not in general satisfy a more fundamental prerequisite of integrability, that of the absence of preimage proliferation [7]. If we consider the "backward" evolution toward diminishing n, then Eq. (1.10), solved for y, leads to multideterminacies. The only mapping of the form (1.10) that has unique preimages is the homographic, discrete Riccati system (1.6).

The natural question that arises from the study of the continuous and discrete Riccati equations is how one can extend these results to higher dimensions (i.e., to higher-order systems). At second order two different approaches exist. The first is based on linearizability, while the second, attributed to Gambier, considers Riccati equations coupled in cascade. Moving to higher-orders, the extension of linearizable systems leads to the consideration of families of the so-called matrix, projective, and conformal Riccati equations. In what follows we shall present our results on coupled Riccati systems with emphasis on the discrete case. We shall thus present the discrete equivalent of the Gambier equations as well as that of the N-dimensional matrix, projective, and conformal Riccati systems. Finally, in the concluding section we shall present some ideas as to how the Gambier approach could be extended to higher-order systems.

2 Brief Review of the Continuous Gambier Equation

In his ground-breaking mémoire [4], Gambier discovered the omissions in the Painlevé classification; filled in the gaps; and produced, among other integrable equations, the one bearing his name:

$$x'' = \frac{n-1}{n}\frac{x'^2}{x} + axx' + bx' - \frac{n-2}{n}\frac{x'}{x} + \frac{na^2}{(n+2)^2}x^3$$
$$+ \frac{n(a'-ab)}{n+2}x^2 + fx - b - \frac{1}{nx}. \tag{2.1}$$

This equation is probably the most complicated that Gambier had to study. It is not one of the transcendental equations, neither is it integrable by quadratures. Rather, it belongs to the class of equations that can be integrated through linearization.

The very essence of the Gambier equation is that it is obtained as a cascade of two Riccati equations [4]. One starts from a first Riccati:

$$y' = -y^2 + by + c, \tag{2.2}$$

where b, c are functions of the independent variable z, and couple y to x through a second Riccati:

$$x' = ax^2 + nyx + \sigma. \tag{2.3}$$

Here, a is a function of z, σ a constant that can be scaled to 1 unless it happens to be 0, and n is an integer. (As we shall see in what follows, this last point is a first requirement for the absence of critical singularities). The quantities a and σ in Eq. (2.3) are related through a duality. Indeed, replacing x by $-1/x$, we find that the equation retains the same form with a and σ interchanged (and $n \to -n$). Then if $a \neq 0$, the new σ can be set to 1 (through $x \to x/a$, and up to a translation of y, the new a becomes $a\sigma$ instead of just σ). In order to study the movable singularities of the coupled Riccati system, we start from the observation that from Eq. (2.2) the dominant behavior of y can only be $y \approx 1/(z - z_0)$. The next terms in the expansion of y can be easily obtained, and involve the functions b, c and their derivatives. In order to study the structure of the singularities of Eq. (2.3), we first remark that since the latter is a Riccati, its movable singularities are poles. However, Eq. (2.3) also has singularities that are owing to the singular behavior of the coefficients of the r.h.s of Eq. (2.3), namely, y. Now, the locations of the singularities of the coefficients are "fixed" as far as (2.3) is concerned. However, from the point of view of the full system (2.2)–(2.3), these singularities are *movable* and thus should be studied. The "fixed" character reflects itself in the fact that -1 is *not* a resonance. (The term *resonance* is used here following the ARS terminology [1] and means the order, in the expansion, where a free coefficient enters. A resonance -1 is related to the arbitrariness of the location of the singularity, and is thus absent when the location of the singularity is determined from the "outside" rather than by the initial conditions.) Because of the pole in y, x has a singular expansion with a resonance different from -1, which may introduce a compatibility condition to be satisfied.

Before proceeding to the examination of the general case (2.3), let us briefly consider the case $a = 0$, whereupon (2.3) becomes linear. If, moreover, we take $\sigma=0$, we find that the behavior of x is obtained through $x'/x = ny$ and thus $x \approx A(z - z_0)^n$. In this case, since n is an integer, (2.3) always has the Painlevé property, and the functions b and c are free. Next, we turn to the case of the general linear (2.3): $x' = nyx + \sigma$ with $\sigma \neq 0$. Let us examine a singularity of the form

$$x \sim \lambda(z - z_0). \tag{2.4}$$

We find $\lambda = \sigma/(1 - n)$ unless $n = 1$. (This last case is excluded as being of non-Painlevé type. Indeed, one easily finds that for $\sigma = 1$, the singular expansion has a logarithmic branching point of the form $x \sim (z - z_0)\log(z - z_0)$.) Thus, excluding the case $n = 1$, we study just the singular behavior (2.4). The resonance associated to the singularity (2.4) turns out to be exactly $n - 1$, which explains the requirement that n be an integer. For every value of $n > 1$ we expect a resonance condition (with complexity increasing with n). For any n, a sufficient condition is $\sigma = 0$ (which, as we have seen above, is also necessary for $n = 1$) and we have a further possibility that, for the first few orders of n, can be written:

$$
\begin{aligned}
n = 2 && 0 &= mb, \\
n = 3 && 0 &= 2b^2 - b' - c, \\
n = 4 && 0 &= 6b^3 - 7bb' + b'' - 8bc + 2c'.
\end{aligned}
\tag{2.5}
$$

For $n < 0$ there is no further constraint. In fact the sufficient condition is $a = 0$, which we have assumed in this paragraph.

We now turn to the case of the full Riccati (2.4) with $a \neq 0$. In this case, thanks to the duality, we can assume $\sigma = 1$ (otherwise interchanging a and σ we are back to the previous case). Again, only the singularity owing to y can lead to trouble. Rewriting (2.3) as $x'/x = ax + ny + \sigma/x$ for $y = 1/(z - z_0) + \ldots$ we remark that unless $n = \pm 1$ a behavior of the form $x \sim (z - z_0)^n$ is impossible when $a\sigma \neq 0$. For $n = 1$, a logarithmic leading behavior will be present for $\sigma \neq 0$, for nonzero a exactly as in the case $a = 0$. Conversely the condition $\sigma = 0$ is necessary and sufficient for the absence of a critical singularity for $n = 1$ even for nonvanishing a. Similarly, in a dual way, for $n = -1$ the necessary and sufficient condition for the absence of a critical singularity is $a = 0$, irrespective of the value of σ.

Next we assume $n \neq \pm 1$, in which case it suffices to study the singularities $x \approx \lambda(z - z_0)$, $(\lambda = \sigma/(1 - n))$ and $x \approx \mu/(z - z_0)$, $(\mu = -(n + 1)/a)$. The first singular behavior $(x \approx \lambda(z - z_0))$ has a resonance at $n - 1$, which is negative for $n < 1$ and thus does not introduce any further condition. For $n > 1$, the resonance condition can be studied at least for the first few values of n. We find that $\sigma = 0$ suffices for the resonance condition to be satisfied even for $a \neq 0$. However, if we demand $\sigma \neq 0$, we find the further possibilities:

$$
\begin{aligned}
n = 2 && 0 &= b, \\
n = 3 && 0 &= 2b^2 - b' - c + \tfrac{1}{2}a\sigma, \\
n = 4 && 0 &= 18b^3 - 21bb' + 3b'' - 24bc + 6c' + 8ab\sigma - 2a'\sigma.
\end{aligned}
\tag{2.6}
$$

The second singular behavior, $x \approx \mu/(z - z_0)$, has a resonance at $-1 - n$. Thus for $n > 0$ this resonance is negative and does not introduce any further condition, while for $n < 0$ a compatibility condition must be satisfied. We find that for every case $n < 0$, $a = 0$ is a sufficient condition for the absence of critical singularity. (This is not in the least astonishing given the duality of a and σ). On the other hand, if we demand $a \neq 0$, then a different resonance condition is obtained, at each value of n. For $n = -1$, whenever $a \neq 0$, a logarithmic singularity of the form $(z - z_0)^{-1} \log(z - z_0)$ appears irrespective of the value of σ. For $n < -1$, we find

$$
\begin{aligned}
n = -2 \quad & 0 = a' - ab, \\
n = -3 \quad & 0 = 4ab^2 - 2ab' - 2ac + a^2\sigma - 6a'b + 2a'', \\
n = -4 \quad & 0 = a(18b^3 - 21bb' + 3b'' - 24bc + 6c' + 8ab\sigma), \\
& \quad + a'(12b' + 12c - 33b^2 - 6a\sigma) + 18ba'' - 3a'''.
\end{aligned}
\tag{2.7}
$$

A remark is in order at this point. The analysis presented above was based explicitly on the assumption that (2.2) is a Riccati. However, if we take $n \to \infty$ in Eq. (2.3) and rescale y (and c) in an appropriate way we find that (2.2) becomes *linear*: $y' = by + c$. Thus, while on the one hand $n \to \infty$ in Eq. (2.7) leads to a resonance condition that should be implemented after an infinite number of steps; on the other hand, we do not have to consider it. Equation (2.2) being now linear, y does not have *any* singularity that could interact with (2.3). The equation for x reads in that case:

$$
x'' = \frac{x'^2}{x} + \left(ax + b - \frac{\sigma}{x}\right)x' + (a' - ab)x^2 + cx - \sigma b, \tag{2.8}
$$

which is a form of the equation XIV in the Painlevé–Gambier classification [2, 4] (not the canonical one but an equivalent, generic one). We recall that the canonical form of the latter is

$$
w'' = \frac{w'^2}{w} + \left(qw + \frac{r}{w}\right)w' + q'w^2 - r'. \tag{2.9}
$$

Equation (2.8) can be integrated to a Riccati:

$$
x' = ax^2 + qx + \sigma, \tag{2.10}
$$

where q is given by $(q\phi)' = c\phi$ and ϕ is an auxiliary function related to b through $b = -\phi'/\phi$.

3   Discrete Analog of the Gambier Equation, Revisited

The discretization of the Gambier equation will be based on the idea of two Riccati equations in cascade. The discrete form of the first is simply

$$
\overline{y} = \frac{by + c}{y + 1}, \tag{3.1}
$$

where the denominator can be generically be brought to this form through a scaling of y and a division by an overall factor. The second equation that contains the coupling can be discretized in several, not necessarily equivalent, ways. In [5] we have proposed the discretization:

$$\bar{x} = \frac{xyd + \sigma}{1 - ax}. \tag{3.2}$$

A different approach could be based on the direct discretization of Eq. (2.3) in the form

$$\bar{x} - x = -fx\bar{x} + (gx + h\bar{x})y + k. \tag{3.3}$$

Instead of choosing *a priori* a particular form, let us start (in the spirit of [6]) from a generic coupling of the form

$$\alpha x\bar{x}y + \beta x\bar{x} + \gamma\bar{x}y + \delta\bar{x} + \varepsilon xy + \zeta x + \eta y + \theta = 0. \tag{3.4}$$

Implementing a homographic transformation on x and y we can generically bring (3.4) under the form

$$x\bar{x} + \gamma\bar{x}y - \varepsilon xy - \eta y - \theta = 0, \tag{3.5}$$

(the sign changes were introduced for future convenience). A choice of different transformations can bring (3.4) to the form (3.3), while (3.2) can be obtained through a special choice of the parameters of (3.4). Note that (3.4) contains an "additive" type coupling $x\bar{x} + \delta\bar{x} + \zeta x + \eta y + \theta = 0$ for special values of its parameters, but the generic form (3.5) is that of a "multiplicative" coupling where γ, ε do not vanish. Solving (3.5) for $\bar{x}$ we obtain the second equation of the discrete Gambier system in the form

$$\bar{x} = \frac{\varepsilon xy + \eta y + \theta}{x + \gamma y}. \tag{3.6}$$

Eliminating y and $\bar{y}$ from (3.1), (3.6) and its upshift, we can obtain a three-point mapping for x alone but the analysis is clearer if we deal with both y and x.

The main tool for the investigation of the integrability of the Gambier mapping will be the singularity confinement criterion. A first remark before implementing the singularity confinement algorithm is that the singularities of a Riccati mapping are automatically confined. Indeed, if we start from $\bar{x} = (ax + \beta)/(\gamma x + \delta)$ and assume that at some step $x = -\delta/\gamma$, we find that $\bar{x}$ diverges but $\bar{\bar{x}}$ and all subsequent x's are finite. Thus, the intrinsic singularities of (3.6) do not play any role. However, the singularities owing to y (obtained from Eq. (3.1)) may cause problems at the level of Eq. (3.6). Whenever y takes a value that corresponds to either of the two roots (v, w) of the equation

$$\varepsilon\gamma y^2 = \eta y + \theta, \tag{3.7}$$

we obtain $\bar{x} = -\varepsilon v$ (or $\bar{x} = -\varepsilon w$) irrespective of the value of x and thus the variable x loses a degree of freedom. On the other hand, once we enter a singularity there is no way to exit it unless y assumes again a special value after a certain number of steps. Thus, if we enter the singularity through, say, $y = v$ we can exit it through $y = w$ after N steps. However, if y were to take the value v again some steps after taking it for the first time, then it would take it periodically and the singularity would be periodic. This is contrary to the requirement that the singularity be movable: a periodic singularity (with fixed period) is "fixed" in our terminology.

Before proceeding further let us choose the scaling of x so as to have $\varepsilon\gamma = 1$. Then the two roots (v, w) of Eq. (3.7) are related through

$$v + w = \eta. \tag{3.8}$$

Assume that v is a root of Eq. (3.7) for some n. Thus x enters a singularity and can only come out of it if after N steps y is identical to w. Thus the first confinement condition reads

$$v(N) + y(N) = \eta(N), \tag{3.9}$$

where $y(N)$ is the Nth iterate of v through the discrete Riccati (3.1). We have for example

$$N = 1 \qquad \frac{av + b}{v + 1} = \bar{\eta} - \bar{v}, \tag{3.10a}$$

$$N = 2 \qquad \frac{\bar{a}(av + b) + \bar{b}(v + 1)}{av + b + v + 1} = \bar{\bar{\eta}} - \bar{\bar{v}}, \tag{3.10b}$$

etc. The equivalent of this requirement in the continuous case is that the resonance be integer. We see here that the discrete condition is much more complicated and while one can easily compute the first few no general expression can be given. Once y passes through the second special value, w, there is a possibility for x to recover its lost degree of freedom through an indeterminate form $0/0$. This is the confinement condition. In full generality (and somewhat abstract form) it reads

$$y(N) + \varepsilon(N)x(N) = 0, \tag{3.11}$$

where $y(N)$ is the same as in condition (3.9) and $x(N)$ is the Nth iterate of x through (3.6). (Notice that (3.11) ensures just the vanishing of the denominator of Eq. (3.6) but because of Eq. (3.7) the numerator also vanishes). In order to implement the confinement condition we must write (3.11) explicitly. We have, for example,

$$N = 1 \qquad \frac{av + b}{v + 1} = -\bar{\varepsilon}\varepsilon v,$$

$$N = 2 \qquad \frac{\bar{a}(av + b) + \bar{b}(v + 1)}{av + b + v + 1} = -\bar{\bar{\varepsilon}}\,\bar{\varepsilon}\frac{\bar{v}^2(v + 1) + \bar{\varepsilon}\varepsilon v(av + b)}{av + b + \bar{\varepsilon}\varepsilon v(v + 1)}.$$

Although the analysis above completes the study of the generic Gambier mapping, there remain several special cases to be examined separately. Some of them correspond to cases where the Riccati for y becomes simply $\bar{y} = a + 1/y$ or even linear $\bar{y} = y + a$. The remaining cases correspond to degenerate forms of the coupling equations (3.6): $\gamma = 0$ or $\varepsilon = 0$ or $\eta = \theta = 0$ or $\eta^2 + 4\varepsilon\gamma\theta = 0$. In fact, it is possible, through the appropriate choices to establish the discrete equivalent of all the linearizable second-order equations of the Painlevé–Gambier classification.

4   Discrete Projective and Matrix Riccati Equations

In this section we shall show how one can discretize in a straightforward way two N-dimensional coupled Riccati systems—projective and conformal Riccati's.

The projective Riccati equations have the form

$$x'_\mu = a_\mu + \sum_\nu b_{\mu\nu} x_\nu + x_\mu \sum_\nu c_\nu x_\nu, \tag{4.1}$$

with $\mu = 1, \ldots, N$. They have been studied in detail in Ref. [2], where they were shown to possess the superposition property. As a matter of fact, the linearization of Eq. (4.1) is straightforward. The idea is to start from a linear system in $(N + 1)$ dimensions:

$$X'_\mu = A_{\mu\nu} X_\nu \quad \mu = 0, 1, \ldots, N, \tag{4.2}$$

and obtain the equations for the quantities

$$x_\mu = X_\mu / X_0 \quad \mu = 1, \ldots, N. \tag{4.3}$$

It turns out that these are precisely projective Riccati equations. Conversely using substitution (4.3) in Eq. (4.1) we can obtain the linearization:

$$X'_0 = \sum_\nu c_\nu X_\nu + X_0.$$

$$X'_\mu = \sum_\nu b_{\mu\nu} X_\nu + a_\mu X_0. \tag{4.4}$$

How can one discretize the projective Riccati equations? To make a long story short we can show that the straightforward discretization:

$$\overline{w}_\mu - w_\mu = a_\mu + \sum_\nu b_{\mu\nu} w_\nu + \overline{w}_\mu \sum_\nu c_\nu w_\nu, \tag{4.5}$$

or equivalently

$$\overline{w}_\mu = \frac{a_\mu + w_\mu + \sum_\nu b_{\mu\nu} w_\nu}{1 - \sum_\nu c_\nu w_\nu}, \tag{4.6}$$

leads to direct linearization. The starting point is the system

$$\overline{X}_\mu = \sum_{\nu=0}^{N} M_{\mu\nu} X_\nu, \quad \mu = 0, 1, \ldots, N. \tag{4.7}$$

We introduce the projective variables $w_\mu = X_\mu/X_0$ and obtain from Eq. (4.7) the equations

$$\overline{w}_\mu = \frac{M_{\mu 0} + \sum_{\nu=1}^{N} M_{\mu\nu} w_\nu}{M_{00} + \sum_{\nu=1}^{N} M_{0\nu} w_\nu}. \tag{4.8}$$

System (4.8) is the general expression of the projective Riccati. (One can easily relate the parameters entering (4.6) to those of Eq. (4.8).)

The case $N = 1$ of the projective Riccati is just the standard homographic mapping. The $N = 2$ case was first obtained in Ref. [9] through a direct linearizability approach. In this special case the mapping can be expressed in terms of a single variable $u = w_1/w_2$ and assumes the form

$$\overline{u} = \frac{(\alpha u + \beta) - u(\gamma u + \delta)}{(\varepsilon u + \zeta) - u(\eta u + \theta)}. \tag{4.9}$$

The quantities $\alpha, \beta, \ldots, \theta$ are expressed in terms of the $M_{00}, \ldots, M_{22}$ of the linear system (as well as their down-shifts, i.e., their values at $n - 1$), which implies some relations among the former quantities. In [9] we have applied the singularity confinement criterion to a mapping of the form of Eq. (4.9). It turned out that the condition for confinement (and thus for integrability) as constraints on the parameters $\alpha, \beta, \ldots, \theta$ was precisely the relations coming from linearizability.

The matrix Riccati equations extend the approach of projective ones in a straightforward way. Let us start with the continuous case. We consider the linear system

$$\begin{pmatrix} X' \\ Y' \end{pmatrix} = \begin{pmatrix} C & A \\ -D & -B \end{pmatrix} \begin{pmatrix} X \\ Y \end{pmatrix}, \tag{4.10}$$

where X, Y, A, B, C, D are matrices of dimensions $N \times L$, $L \times L$, $N \times L$, $L \times L$, $N \times N$, $L \times N$, respectively. We introduce the variable $W = XY^{-1}$ (W is thus a $N \times L$ matrix) and obtain the matrix Riccati equations:

$$W' = A + WB + CW + WDW. \tag{4.11}$$

The discretization of Eq. (4.11) is straightforward. We start with the matricial linear mapping:

$$\begin{pmatrix} \overline{X} \\ \overline{Y} \end{pmatrix} = \begin{pmatrix} C & A \\ D & B \end{pmatrix} \begin{pmatrix} X \\ Y \end{pmatrix}, \tag{4.12}$$

introduce again the variable $W = XY^{-1}$, and find the equation

$$\overline{W} = (CW + A)(DW + B)^{-1}. \tag{4.13}$$

Equation (4.13) is the discrete matrix Riccati system.

5   Discrete Conformal Riccati Equations

We now turn to the case of the conformal Riccati's. We start with the case $1 - N - 1$ associated to the metric:

$$K = \begin{pmatrix} 0 & 0 & 1 \\ 0 & \mathbb{I}_N & 0 \\ 1 & 0 & 0 \end{pmatrix}, \tag{5.1}$$

where $\mathbb{I}_N$ is the $N \times N$ unit matrix. We introduce the coefficient matrix (where the symbol $M^\dagger$ denotes the transpose of the matrix M):

$$A = \begin{pmatrix} a & \alpha^\dagger & 0 \\ \beta & B & -\alpha \\ 0 & -\beta^\dagger & -a \end{pmatrix}, \tag{5.2}$$

where B is an antisymmetric $(B + B^\dagger = 0)$ $N \times N$ matrix. Given the form of A we have $AK + KA^\dagger = 0$. Next, we introduce the linear system

$$V' = AV, \tag{5.3}$$

with $V^\dagger = (x, y^\dagger, z)$ (where y is a N-dimensional vector) and impose the constraint

$$V^\dagger K V = 0. \tag{5.4}$$

Introducing the new variables $u = x/z$ and $w = y/z$ we obtain from Eq. (5.4) the relation $u = -w^\dagger w/2$. From Eq. (5.3) we obtain the equations of motion for u (which turns out to be identically satisfied) and for w, which is precisely the conformal Riccati. We obtain, in matrix notation

$$w' = -\alpha + (a + B)w + w(\beta^\dagger w) - \tfrac{1}{2}\beta w^\dagger w. \tag{5.5}$$

The discretization of the conformal Riccati turns out to be straightforward. Starting with the same K we introduce the coefficient matrix M through the relation

$$M^\dagger K M = K \tag{5.6}$$

(where $M^\dagger$ denotes the transpose of the matrix M). This is a nontrivial condition but the practical construction of M lies beyond the scope of the present work. We simply parametrize M as

$$M = \begin{pmatrix} a & B^\dagger & c \\ 2D & E & F \\ 2g & -H^\dagger & j \end{pmatrix}, \tag{5.7}$$

where B, D, F, H are (N component) column vectors and E is a $N \times N$ antisymmetric matrix. Next, we introduce the linear problem

$$\overline{X} = MX, \tag{5.8}$$

where $X^\dagger = (x, y^\dagger, z)$ (with scalar x, z and a vector y) and the (conformality) constraint

$$X^\dagger K X = 0. \tag{5.9}$$

It is straightforward to show that this condition is conserved under the evolution (5.8). Indeed, $\overline{X}^\dagger K \overline{X} = X^\dagger M^\dagger K M X = X^\dagger K X = 0$. We introduce the variables: $u = x/z$ and $w = y/z$. The condition (5.9) writes $2xz + y^\dagger y = 0$ and becomes $u = -w^\dagger w/2$. We obtain thus the conformal Riccati mapping

$$\overline{w} = \frac{Ew + F - w^\dagger w D}{H^\dagger w + j - g w^\dagger w}, \tag{5.10}$$

(while we have automatically $\overline{u} = -\overline{w}^\dagger \overline{w}/2$, since $\overline{X}^\dagger K \overline{X} = 0$).

The extension of the conformal Riccati mapping to the $L + N + L$ case does not present any difficulties. We start from a metric:

$$K = \begin{pmatrix} 0 & 0 & \mathbb{I}_L \\ 0 & \mathbb{I}_N & 0 \\ \mathbb{I}_L & 0 & 0 \end{pmatrix}. \tag{5.11}$$

The coefficient matrix is now

$$M = \begin{pmatrix} A & B & C \\ D & E & F \\ G & H & J \end{pmatrix}, \tag{5.12}$$

where we require $M^\dagger K M = 0$. Again, we will not elaborate on the practical implementation of this condition. The linear problem is again

$$\overline{X} = MX, \tag{5.13}$$

where $X^\dagger = (x, y^\dagger, z)$ and x, y, z are matrices of dimensions $L \times L$, $N \times L$, and $L \times L$, respectively. The (conformality) condition reads again

$$X^\dagger K X = 0, \tag{5.14}$$

and is automatically conserved under the evolution (5.13). We expand (5.14) to

$$z^\dagger x + y^\dagger y + x^\dagger z = 0. \tag{5.15}$$

Next, we introduce the variables $v = xz^{-1}$ and $u = yz^{-1}$ and rewrite condition (5.15) as

$$v + v^\dagger + u^\dagger u = 0, \tag{5.16}$$

(and we have, by construction $\bar{v} + \bar{v}^\dagger + \bar{u}^\dagger \bar{u} = 0$). The equations of motion become

$$\bar{v} = (Av + Bu + C)(Gv + Hu + J)^{-1}, \qquad (5.17a)$$
$$\bar{u} = (Dv + Eu + F)(Gv + Hu + J)^{-1}. \qquad (5.17b)$$

We introduce the antisymmetric part of v through $w = (v - v^\dagger)/2$ and we have $v = w - u^\dagger u/2$, $v^\dagger = -w - u^\dagger u/2$. Substituting into (5.17b), we obtain an equation for $\bar{u}$, in terms of w and u, while using (5.17a) and its transpose we can express the evolution of w. These equations are the discrete conformal Riccati equations in $L + N + L$ dimensions. Starting from this general case one can construct various interesting subcases.

6 Conclusions and Outlook

In the previous sections we have presented extensions of the Riccati equations to higher dimensions with emphasis on their discrete forms. In particular we have obtained the discrete forms of the N-dimensional linearizable Riccati systems that are known as projective, matrix, and conformal Riccati. An exhaustive treatment of discrete linearizable systems can be found in our work [8], in collaboration with P. Winternitz.

One interesting question that remains essentially open at present is that of extending the approach of Gambier to higher-order equations. The structure of the singularities of more than two Riccati's in cascade is rather complicated and would necessitate an extension of the "standard" Painlevé method. On the other hand, the essence of the Gambier approach was to couple an integrable first-order equation to another integrable first-order equation. Thus the first extension of this approach would be to couple an integrable first-order equation (Riccati or linear) to an integrable second-order one. Although the coupling of the Gambier equation to a Riccati seems excessively difficult, the coupling of (some of) the Painlevé equations to a Riccati does not present major difficulties. (As a matter of fact the coupling of P_I to a Riccati was already studied by Chazy [3]). We hope in the near future to be able to present results on this problem in both the continuous and discrete case.

Acknowledgments: The authors wish to express their sincerest thanks to P. Winternitz for a collaboration that started 15 years ago and is still going strong.

REFERENCES

1. M.J. Ablowitz, A. Ramani, and H. Segur, *Nonlinear evolution equations and ordinary differential equations of Painlevé-type*, Nuovo Cimento **23** (1978), 333–338.

2. R.L. Anderson, J. Harnad, and P. Winternitz, *Systems of ordinary differential equations with nonlinear superposition principles*, Phys. D **4** (1982), 164–182.

3. J. Chazy, *Sur les équations différentielles du troisième ordre et d'ordre supérieur dont l'intégrale générale a ses points critiques fixes*, Acta Math. **34** (1910), 317–385.

4. B. Gambier, *Sur les équations différentielles du second ordre et du premier degré dont l'intégrale générale est à points critiques fixes*, Acta Math. **33** (1910), 1–55.

5. B. Grammaticos and A. Ramani, *The Gambier mapping*, Phys. A **223** (1995), 125–136.

6. B. Grammaticos and A. Ramani, *Retracing the Painlevé–Gambier classification for discrete systems*, Methods Appl. Anal. **4** (1997), no. 2, 196–211.

7. B. Grammaticos, A. Ramani and K.M. Tamizhmani, *Nonproliferation of preimages in integrable mappings*, J. Phys. A **7** (1994), 559–566.

8. B. Grammaticos, A. Ramani, and P. Winternitz, *Dicretizing families of linearizable equations*, Phys. Lett. A **245** (1998), no. 5, 382–388.

9. A. Ramani, B. Grammaticos, and G. Karra, *Linearizable mappings*, Phys. A **181** (1992), 115–127.

7

Superintegrability on Two-Dimensional Complex Euclidean Space

E.G. Kalnins, W. Miller, Jr., and G.S. Pogosyan

ABSTRACT In this work we examine the basis functions for those quantum mechanical systems on two-dimensional complex Euclidean space, with nonzero potential, that admit separation in at least two coordinate systems. We present all of these cases from a unified point of view. In particular, all of the polynomial special functions that arise via variable separation have their essential features expressed in terms of their zeros.

1 Introduction

In a previous work [12] we examined the basis functions for those classical and quantum mechanical systems in real two-dimensional Euclidean space that admit a solution via separation of variables in more than one coordinate system. (We recall that the notion of a *superintegrable system* relates to a potential for which the solution via separation of variables is possible in more than one coordinate system, [4, 5, 16]. The interest in these systems results from the fact that they have extra integrals of the motion that can be used to break the degeneracy of the energy eigenstates.) The methodical search for such potentials was initiated by Smorodinsky, Winternitz *et al.* in two and three dimensions [1, 6, 7]. (For some related work, see [2, 9, 10, 13].) In this paper we look at the same problem for the case of complex Euclidean space. In so doing we show that there are a number of additional potentials possible other than those of the real case alone. These additional potentials are complexifications of potentials for the Klein Gordan equation on real flat space with indefinite metric.

Our basic problem is to solve Schrödinger's equation

$$H\Psi = -\tfrac{1}{2}\Delta\Psi + V(x,y)\Psi = -\tfrac{1}{2}(\partial_x^2 + \partial_y^2)\Psi + V(x,y)\Psi = E\Psi, \quad (1.1)$$

via an ansatz of the form $\Psi = \psi_1(u_1)\psi_2(u_2)$ in some suitable coordinate system $x = f(u_1, u_2), y = g(u_1, u_2)$. We are interested in potentials $V(x,y)$ for which this is possible for two or more distinct coordinate systems. For the case of real Euclidean space there are exactly four such potentials, and a detailed analysis of their finite solutions or bound states has been given previously, e.g., [2, 10, 12]. We summarize the results of previous work that we need in order to give a readable presentation. In addition to the four potentials possible for real Euclidean space we find in the complex case three more potentials. For one of the potentials there is the possibility of finite polynomial solutions. We give the details of the eigenfunction solutions and the structure of the quadratic algebra associated with each system. The quadratic algebra yields essential information for the expansion of the basis functions of one separable system in terms of the basis functions for a second system.

There are seven possible potentials. The four potentials already treated in the real Euclidean case are

$$V_1(x, y) = \frac{1}{2}\omega^2(x^2 + y^2) + \frac{k_1^2 - 1/4}{x^2} + \frac{k_2^2 - 1/4}{y^2}, \tag{1.2}$$

$$V_2(x, y) = \frac{1}{2}\omega^2(4x^2 + y^2) + \frac{k_2^2 - 1/4}{y^2}, \tag{1.3}$$

$$V_3(x, y) = \frac{-\alpha}{\sqrt{x^2 + y^2}} \tag{1.4}$$

$$+ \frac{k_1^2 - 1/4}{\sqrt{x^2 + y^2}[\sqrt{x^2 + y^2} + x]} + \frac{k_2^2 - 1/4}{\sqrt{x^2 + y^2}[\sqrt{x^2 + y^2} - x]}, \tag{1.5}$$

$$V_4(x, y) = \frac{-\alpha}{\sqrt{x^2 + y^2}} + B_1 \frac{\sqrt{\sqrt{x^2 + y^2} + x}}{\sqrt{x^2 + y^2}} + B_2 \frac{\sqrt{\sqrt{x^2 + y^2} - x}}{\sqrt{x^2 + y^2}}. \tag{1.6}$$

These potentials are thoroughly treated in our previous article. We now turn our attention to the *new* potentials that occur in the complex case. The possible orthogonal separable coordinate systems in complex Euclidean space, together with their defining symmetry operators for the zero potential case [14], are listed in Table 7.1 and the remaining potentials are listed in Table 7.2. Here,

$$M = y\partial_x - x\partial_y, \quad P_1 = \partial_x, \quad P_2 = \partial_y.$$

2 Potential V_5

The Schrödinger equation for potential V_5 admits separable solutions in three distinct coordinate systems. In cartesian coordinates this equation has the form

$$-\tfrac{1}{2}(\partial_x^2 + \partial_y^2)\Psi + \tfrac{1}{2}B(x - iy)\Psi = E\Psi. \tag{2.1}$$

TABLE 7.1. Separable coordinate systems on two-dimensional complex Euclidean space.

Coordinate System	Integrals of Motion	Coordinates
I. Polar	$I = M^2$	$x = r \cos \theta$
$r > 0, \ \theta \in [0, \pi]$		$y = r \sin \theta$
II. Cartesian	$I = P_1^2$	x, y
$x, y \in R$		
III. Light Cone	$I = (P_1 + iP_2)^2$	$\widehat{x} = x - iy$
$x, y \in R$		$\widehat{y} = x + iy$
IV. Elliptic	$I = M^2 - a^2 P_2^2,$	$x = a \cosh \alpha \cos \beta$
$\alpha > 0, \beta \in [0, 2\pi]$	$a \neq 0$	$y = a \sinh \alpha \sin \beta$
V. Parabolic	$I = \{M, P_2\}$	$x = \frac{1}{2}(\xi^2 - \eta^2)$
$\xi, \eta > 0$		$y = \xi \eta$
VI. Hyperbolic	$I = M^2 + (P_1 + iP_2)^2$	$x = (u^2 + u^2 v^2 + v^2)/2vu$
.		$y = i(u^2 - u^2 v^2 + v^2)/2vu$
VII. Semi-Hyperbolic	$I = \{M, P_1 + iP_2\}$	$x = -\frac{1}{4}(w - z)^2$
	$+ (P_1 - iP_2)^2$	$+\frac{1}{2}(w + z)$
		$iy = -\frac{1}{4}(w - z)^2$
		$-\frac{1}{2}(w + z)$

TABLE 7.2. Complex superintegrable systems.

Potential $V(\vec{x})$	Coordinate System
$V_5 = \frac{B}{2}(x - iy)$	Cartesian
	Light Cone
	Semi-Hyperbolic
$V_6 = \frac{\alpha}{2\sqrt{x+iy}}$	Parabolic
	Semi-Hyperbolic
	Light Cone
$V_7 = \frac{1}{2}\left[\frac{\alpha(x^2+y^2)}{(x+iy)^4} + \frac{\beta}{(x+iy)^2} + \gamma(x^2 + y^2) \right]$	Polar
	Hyperbolic

It admits separable solutions $\Psi = \psi(x)\varphi(y)$ that satisfy

$$\partial_x^2 \psi(x) + Bx\psi(x) = (2E + \lambda)\psi(x), \quad \partial_y^2 \varphi(y) - iyB\varphi(y) = (2E - \lambda)\varphi(y).$$

The solutions obtained in this way are products of Airy functions [3], typically

$$\psi(x) = Ai\left((-B)^{1/3}x - \frac{2E + \lambda}{B}\right), \quad \varphi(y) = Ai\left((iB)^{1/3}y - \frac{2E - \lambda}{B}\right).$$

The operator K, whose eigenvalue is the separation constant λ, is defined by

$$K\Psi = (-\partial_x^2 + \partial_y^2)\Psi + B(x + iy)\Psi.$$

In light cone coordinates Schrödinger's equation has the form

$$-4\partial_{\widehat{x}}\partial_{\widehat{y}}\Psi + B\widehat{y}\Psi = 2E\Psi, \tag{2.2}$$

which admits a solution of the form $\Psi = h(\widehat{x})k(\widehat{y})$, where

$$\partial_{\widehat{x}}h(\widehat{x}) = \rho h(\widehat{x}), \quad -4\rho\partial_{\widehat{y}}k(\widehat{y}) + (B\widehat{y} - 2E)k(\widehat{y}) = 0.$$

These equations have the solution

$$\Psi = \exp\left(\rho\widehat{x} + \frac{E}{2\rho}\widehat{y} - \frac{B}{8\rho}\widehat{y}^2\right).$$

The solutions are clearly eigenfunctions of the operator $N = \partial_x + i\partial_y = 2\partial_{\widehat{x}}$.

Finally, if we write the Schrödinger equation in semi-hyperbolic coordinates, it becomes

$$(\partial_w^2 - \partial_z^2)\Psi + B(w^2 - z^2)\Psi = 2E(w - z)\Psi. \tag{2.3}$$

The solutions of this equation can be obtained via the separation of variables ansatz $\Psi = W(w)Z(z)$. The separation equations are

$$\partial_x^2 X + Bx^2 X = (2Ex + \lambda)X,$$

where $X = W, Z$ when $x = w, z$, respectively. The solutions of these equations are typically

$$X(x) = D_\nu\left(\sqrt{2}(-B)^{1/4}\left(x - \frac{E}{B}\right)\right),$$

where $\nu = (-1/2\sqrt{-B})(-\lambda - (E/B)^2) - 1/2$, and $D_\nu(z)$ is a parabolic cylinder function [3]. The operator, whose eigenvalue is the separation parameter, is

$$L = \frac{1}{z - w}(z\partial_w^2 - w\partial_z^2) + Bzw.$$

The operators K, L, N, and H that define these bases satisfy the following quadratic algebra relations [8, 12] :

$$[N, K] = -\tfrac{1}{2}B, \quad [N, L] = -N^2 - B,$$
$$[K, L] = \tfrac{1}{2}N^3 + \{N, K\} + \tfrac{1}{4}BN, \tag{2.4}$$
$$4H^2 + N^4 + 4\{N^2, K\} - 4BL - 8K - BN^2 = 0.$$

Here, $[A, B] = AB - BA$, $\{A, B\} = AB + BA$. There are no solutions that are the analogue of bound states in this case.

3   Potential V_6

For potential V_6 the Schrödinger equation admits separable solutions in three coordinate systems.

In parabolic coordinates the Schrödinger equation has the form

$$(\partial_\xi^2 + \partial_\eta^2)\Psi - \alpha\sqrt{2}(\xi - i\eta)\Psi = 2E(\xi^2 + \eta^2)\Psi. \tag{3.1}$$

The separation equations are

$$\partial_\xi^2 X - \alpha\sqrt{2}\xi X = (2E\xi^2 + \lambda)X, \quad \partial_\eta^2 Y + \alpha\sqrt{2}i\eta Y = (2E\eta^2 - \lambda)Y,$$

where $\Psi = X(\xi)Y(\eta)$. Solutions to these equations are typically

$$X(\xi) = D_\mu\left((8E)^{1/4}\left(\xi + \frac{\alpha}{\sqrt{2E}}\right)\right), \quad Y(\eta) = D_\kappa\left((8E)^{1/4}\left(\xi + \frac{i\alpha}{\sqrt{2E}}\right)\right),$$

where

$$\mu = \frac{1}{\sqrt{8E}}\left(-\lambda + \frac{\alpha^2}{4E}\right) - \frac{1}{2}, \quad \kappa = \frac{1}{\sqrt{8E}}\left(\lambda - \frac{\alpha^2}{4E}\right) - \frac{1}{2},$$

and $D_\mu(x)$ is a parabolic cylinder function [3]. The operator whose eigenvalue is λ is

$$L = \frac{1}{\xi^2 + \eta^2}(\xi^2\partial_\xi^2 - \eta^2\partial_\eta^2) + \frac{i\sqrt{2}\alpha}{\xi + i\eta}.$$

In light cone coordinates the Schrödinger equation has the form

$$-4\partial_{\widehat{x}}\partial_{\widehat{y}}\Psi + \frac{\alpha}{\sqrt{\widehat{y}}}\Psi = 2E\Psi, \tag{3.2}$$

which admits a solution $\Psi = p(\widehat{x})q(\widehat{y})$ where

$$\partial_{\widehat{x}}p(\widehat{x}) = \rho p(\widehat{x}), \quad -4\rho\partial_{\widehat{y}}q(\widehat{y}) + \left(\frac{\alpha}{\sqrt{\widehat{y}}} - 2E\right)q(\widehat{y}) = 0.$$

These equations have the solution

$$\Psi = \exp\left(\rho\widehat{x} + \frac{E}{2\rho}\widehat{y} - \frac{\alpha}{2\rho}\sqrt{\widehat{y}}\right).$$

The solutions are clearly eigenfunctions of the operator $N = \partial_x + i\partial_y = 2\partial_{\widehat{x}}$.

For semi-hyperbolic coordinates the Schrödinger equation has the form

$$(\partial_w^2 - \partial_z^2)\Psi - i\alpha\sqrt{2}\Psi = 2E(w - z). \tag{3.3}$$

The separation equations are

$$\partial_w^2 R(w) + \left(-2Ew + \rho - \frac{i\alpha}{\sqrt{2}}\right)R(w) = 0,$$

$$\partial_z^2 S(z) + \left(-2Ez + \rho + \frac{i\alpha}{\sqrt{2}}\right)S(z) = 0,$$

where $\Psi = R(w)S(z)$. These equations have Airy function solutions

$$Ai\left((2E)^{1/3}w - \frac{\rho - i(\alpha)/(\sqrt{2})}{2E}\right), \quad Ai\left((2E)^{1/3}w - \frac{\rho + i(\alpha)/(\sqrt{2})}{2E}\right).$$

These solutions are eigenfunctions of the operator

$$M = \frac{1}{w-z}(z\partial_w^2 - w\partial_z^2 - \frac{i\alpha}{\sqrt{2}}(w+z)).$$

The basis elements L, M, T, and H satisfy the quadratic algebra relations [8, 12]

$$[L, N] = H + N^2, \quad [M, N] = -\tfrac{1}{2}H,$$
$$[M, L] = \tfrac{1}{4}N^3 + \{N, M\} + \tfrac{1}{8}HN, \qquad (3.4)$$
$$HL + 2HM + \tfrac{1}{8}N^2H + \{N^2, M\} + \tfrac{1}{8}N^4 - \tfrac{1}{4}\alpha^2 = 0.$$

4   Potential V_7

The Schrödinger equation for potential V_7 admits separable solutions in two coordinate systems.

In polar coordinates the equation has the form

$$\left(\partial_r^2 + \frac{1}{r}\partial_r + \frac{1}{r^2}\partial_\theta^2 - \frac{\alpha}{r^2}e^{-4i\theta} - \frac{\beta}{r^2}e^{-2i\theta} - \gamma^2r^2\right)\Psi = -2E\Psi. \qquad (4.1)$$

The θ separation equation has the form

$$(\partial_\theta^2 - \alpha e^{-4i\theta} - \beta e^{-2i\theta} - \lambda)\Lambda(\theta) = 0,$$

which has solutions of the form

$$\Lambda(\theta) = e^{(\alpha z)/2}z^{-1/2-\beta/(4\alpha)-p}L_p^{2p+\beta/(2\alpha)-1}(-\alpha z),$$

where $z = e^{-2i\theta}$, $\lambda = -[2p+1+\beta/(2\alpha)]^2$ and $L_p^q(x)$ is a Laguerre polynomial with p an integer [3]. The corresponding separation equation in r is

$$\left(\partial_r^2 + \frac{1}{r}\partial_r + \frac{\lambda}{r^2} - \gamma^2r^2 + 2E\right)R(r) = 0,$$

which has solutions

$$R(r) = e^{-\gamma r^2/2}r^{1/2+\beta/(4\alpha)+p}L_m^{1/2+\beta/(4\alpha)+p}(\gamma r^2),$$

for which $E = \gamma(2(m+p+1) + \beta/(2\alpha))$. The operator that corresponds to the eigenvalue λ is

$$M = \partial_\theta^2 - \alpha e^{-4i\theta} - \beta e^{-2i\theta}.$$

For hyperbolic coordinates the Schrödinger equation has the form

$$\left[\frac{1}{u^2 - v^2}\left(v\partial_v(v\partial_v) - u\partial_u(u\partial_u)\right) + \frac{\beta}{u^2 v^2} - \alpha^2\frac{u^2 + v^2}{u^4 v^4} + \gamma^2(u^2 + v^2)\right]\Psi$$
$$= 2E\Psi. \quad (4.2)$$

The separation equations are

$$\left[t\partial_t(t\partial_t) - \frac{\alpha^2}{t^4} + \frac{\beta}{t^2} - \gamma^2 t^4 + 2Et^2 + \mu\right]T(t) = 0,$$

where $\Psi = U(u)V(v)$ and $T = U$, V when $t = u$, v, respectively. The operator that characterizes this separation is

$$L = \frac{1}{u^2 - v^2}\left(v^2 u\partial_u(u\partial_u) - u^2 v\partial_v(v\partial_v)\right)$$

$$+ \alpha^2\left(\frac{1}{u^2 v^2} + \frac{1}{u^4} + \frac{1}{v^4}\right) - \beta\left(\frac{1}{u^2} + \frac{1}{v^2}\right) - \gamma^2 u^2 v^2.$$

To obtain polynomial eigenfunctions we can look for solutions of the form [11]

$$\Psi = \exp\left(\frac{\alpha}{2}\left[\frac{1}{u^2} + \frac{1}{v^2}\right] - \frac{1}{2}\gamma(u^2 + v^2)\right)$$

$$\times \prod_{i=1}^{n}\left(\frac{1}{u^2} - \theta_i\right)\left(\frac{1}{v^2} - \theta_i\right)(uv)^{1+n+\beta/2\alpha},$$

where the zeros satisfy

$$2\sum_{j\neq i}\frac{\theta_i^2}{\theta_i - \theta_j} - \left(\frac{\beta}{2\alpha} + 2n + 2\right)\theta_i + \gamma + \alpha^2\theta_i^2 = 0, \quad i = 1, \ldots, n.$$

The energy eigenvalues and the separation constant are

$$E = \gamma\left(2(n + 1) + \frac{\beta}{2\alpha}\right), \quad \lambda = -\left[2n + 1 + \frac{\beta}{2\alpha}\right]^2 - 2\alpha\gamma + 4\gamma\sum_{\kappa=1}^{n}\frac{1}{\theta_\kappa}.$$

The relations of the corresponding quadratic algebra are

$$[R, L] = -8L^2 - 24M^2 - 16\{L, M\} - 8\beta H - 32\alpha\gamma, \quad (4.3)$$

$$[R, M] = 16M^2 - 8M - 8L + 8\beta H + 8\{L, M\}, \quad (4.4)$$

$$R^2 = 16M^2 + \frac{16}{3}(L^2 M + ML^2 + LML) - \frac{176}{3}L^2 + \frac{80}{3}M^2 \quad (4.5)$$

$$- 16\alpha H^2 + 16\beta(HL + HM) - 16\{L, M\} + \left(64\gamma\alpha - \frac{256}{3}\right)M$$

$$- \frac{256}{3}L + \frac{128}{3}\beta H + \frac{64}{3}\gamma\alpha + 16\gamma\beta^2,$$

where $R = [L, M]$.

Acknowledgments: Two of the authors of this paper, Kalnins and Miller, met at the CRM in 1973 as a result of invitations from Patera and Winternitz, became involved in the work on variable separation being done at that time, e.g., [15], and have continued collaborating until this day. We are grateful to Pavel and George for stimulating this interaction and wish them the best on their 60th birthdays.

REFERENCES

1. C.P. Boyer, E.G. Kalnins, and P. Winternitz, *Completely integrable relativistic Hamiltonian systems and separation of variables in Hermitian hyperbolic spaces*, J. Math. Phys. **24** (1983), 2022–2034.

2. L.P. Eisenhart, *Enumeration of potentials for which one-particle Schrödinger equations are separable*, Physical Rev. (2) **74** (1948), 87–89.

3. A. Erdélyi, W. Magnus, F. Oberhettinger, and F.G. Tricomi (eds.), *Higher transcendental functions*, Vols. I–II, McGraw Hill, New York, 1953.

4. N.W. Evans, *Superintegrability in classical mechanics*, Phys. Rev. A (3) **41** (1990), 5666–5676; *Group theory of the Smorodinsky–Winternitz system*, J. Math. Phys. **32** (1991), 3369–3375.

5. N.W. Evans, *Super-integrability of the Winternitz system*, Phys. Lett. A **147** (1990), 483–486.

6. J. Friš, V. Mandrosov, Ya. A. Smorodinsky, M. Uhlir, and P. Winternitz, *On higher symmetries in quantum mechanics*, Phys. Lett. **16** (1965), 354–356.

7. J. Friš, Ya. A. Smorodinskii, M. Uhlír, and P. Winternitz, *Symmetry groups in classical and quantum mechanics*, Soviet J. Nuclear Phys. **4** (1967), 444–450.

8. Ya.A. Granovsky, A.S. Zhedanov, and I.M. Lutzenko, *Quadratic algebra as a "hidden" symmetry of the Hartmann potential*, J. Phys. A **24** (1991), 3887–3894.

9. C. Grosche, G.S. Pogosyan, and A.N. Sissakian, *Path integral approach to superintegrable potentials. The two-dimensional hyperboloid*, Phys. Particles Nuclei **27** (1996), 244.

10. C. Grosche, G.S. Pogosyan, and A.N. Sissakian, *Path integral discussion for Smorodinsky—Winternitz potentials. I. Two- and three-dimensional Euclidean space*, Fortschr. Phys. **43** (1995), 453–521.

11. E.G. Kalnins and W. Miller, *Separable coordinates, integrability and the Niven equations*, J. Phys. A **25** (1992), 5663.

12. E.G. Kalnins, W. Miller Jr., and G.S. Pogosyan, *Superintegrability and associated polynomial solutions. Euclidean space and the sphere in two dimensions* J. Math. Phys. **37** (1996), 6439–6467.

13. P. Letourneau and L. Vinet, *Superintegrable systems: Polynomial algebras and quasiexactly solvable Hamiltonians*, Ann. Physics **243** (1995), 144–168.

14. W. Miller, Jr, *Symmetry and separation of variables*, Addison-Wesley Publishing Company, Providence, Rhode Island, 1977.

15. J. Patera and P. Winternitz, *A new basis for the representation of the rotation group. Lamé and Heun polynomials*, J. Math. Phys. **14** (1973), 1130.

16. S. Wojciechovski, *Superintegrability of the Calogero–Moser System*, Phys. Lett. A **95** (1983), 279.

8

Hydrodynamic Systems and the Higher-Dimensional Laplace Transformations of Cartan Submanifolds

N. Kamran and K. Tenenblat

Dedicated to Jiři Patera and Pavel Winternitz
on the occasion of their 60th birthday

1 Introduction

Our purpose in this paper is to show how the recently discovered higher-dimensional Laplace transformation [5] for overdetermined systems of partial differential equations can be applied to the study of the class of systems of hydrodynamic type that are rich in conservation laws [6], also known as semi-Hamiltonian hydrodynamic systems [7]. This class includes such important examples as the electrophoresis system, the equations of ideal chromatography, the Benney system, and the system describing nonlinear electromagnetic plane waves [6, 7]. We will demonstrate that the higher-dimensional Laplace transformation, induced by the geometric Laplace transformation of Cartan manifolds [1, 2], is an effective tool for computing in closed form the infinitely many conserved densities admitted by the hydrodynamic systems belonging to this class. We will also show that there is a transformation of hydrodynamic systems that is induced by the higher-dimensional Laplace transformation. (A similar result has been announced in a preprint of Ferapontov [3].) As such, this transformation is a powerful tool for generating new systems that are rich in conservation laws from known ones.

Our paper is organized as follows. In Section 2, we recall briefly the basic definitions and fundamental properties of the class of hydrodynamic systems under consideration. In Section 3, we show how the higher-dimensional Laplace transformation of Cartan manifolds gives rise to a transformation mapping hydrodynamic systems that are rich in conservation laws to other such systems. We also characterize those hydrodynamic systems whose conserved densities are governed by an overdetermined system with vanishing Laplace invariants, so that their conserved densities may be determined by quadratures from those of a lower-dimensional system of the same type.

2 Hydrodynamic Systems Rich in Conservation Laws

Our goal in this section is to recall some fundamental results of D. Serre [6] and Tsarev [7] concerning hydrodynamic systems which are rich in conservation laws. We shall see in Section 3 that there is a well-defined action of the higher-dimensional Laplace transformation on these systems that is induced by the transformation of their conserved densities.

By a *hydrodynamic system*, we mean as usual a determined system of n first-order quasilinear partial differential equations of the form

$$u^i_{,t} = \sum_{i=1}^{n} v^i_j(u^1, \ldots, u^n) u^j_{,x}, \tag{2.1}$$

for n functions $u^i(x,t)$, $1 \leq i \leq n$. The coefficients v^i_j, $1 \leq i, j \leq n$ thus transform like the components of a $(1,1)$-tensor under smooth and invertible changes of variables $\bar{u}^i = f^i(u^1, \ldots, u^n)$, $1 \leq i \leq n$.

We shall consider only those hydrodynamic systems that are *diagonalizable* and *strongly hyperbolic*, meaning that there exist coordinates $\bar{u}^1, \ldots, \bar{u}^n$, usually called Riemann invariants, in which the $(1,1)$-tensor (v^i_j) is diagonal, with distinct eigenvalues $\bar{v}^i(u^1, \ldots, u^n)$, $1 \leq i \leq n$. Since we will always work in a coordinate system composed of Riemann invariants, we will drop the overbars and write our hydrodynamic system in diagonal form as

$$u^i_{,t} = v^i(u^1, \ldots, u^n) u^i_{,x}, \quad 1 \leq i \leq n. \tag{2.2}$$

A *conservation law*, or an *entropy-flux pair*, for a hydrodynamic system (2.1) consists in a pair of functions $\big(P(u^1, \ldots, u^n), Q(u^1, \ldots, u^n)\big)$, such that

$$\frac{DP}{Dt} + \frac{DQ}{Dx} = 0, \tag{2.3}$$

whenever $(u^1, \ldots, u^n)$ is a solution of Eq. (2.1). If the hydrodynamic system is diagonalizable and strongly hyperbolic, then it is straightforward to show that P must satisfy the overdetermined system of $n(n-1)/2$ second-order equations given by

$$P_{,ij} - \frac{v^i_{,j}}{v^j - v^i} P_{,i} - \frac{v^j_{,i}}{v^i - v^j} P_{,j} = 0, \quad 1 \leq i \neq j \leq n, \tag{2.4}$$

where $P_{,i} = \partial/(\partial u^i)$, etc. We will refer to the solutions P of Eq. (2.4) as *conserved densities* for the hydrodynamic system (2.2). If (2.2) is to have any nontrivial conserved densities, then the coefficients v^i, $1 \le i \le n$, of (2.2) must satisfy the constraints

$$\left(\frac{v^i_{,j}}{v^j - v^i}\right)_{,k} = \left(\frac{v^i_{,k}}{v^k - v^i}\right)_{,j},$$

(2.5)

for all $1 \le i \ne j \ne k \le n$. It is a nontrivial result, proved independently by D. Serre [6] and Tsarev [7] that the conditions (2.5) are also sufficient for the existence of infinitely many conserved densities, which form a complete set in the periodic and rapidly decreasing cases. We shall limit ourselves to stating the existence theorem, which appears already in the lectures of Darboux [7] on orthogonal coordinate systems.

Theorem 2.1. *A strongly hyperbolic hydrodynamic system* (2.2) *whose coefficients satisfy the conditions* (2.5) *has infinitely many linearly independent conserved densities depending on n arbitrary functions of one variable.*

In view of Theorem 2.1, the hydrodynamic systems whose coefficients satisfy (2.5) are called *rich in conservation laws* by D. Serre [6]. They are also called semi-Hamiltonian by Tsarev [7]. It has been conjectured by Ferapontov [4] that all such systems can be written in Hamiltonian form,

$$u^i_{,t} = \sum_{j=1}^{n} A^{ij} \frac{\delta H}{\delta u^j},$$

for some Hamiltonian $(0, 2)$-tensor-valued operator A^{ij}, which will generally be *nonlocal* and some Hamiltonian density H which will always be *local*.

An important technical tool in the proof of Theorem 2.1 and in the application of the higher-dimensional Laplace transformation to systems of hydrodynamic type is provided by the following existence theorem, attributed also to Darboux [7].

Theorem 2.2. *Let* $\Gamma^i_{ij}(u^1, \ldots, u^n)$, $1 \le i \ne j \le n$, *be a collection of* $n(n-1)$ *smooth functions, which are commutative in the lower indices, satisfying,*

$$\Gamma^i_{il,k} = \Gamma^i_{ik,l}, \qquad\qquad 1 \le i \ne k \ne l \le n, \qquad (2.6)$$
$$\Gamma^k_{ik,l} + \Gamma^k_{ik}\Gamma^k_{kl} - \Gamma^i_{il}\Gamma^k_{ik} - \Gamma^l_{il}\Gamma^k_{lk} = 0, \quad 1 \le i \ne k \ne l \le n. \qquad (2.7)$$

Then the linear first-order system

$$\frac{w^i_{,j}}{w^j - w^i} = \Gamma^i_{ij}, \quad 1 \le i \ne j \le n, \qquad (2.8)$$

(which is overdetermined for $n > 2$) admits infinitely many linearly independent solutions depending on n arbitrary functions of one variable.

Corollary 2.1. *If $w^i(u^1, \ldots, w^n)$ is a solution of (2.8), then the hydrodynamic system*

$$u^i_{,t} = w^i(u^1, \ldots, u^n)u^i_{,x}, \quad 1 \le i \le n, \tag{2.9}$$

will automatically be rich in conservation laws.

In fact Tsarev [7] proved that the flows that are solutions of the systems (2.9) corresponding to different solutions of (2.8) are commuting flows.

We conclude this section by listing some well-known examples of hydrodynamic systems that are rich in conservation laws:

(i) The electrophoresis system [6]:

$$u^1_{,t} = 0, \quad u^i_{,t} = u^1 \left(\prod_{j=2}^{n} u^j \right) u^i_{,x}, \quad 2 \le i \le n. \tag{2.10}$$

(ii) The ideal chromatography system [7]:

$$u^i_{,t} = u^i \left(\prod_{j=1}^{n} u^j \right) u^i_{,x}, \quad 1 \le i \le n. \tag{2.11}$$

3   Applications of the Higher-Dimensional Laplace Transformation to Hydrodynamic Systems that are Rich in Conservation Laws

In this section, we will show that the higher-dimensional Laplace transformation of n-dimensional Cartan submanifolds of $\mathbb{R}^{2n}$, which was developed in Ref. [5], induces naturally a new kind of transformation mapping hydrodynamic systems that are rich in conservation laws into other such systems. Moreover, we will show that the Laplace transformation can be used in order to obtain conserved densities for such systems.

Laplace transformations for Cartan submanifolds were first considered by Chern [1, 2] in projective space. In this section, we state the Euclidean version of this transformation so as to set up the basic notation. We refer the reader to [5] for all the proofs and details.

Definition. A Riemannian n-dimensional manifold M^n isometrically immersed in $\mathbb{R}^{2n}$ is said to be a *Cartan submanifold* if there exist local coordinates $(u^1, \ldots, u^n)$ such that the net of coordinate curves is conjugate and the osculating space is $2n$-dimensional.

It follows from this definition that given a parametrization of a Cartan manifold by conjugate curves, $X: U \subset \mathbb{R}^n \to \mathbb{R}^{2n}$, the second fundamental forms are simultaneously diagonalized. If we denote as usual by Γ^k_{ij} the Christoffel symbols for this parametrization we see (see lemma in Ref. [5])

that for each i, j with $i \neq j$ the vector field X_{ij} lies in the space spanned by X_i and X_j; i.e.,

$$X_{,ij} = \Gamma^i_{ij} X_{,i} + \Gamma^j_{ij} X_{,j}, \quad i \neq j. \tag{3.1}$$

Moreover, the Christoffel symbols are rather strongly constrained since they satisfy (2.6) and (2.7).

It has been shown that to each n-dimensional Cartan submanifold X, one can associate in general a family of $n(n-1)$ submanifolds that will also be Cartan submanifolds. The construction can be summarized as follows (we refer the reader to [5] for the details). For a fixed ordered pair (i,j), $1 \leq i \neq j \leq n$, such that $\Gamma^i_{ij} \neq 0$, consider the edge of regression of the ruled submanifold constructed from X by taking the envelope of the $n-1$-parameter family of n-planes tangent to X along the coordinate $n-1$-dimensional coordinate hypersurface $u^j = u^j_0$. By letting u^j_0 vary, we obtain the map

$$Y(u^1, \ldots, u^n)) = X - X_{,j}/\Gamma^i_{ij}, \quad i \neq j,$$

which will be called the (i,j)-*Laplace Transform* of X. A Laplace Transform of a Cartan submanifold is generically also a Cartan submanifold. In order to state the theorem, we need to introduce some notation. For each ordered pair (i,j), $i \neq j$ for which Γ^i_{ij} is nonzero, we define the functions

$$M_{ij} = \frac{\Gamma^i_{ij,i}}{\Gamma^i_{ij}} - \Gamma^j_{ij},$$

$$M_{ijl} = \Gamma^i_{ij} - \Gamma^l_{jl}, \quad \forall l, l \neq j. \tag{3.2}$$

The functions M_{ij} and M_{ijl} are called the (i,j)-*Laplace invariants* of the Cartan submanifold X. We observe that in order to decrease the number of indices used to denote the invariants associated to distinct pairs (i,j), the notation introduced in Eq. (3.2) is slightly different from the one used in Ref. [5].

Theorem 3.1. *Let* $X(u^1, \ldots, u^n)$ *be a Cartan submanifold of* $\mathbb{R}^{2n}$ *parametrized by conjugate coordinates. Consider an ordered pair* (i,j), $i \neq j$ *such that* $\Gamma^i_{ij} \neq 0$. *If the* (i,j)-*Laplace invariants* M_{ij} *and* M_{ijl} *are nonzero, then the map*

$$Y = X - \frac{1}{\Gamma^i_{ij}} X_{,j},$$

defines a Cartan manifold; that is, Y *satisfies*

$$Y_{kl} = \tilde{\Gamma}^k_{kl} Y_{,k} + \tilde{\Gamma}^l_{kl} Y_{,l}, \quad 1 \leq k \neq l \leq n,$$

where the coefficients $\tilde{\Gamma}^k_{kl}$, $1 \leq k \neq l \leq n$ *are given by the equations*

$$Y_{,ij} = \left(\frac{M_{ij,j}}{M_{ij}} + \Gamma^i_{ij} \right) Y_{,i} - M_{ij} Y_{,j},$$

$$Y_{,ik} = \left(M_{ij,k} + \Gamma_{ik}^i\right)Y_{,i} + \frac{M_{ij}}{M_{ijk}}\Gamma_{jk}^k Y_{,k},$$

$$Y_{,kj} = \left[\left(\frac{M_{ijk}}{\Gamma_{ij}^i}\right)_{,j}\frac{\Gamma_{ij}^i}{M_{ijk}} + \Gamma_{kj}^k\right]Y_{,k} + \frac{\Gamma_{ik}^i}{\Gamma_{ij}^i}M_{ijk}Y_{,j},$$

$$(3.3)$$

$$Y_{,k\ell} = \left[\left(M_{ijk}\Gamma_{ij}^i\right)_{,\ell}\frac{\Gamma_{ij}^i M_{ijk}}{+}\Gamma_{k\ell}^k\right]Y_{,k} + \frac{M_{ijk}}{M_{ij\ell}}\left(\Gamma_{k\ell}^\ell - \frac{\Gamma_{ik}^i\Gamma_{j\ell}^\ell}{\Gamma_{ij}^i}\right)Y_{,\ell}.$$

It follows from the definition there are at most $n(n-1)$ Laplace transforms for a given Cartan manifold. It is easy to see that generically the Laplace transformation is invertible.

Proposition 3.1. *If $M_{ij} \neq 0$, the inverse of the (i,j)-Laplace transform exists and it is given by the (j,i)-Laplace transform; i.e.,*

$$X = Y + Y_{,i}/M_{ij}. \qquad (3.4)$$

The vanishing of the higher-dimensional Laplace invariants has a simple geometric interpretation given in the following proposition:

Proposition 3.2. *The (i,j)-Laplace transform of a Cartan submanifold X reduces to a curve if and only if*

$$M_{ij} = M_{ijk} = 0, \quad \forall k, \ k \neq i, \ k \neq j.$$

At the analytic level, the vanishing of the higher-dimensional Laplace invariants leads to an interesting reduction theorem that will play a significant role in the applications of the higher-dimensional Laplace transformation to hydrodynamic systems.

Theorem 3.2. *Suppose that the immersion $X \colon U \subset \mathbb{R}^n \to \mathbb{R}^{2n}$ defines a Cartan submanifold parametrized by conjugate coordinates, so that we have*

$$X_{,ij} = \Gamma_{ij}^i X_{,i} + \Gamma_{ij}^j X_{,j}, \quad i \neq j.$$

If for some ordered pair (i,j), the higher-dimensional Laplace invariants M_{ij} and M_{ijk}, $1 \leq k \leq n$, $k \neq i$, $k \neq j$, are identically zero, then X is given by

$$X = Q + e^J G(\widehat{u^j}),$$

where

$$Q = e^J \int e^{I-J} F(u^j)\, du^j, \quad I = \int \Gamma_{ij}^j\, du^i, \quad J = \int \Gamma_{ij}^i\, du^j,$$

where F is an arbitrary function of x_j, $G(u^1, \ldots, \widehat{u^j}, \ldots, u^n)$ does not depend on x_j and where the antiderivative I is such that $I_{,k} = -\Gamma_{jk}^j$ for

$k \neq i$, $k \neq j$. Then G satisfies a linear system in $n-1$ independent variables $u^1, \ldots, \widehat{u^j}, \ldots, u^n$ of the form

$$G_{,kl} + g_{kl}^k G_{,k} + g_{kl}^\ell G_{,\ell} + b_{kl} G = 0, \quad k \neq l \text{ distinct from } j,$$

where

$$g_{lk}^l = \Gamma_{lk}^l - J_{,k}, \quad g_{lk}^k = \Gamma_{lk}^k - J_{,l},$$
$$b_{lk} = J_{,k} J_{,l} - J_{,lk} - \Gamma_{lk}^l J_{,l} - \Gamma_{lk}^k J_{,k}.$$

We are now ready to apply the above results concerning Cartan submanifolds to the class of hydrodynamic systems that are rich in conservation laws and to their conserved densities. We begin with some general theoretical results that follow naturally from the transformation theorems. These theoretical results are similar to those recently announced by Ferapontov [3]. Our approach is based on the Laplace transformations of Cartan submanifolds given in this section. We shall also make use of the reduction theorem (Theorem 3.2) and Theorem 3.1 in the computation of the conserved densities for concrete examples.

From Theorem 2.1 and Corollary 2.1, we see that if $X : U \subset \mathbb{R}^n \to \mathbb{R}^{2n}$ is a Cartan submanifold parametrized by conjugate coordinates, then every component of the $\mathbb{R}^{2n}$-valued function X is a conserved density for an infinite number of hydrodynamic systems that are rich in conservation laws and whose flows are commuting. The higher-dimensional Laplace transformation will therefore induce a transformation at the level of these hydrodynamic systems. We have the following result:

Theorem 3.3. *Let*

$$u_{,t}^i = v^i(u^1, \ldots, u^n) u_{,x}^i, \quad 1 \leq i \leq n,$$

be a strongly hyperbolic hydrodynamic system that is rich in conservation laws, meaning that

$$v^i \neq v^j, \quad 1 \leq i \neq j \leq n,$$
$$\left(\frac{v_{,j}^i}{v^j - v^i} \right)_{,k} = \left(\frac{v_{,k}^i}{v^k - v^i} \right)_{,j},$$

in some open set U of $\mathbb{R}^n$ with coordinates $(u^1, \ldots, u^n)$. Suppose that for some ordered pair (i,j), $1 \leq i \neq j \leq n$, the higher-dimensional Laplace invariants of the corresponding system (2.4) are nonzero in U,

$$M_{ij} = \frac{v_{,ij}^i}{v_{,j}^i} + \frac{v_{,i}^i}{v^j - v^i} \neq 0,$$

$$M_{ijk} = -\frac{v_{,j}^k}{v^j - v^k} + \frac{v_{,j}^i}{v^j - v^i} \neq 0,$$

for all $1 \leq k \leq n$, $k \neq i$, $k \neq j$. Consider now the coefficients $\widetilde{\Gamma}^k_{kl}$ of the (i,j)-transformed system, given by (3.3). Then we have—

(i) *The system*

$$\frac{\widetilde{v}^k_{,l}}{\widetilde{v}^l - \widetilde{v}^k} = \widetilde{\Gamma}^k_{kl}(u^1,\ldots,u^n), \quad 1 \leq k \neq l \leq n, \tag{3.5}$$

admits infinitely many solutions $\widetilde{v}^k(u^1,\ldots,u^n)$ depending on n functions of one variable.

(ii) *The corresponding hydrodynamic systems*

$$u^i_{,t} = \widetilde{v}^i(u^1,\ldots,u^n)u^i_{,x}, \quad 1 \leq i \leq n,$$

are rich in conservation laws. The flows corresponding to different solutions of (3.5) are commuting.

Ferapontov [3] gives an explicit formula of $\widetilde{v}$ in terms of v that provides a particular solution of (3.5) (see our last example in this paper).

The following result shows that the higher-dimensional Laplace transformation induces a one-to-one correspondence between the conserved densities for any pair of hydrodynamic systems related as in Theorem 3.3.

Theorem 3.4. *Suppose that we have two strongly hyperbolic hydrodynamic systems that are rich in conservation laws and related as in Theorem 3.3,*

$$u^i_{,t} = v^i(u^1,\ldots,u^n)u^i_{,x}, \quad 1 \leq i \leq n, \tag{3.6}$$
$$u^i_{,t} = \widetilde{v}^i(u^1,\ldots,u^n)u^i_{,x}, \quad 1 \leq i \leq n. \tag{3.7}$$

Then there is a one-to-one correspondence between the conserved densities $P(u^1,\ldots,u^n)$ for (3.6) and $\widetilde{P}(u^1,\ldots,u^n)$ for (3.7), given by

$$\widetilde{P} = P - P_{,j}/\Gamma^i_{ij}, \quad P = \widetilde{P} + \widetilde{P}_{,i}/M_{ij}. \tag{3.8}$$

where M_{ij} is given by (3.2).

In what follows, we will show how the Laplace transformation can be used in order to obtain conserved densities for hydrodynamic systems. We will denote the (i,j)-Laplace transform of X by $\mathcal{L}_{(i,j)}(X)$. We will first consider the ideal chromatography system.

Proposition 3.3. *The conserved densities $P(u^1,\ldots,u^N)$, $N \geq 2$ for the chromatography system*

$$u^i_{,t} = u^i \prod_{\ell=1}^{N} u^\ell u^i_{,x}, \quad 1 \leq i \leq N,$$

are given inductively as follows:

$$P^{(N)} = \frac{1}{u^1} \sum_{k=2,i=0}^{N} (-1)^i \frac{\partial^i}{\partial u^k} \left((u^1 - u^k)^N \right) \frac{\partial^{N-i} U_k}{\partial u^k} + U_1, \tag{3.9}$$

$$P^{(n)} = P^{(n+1)} - \frac{u^1(u^1 - u^{n+1})}{(n+1)u^{n+1}} P_{,1}^{(n+1)}, \quad 2 \le n \le N - 1, \text{ if } N \ge 3,$$

$$P = P^{(2)} - \frac{u^1(u^1 - u^2)}{2u^2} P_{,1}^{(2)}, \tag{3.10}$$

where U_ℓ are arbitrary differentiable function of u^ℓ.

Proof. We will prove this result in three steps: We will first consider the system satisfied by the conserved densities P for the chromatography system. Then we will apply to P the composition of Laplace transformations

$$P^{(N)} = \mathcal{L}_{(1,N)} \circ \cdots \circ \mathcal{L}_{(1,2)}(P),$$

and we will solve the system satisfied by $P^{(N)}$. Finally, we will obtain P by inverting the Laplace transformations.

We start observing that it follows from (2.4) that the conserved densities for the chromatography system must satisfy the overdetermined system given by

$$P_{,ij} = \Gamma_{ij}^i P_{,i} + \Gamma_{ij}^j P_{,j} \quad \text{where} \quad \Gamma_{ij}^i = \frac{u^i}{u^j(u^j - u^i)},$$

$$1 \le i \ne j \le N. \tag{3.11}$$

We will show that the general solution P for this system is given by Eq. (3.10).

We first compute the (i, j)-Laplace invariants of the system (3.11). For a fixed pair (i, j), $i \ne j$, it follows from (3.5) that

$$M_{ij} = -2\Gamma_{ij}^j, \quad M_{ijk} = \Gamma_{ij}^i - \Gamma_{kj}^k, \quad k \text{ distinct from } i \text{ and } j. \quad \square \tag{3.12}$$

Claim. If $N \ge 3$, then for any integer n, $2 \le n < N$, the composition of Laplace transformations given by

$$P^{(n)} = \mathcal{L}_{(1,n)} \circ \cdots \circ \mathcal{L}_{(1,3)} \circ \mathcal{L}_{(1,2)}(P),$$

satisfies the system of equations

$$P_{,1k}^{(n)} = \begin{cases} n\Gamma_{1k}^k P_{,k}^{(n)} & \text{if } 2 \le k \le n; \\ \Gamma_{1k}^1 P_{,1}^{(n)} + n\Gamma_{1k}^k P_{,k}^{(n)} & \text{if } n + 1 \le k \le N. \end{cases} \tag{3.13}$$

$$P_{,\ell k}^{(n)} = \begin{cases} 0 & \text{if } 2 \le \ell \ne k \le n; \\ \Gamma_{\ell k}^\ell P_{,\ell}^{(n)} & \text{if } 2 \le \ell \le n, \quad n + 1 \le k \le N; \\ \Gamma_{\ell k}^\ell P_{,\ell}^{(n)} + \Gamma_{\ell k}^k P_{,k}^{(n)} & \text{if } n + 1 \le \ell \ne k \le N. \end{cases} \tag{3.14}$$

The claim will be proved by induction on n. First we show that our claim holds for $n = 2$. In fact, we consider $P^{(2)}$ defined by

$$P^{(2)} = \mathcal{L}_{(1,2)}(P).$$

We will use (3.3) where $i = 1$ and $j = 2$. From (3.12) and (3.11) we observe that

$$\frac{\partial}{\partial u^2}(\log M_{12}) + \Gamma_{12}^1 = 0.$$

Hence, it follows from the first equation of (3.3) that

$$P_{,12}^{(2)} = 2\Gamma_{12}^2 P_{,2}^{(2)}.$$

From the second equation of (3.3), using (3.12) and (3.11) we get

$$P_{,1k}^{(2)} = \frac{\Gamma_{12}^1 \Gamma_{2k}^2}{\Gamma_{12}^1 - \Gamma_{k2}^k} P_{,1}^{(2)} - \frac{2\Gamma_{12}^2 \Gamma_{2k}^k}{\Gamma_{12}^1 - \Gamma_{k2}^k} P_{,k}^{(2)},$$

which reduces to

$$P_{,1k}^{(2)} = \Gamma_{1k}^1 P_{,1}^{(2)} + 2\Gamma_{1k}^k P_{,k}^{(2)}.$$

Similarly, from the third equation of (3.3) we get that

$$P_{,2k}^{(2)} = \Gamma_{2k}^2 P_{,2}^{(2)}, \quad 3 \le k \le N.$$

Finally, from the last equation of (3.3) we obtain

$$P_{,\ell k}^{(2)} = \Gamma_{k\ell}^k P_{,k}^{(2)} + \frac{1}{M_{12\ell}}(-\Gamma_{2\ell}^\ell \Gamma_{2k}^2 + M_{12k}\Gamma_{k\ell}^\ell)P_{,\ell}^{(2)}, \quad \text{for} \quad 3 \le k \ne \ell \le N,$$

which reduces to

$$P_{,\ell k}^{(2)} = \Gamma_{k\ell}^k P_{,k}^{(2)} + \Gamma_{k\ell}^\ell P_{,\ell}^{(2)},$$

when we use equations (3.12) and (3.11). This concludes the first step of our claim. Now assuming the induction hypothesis (3.13) and (3.14) we will prove that it also holds for $n + 1$. We consider

$$P^{(n+1)} = \mathcal{L}_{(1,n+1)}(P^{(n)}).$$

We observe that the $(1, n+1)$-Laplace invariants for the system (3.13) and (3.14) are

$$\overset{(n)}{M}_{1n+1} = -(n+1)\Gamma_{1n+1}^{n+1},$$

$$\overset{(n)}{M}_{1n+1k} = \Gamma_{1n+1}^1 - \Gamma_{kn+1}^k,$$

(3.15)

where k is distinct from 1 and $n+1$. In fact, this follows from the fact that

$$\overset{(n)}{M}_{1n+1} = \frac{\partial}{\partial u^1}(\log \Gamma_{1n+1}^1) - n\Gamma_{1n+1}^{n+1}.$$

Now we will obtain the system of equations that are satisfied by $P^{(n+1)}$ by applying (3.3), with $i = 1$ and $j = n+1$, to the system (3.13) and (3.14) From the first equation of (3.3), we obtain

$$P_{,1n+1}^{(n+1)} = (n+1)\Gamma_{1n+1}^{n+1}P_{,n+1}^{(n+1)}.$$

For k distinct from 1 and $n+1$, it follows from the second equation of (3.3) and from (3.11) that

$$P_{,1k}^{(n+1)} = \begin{cases} (n+1)\Gamma_{1k}^k P_{,k}^{(n+1)} & \text{if } 2 \leq k \leq n, \\ \Gamma_{1k}^1 P_{,1}^{(n+1)} + (n+1)\Gamma_{1k}^k P_{,k}^{(n+1)} & \text{if } n+1 < k \leq N. \end{cases} \tag{3.16}$$

From the third equation of (3.11), we get

$$P_{,n+1k}^{(n+1)} = \begin{cases} 0 & \text{if } 2 \leq k < n+2, \\ \Gamma_{u+1k}^{n+1} P_{,n+1}^{(n+1)} & \text{if } n+2 \leq k \leq N. \end{cases}$$

Finally, it follows from the last equation of (3.3), that for k and ℓ distinct from 1 and $n+1$

$$P_{,k\ell}^{(n+1)} = \begin{cases} 0 & \text{if } 2 \leq k \neq \ell < n+1, \\ \Gamma_{k\ell}^\ell P_{,\ell}^{(n+1)} & \text{if } 2 \leq \ell < n+1, \\ & \quad n+2 \leq k \leq N, \\ \Gamma_{k\ell}^k P_{,k}^{(n+1)} + \Gamma_{k\ell}^\ell P_{,\ell}^{(n+1)} & \text{if } n+2 \leq \ell \neq k \leq N. \end{cases} \tag{3.17}$$

This concludes the proof of our claim.
As a consequence we get that

$$P^{N-1} = \mathcal{L}_{(1,N-1)} \circ \cdots \circ \mathcal{L}_{(1,2)}(P),$$

satisfies the system of equations

$$P_{,1k}^{(N-1)} = \begin{cases} (N-1)\Gamma_{1k}^k P_{,k}^{(N-1)} & 2 \leq k \leq N-1, \\ \Gamma_{1N}^1 P_{,1}^{(N-1)} + (N-1)\Gamma_{1N}^N P_{,k}^{(N-1)} & \text{if } k = N, \end{cases}$$

$$P_{,\ell k}^{(N-1)} = \begin{cases} 0 & \text{if } 2 \leq \ell \neq k \leq N-1, \\ \Gamma_{\ell N}^\ell P_{,\ell}^{(N-1)} & \text{if } 2 \leq \ell \leq N-1, \ k = N. \end{cases}$$

Hence

$$P^{(N)} = \mathcal{L}_{(1,N)} \circ \cdots \circ \mathcal{L}_{(1,2)}(P),$$

satisfies the system of equations

$$P_{,1k}^{(N)} = N\Gamma_{1k}^k P_{,k}^{(N)} \quad 2 \leq k \leq N,$$
$$P_{,\ell k}^{(N)} = 0 \quad \text{if } 2 \leq \ell \neq k \leq N.$$

It follows that

$$P^{(N)} = \sum_{k=2}^{N} \int \left(1 - \frac{u^k}{u^1}\right)^N U_k(u^k)\, du^k + U_1(u^1),$$

where U_i are arbitrary functions of u^i. This last expression reduces to (3.9).

Now using the inversion of the Laplace transformation (3.4), we obtain from (3.15) that

$$P^{(N-1)} = P^N - \frac{1}{N\Gamma_{1N}^N} P_{,1}^N.$$

Inductively, we obtain

$$P^{(n)} = P^{(n+1)} - \frac{1}{(n+1)\Gamma_{1n+1}^{n+1}} P_{,1}^{(n+1)}, \quad 2 \le n \le N-1,$$

and

$$P = P^{(2)} - \frac{1}{2\Gamma_{12}^2} P_1^{(2)},$$

which gives the general solution of the system (3.11).   □

As a consequence of Theorem 3.2 and the previous proposition we will obtain the conserved densities for another system of hydrodynamic type.

Proposition 3.4. *The conserved densities $P(u^1,\ldots,u^N)$, $N \ge 2$ for the electrophoresis system*

$$u_{,t}^1 = 0,$$

$$u_{,t}^i = -u^i \prod_{\ell=1}^{N} u^\ell u_{,x}^i, \quad 2 \le i \le N,$$

are given by

$$P = U_1(u^1) + \frac{1}{u^1} G(u^2,\ldots,u_N), \tag{3.18}$$

where G is defined inductively as follows: Let

$$G^{(N-1)} = \frac{1}{u^2} \sum_{k=3,i=0}^{N} (-1)^i \frac{\partial^i}{\partial u^k} \left((u^2 - u^k)^{N-1}\right) \frac{\partial^{N-1-i} U_k}{\partial u^k} + U_2,$$

$$G^{(n-1)} = G^{(n)} - \frac{u^2(u^2 - u^{n+1})}{n\, u^{n+1}} G_{,2}^{(n)}, \quad 2 \le n-1 \le N-2 \quad if \ \ N \ge 4,$$

$$G = G^{(2)} - \frac{u^2(u^2 - u^3)}{2u^3} G_{,2}^{(2)},$$

where $U_i(u^i)$ are arbitrary differentiable functions of u^i.

Proof. It follows from (2.4) that the conserved densities $P(u^1, \ldots, u_N)$ must satisfy the system of equations

$$P_{,ij} = \Gamma^i_{ij}P_{,i} + \Gamma^j_{ij}P_{,j}, \quad i \neq j, \ 1 \leq i \neq j \leq N, \qquad (3.19)$$

where

$$\Gamma^1_{1k} = 0, \quad \Gamma^k_{k1} = -\frac{1}{u^1}, \quad \Gamma^k_{kj} = \frac{u^k}{u^j(u^j - u^k)}, \quad 2 \leq k \neq j \leq N. \quad (3.20)$$

We want to obtain the general solution for this system.

We consider the (2,1)-Laplace invariants for the system (3.19) and (3.20). It follows from (3.20) that

$$M_{21} = M_{21k} = 0 \quad \forall k, \ 3 \leq k \leq N.$$

Applying Theorem 3.2, we get that its general solution P is given by (3.18) and $G(u_2, \ldots, u_N)$ satisfies the differential system

$$G_{,ij} = \Gamma^i_{ij}G_{,i} + \Gamma^j_{ij}G_{,j}, \quad 2 \leq i \neq j \leq N,$$

where

$$\Gamma^i_{ij} = \frac{u^i}{u^j(u^j - u^i)}.$$

In order to apply Proposition 3.3 we change our notation by considering

$$G(v^1, \ldots, v^{N-1}),$$

where $v^i = u^{i+1}$ for $1 \leq i \leq N-1$, and

$$\widetilde{\Gamma}^i_{ij} = \frac{v^i}{v^j(v^j - v^i)} = \Gamma^{i+1}_{i+1\,j+1}, \quad i \neq j, \ 1 \leq i \neq j \leq N-1.$$

It follows from Proposition 3.3 that G is defined inductively as follows:

$$G^{(N-1)} = \sum_{k=2}^{N-1}\left(1 - \frac{v^k}{v^1}\right)^{N-1} V_\ell(v^\ell)dv^\ell + V_1(v^1),$$

$$G^{(n-1)} = G^{(n)} - \frac{1}{n\widetilde{\Gamma}^n_{1n}}, \quad 2 \leq n-1 \leq N-2 \quad \text{if} \quad N \geq 4,$$

$$G = G^{(2)} - \frac{1}{2\widetilde{\Gamma}^2_{12}}G^{(2)}_{,v^1}.$$

Replacing v^ℓ and $\widetilde{\Gamma}$ we obtain

$$G^{(N-1)} = \sum_{k=2}^{N-1}\int\left(1 - \frac{u^{k+1}}{u^2}\right)^{N-1} U_{\ell+1}(u^{\ell+1})\,du^{\ell+1} + U_2(u^2),$$

$$G^{(n-1)} = G^{(n)} - \frac{1}{n\Gamma^{n+1}_{2n+1}}G^{(n)}_{,2}, \quad 2 \leq n-1 \leq N-2 \quad \text{if} \quad N \geq 4,$$

$$G = G^{(2)} - \frac{1}{2\Gamma^3_{23}}G^{(2)}_{,2},$$

which concludes the proof of Proposition 3.4.   □

Similar arguments will show—

Proposition 3.5. *The conserved quantities P for the system*

$$u^i_{,x} = -u^i \prod_{\ell=1}^{N} u^\ell u^i_{,t}, \quad 1 \leq i \leq N,$$

are given by the system of equations

$$P_{,ij} = \frac{1}{u^j - u^i}(P_{,i} - P_{,j}),$$

whose general solution is given inductively by

$$P^{(N)} = \sum_{k=2,i=0}^{N} (-1)^i \frac{\partial^i}{\partial u^k}((u^1 - u^k)^N)\frac{\partial^{N-i}U_k}{\partial u^k} + U_1,$$

$$P^{(n)} = P^{(n+1)} - \frac{u^1 - u^{n+1}}{(n+1)}P_1^{(n+1)}, \quad 2 \leq n \leq N-1, \quad if \quad N \geq 3,$$

$$P = P^{(2)} - \frac{u^1 - u^2}{2}P_1^{(2)},$$

where U_ℓ are arbitrary differentiable function of u^ℓ.

As an application of Theorem 3.4 we provide the following example of hydrodynamic systems related by a Laplace transformation.

Example. Consider the systems

$$u^1_{,t} = 0, \tag{3.21}$$

$$u^i_{,t} = u^i \left(\prod_{j=1}^{3} u^j\right)u^i_{,x}, \quad 2 \leq i \leq 3,$$

and

$$\begin{aligned} u^1_{,t} &= -(u^1)^2 U_{1,1}u^1_{,x}, \\ u^2_{,t} &= u^1(U_1 + U_2)u^2_{,x}, \\ u^3_{,t} &= u^1(U_1 + f(u^2, u^3))u^3_{,x}, \end{aligned} \tag{3.22}$$

where

$$f = \left(\frac{u^2}{u^2 - u^3}\right)^2\left[2\int \frac{u^3(u^2 - u^3)}{(u^2)^3}U_2\,du^2 + U_3\right],$$

and U_i are arbitrary differentiable functions of u^i. Then the electrophoresis system (3.21) and system (3.22) are related as in Theorem 3.3 by the (2, 3)-Laplace transformation. Therefore, the conserved densities $\widetilde{P}$ of (3.22) are obtained from the conserved densities P of (3.21) by the relation

$$\widetilde{P} = P - \frac{u^3(u^3 - u^2)}{u^2}P_{,3}.$$

In fact, the conserved densities $P(u^1, u^2, u^3)$ for (3.21) must satisfy (3.19), (3.20), where $N = 3$. Since the $(2,3)$-Laplace invariants $M_{23} = -2\Gamma^3_{23}$ and $M_{231} = \Gamma^2_{23}$ are nonzero, it follows from Theorem 3.1 that the $(2,3)$-Laplace transform of the system 3.19 is given by

$$\widetilde{P}_{k\ell} = \widetilde{\Gamma}^k_{k\ell}\widetilde{P}_{,k} + \widetilde{\Gamma}^\ell_{k\ell}\widetilde{P}_{,\ell}, \quad 1 \leq k \neq \ell \leq 3, \tag{3.23}$$

where the coefficients are given by $\widetilde{\Gamma}^1_{1\ell} = 0$, $\widetilde{\Gamma}^\ell_{1\ell} = \Gamma^1_{1\ell}$ for $\ell \neq 1$, $\widetilde{\Gamma}^2_{23} = 0$, and $\widetilde{\Gamma}^3_{23} = 2\Gamma^3_{23}$. Applying Theorem 3.3, by solving the system

$$\frac{\widetilde{v}^k_{,l}}{\widetilde{v}^l - \widetilde{v}^k} = \widetilde{\Gamma}^k_{kl}, \quad 1 \leq k \neq l \leq 3,$$

we obtain (3.22). The relation between the conserved densities is given by Theorem 3.4 as in Ref. (3.23). We conclude, observing that by choosing the particular functions $U_1 = 0$, $U_2 = (u^2)^3$ and $U_3 = (u^3)^3$ in Ref. (3.22), we obtain the transformed system given in Ref. [3].

Acknowledgments: We thank E. Ferapontov for some helpful observations on an earlier version of this paper.

REFERENCES

1. S.S. Chern, *Laplace transforms of a class of higher-dimensional varieties in projective space of n dimensions*, Proc. Nat. Acad. Sci. **30** (1944), 95–97.

2. S.S. Chern, *Sur une classe remarquable de variétés dans l'espace projectif a n dimensions*, Science Reports Tsing-Hua Univ. **4** (1947), 328–336; reprinted in Selected papers of S.S. Chern, Vol. 1, Springer-Verlag, 1978, pp. 138–146.

3. E. Ferapontov, *Laplace transformations of hydrodynamic-type systems in Riemann invariants* Teoret. Mat. Fiz. **110** (1997), No. 1, 86–97 (Russian); transl. Theoret. and Math. Phys. **110** (1997), No. 1, 68–77.

4. E. Ferapontov, *Differential geometry of nonlocal Hamiltonian operators of hydrodynamic type*, Funct. Anal. Appl. **25** (1991), 37–49.

5. N. Kamran and K. Tenenblat, *Laplace transformation in higher dimensions*, Duke Math. J. **80** (1996), 237–266.

6. D. Serre, *Systèmes hyperboliques riches de lois de conservation*, Nonlinear Partial Differential Equations and Their Applications (Paris, 1989–1991), Collège de France Seminar, Vol. XI, Pitman Res. Notes Math. Ser. 299, Longman, Harlow, 1994, pp. 248–281.

7. S.P. Tsarev, *The geometry of Hamiltonian systems of hydrodynamic type*, The generalized hodograph method (Russian) Izv. Akad. Nauk SSSR Ser. Mat. **54** (1990), No. 5, 1048–1068; transl. Math. USSR-Izv. **37** (1991), No. 2, 397–419.

9

Branching Rules and Weight Multiplicities for Simple and Affine Lie Algebras

Ronald C. King

The author is delighted to dedicate this paper to Jiři Patera and Pavel Winternitz in celebration of their 60th birthdays. He has greatly valued their friendship over many years and wishes to express his appreciation both of the warmth of their hospitality on numerous visits to Montreal and of the inspiration he has derived from their work that has so significantly enriched both mathematics and theoretical physics.

ABSTRACT It is pointed out that the tabulations of branching rule and weight multiplicity data provided by Patera and his colleagues [3, 6, 10] suggest that certain of these multiplicities are either rank-independent or polynomial functions of the rank. Character formulae and Young diagram methods are then used to confirm the validity of these observations in the case of both simple and affine Lie algebras. Examples are given pertaining to the rank-independent branching rules for both $A_\ell \supset A_{\ell-1}$ and $A_\ell^{(1)} \supset A_\ell$ and to the polynomial rank-dependent weight multiplicities of both A_ℓ and $A_\ell^{(1)}$.

1 Introduction

The aim here is to describe methods of calculating the branching rule and weight multiplicities of irreducible representations V^λ of both simple and affine Lie algebras as functions of their rank ℓ. The results are complementary to those developed by Jiři Patera and his collaborators, who have produced definitive tabulations appropriate to all simple Lie algebras of rank $\ell \leq 12$ and all untwisted affine Lie algebras of rank $\ell \leq 8$. These tabulations are to be found in Refs. [3, 6, 10]

These texts are a rich source of data from which can be extracted both a wealth of specific results ripe for exploitation in numerous applications of Lie algebras to problems of theoretical physics and, as will be emphasized here, a range of conjectures regarding the independence or dependence of various branching rule and weight multiplicities on the rank of the appropriate Lie algebra.

2   Simple and Affine Lie Algebras

A matrix $A = (a_{ij})_{i,j \in I}$ is said to be a generalized Cartan matrix (GCM) if

(i) $a_{ii} = 2$;

(ii) $a_{ij} \leq 0$ for $i \neq j$;

(iii) $a_{ij} = 0$ implies $a_{ji} = 0$;

(iv) $a_{ij} \in \mathbb{Z}$.

The associated Kac–Moody algebra $\mathbf{g}(A)$ is a complex Lie algebra with generators e_i, f_i, h_i for $i \in I$ satisfying the relations

$$[h_i, h_j] = 0, \quad [e_i, f_j] = \delta_{ij} h_i,$$
$$[h_i, e_j] = a_{ij} e_j, \quad [h_i, f_j] = -a_{ij} f_j,$$
$$(\operatorname{ad} e_i)^{1-a_{ij}} e_j = 0 \quad \text{and} \quad (\operatorname{ad} f_i)^{1-a_{ij}} f_j = 0 \quad \text{if} \quad i \neq j,$$

where $(\operatorname{ad} x) y = [x, y]$ for all $x, y \in \mathbf{g}(A)$.

If A is indecomposable and $\det A > 0$ then $\mathbf{g}(A)$ is finite-dimensional and isomorphic to one of the simple Lie algebras $\mathbf{g}_0$. On the other hand, if A is indecomposable and there exists δ such that $A\delta = 0$, then $\mathbf{g}(A)$ is infinite-dimensional and isomorphic to one of the affine Lie algebras $\mathbf{g}$.

The list of simple Lie algebras $\mathbf{g}_0$ owing to Cartan and Killing takes the form

$$A_\ell, \ B_\ell, \ C_\ell, \ D_\ell, \ E_6, \ E_7, \ E_8, \ F_4, G_2,$$

while the corresponding list of affine Lie algebras $\mathbf{g}$ owing to Kac and Moody is given by

$$A_\ell^{(1)}, \ B_\ell^{(1)}, \ C_\ell^{(1)}, \ D_\ell^{(1)}, \ A_{2\ell}^{(2)}, \ A_{2\ell-1}^{(2)}, \ D_{\ell+1}^{(2)},$$
$$E_6^{(1)}, \ E_7^{(1)}, \ E_8^{(1)}, \ F_4^{(1)}, \ G_2^{(1)}, \ E_6^{(2)}, \ D_4^{(3)}.$$

Thus there are four infinite sequences of simple Lie algebras $\mathbf{g}_0$ and seven infinite sequences of affine Lie algebras $\mathbf{g}$, each indexed by their rank ℓ.

Each simple Lie algebra $\mathbf{g}_0$ is embedded naturally in at least one affine Lie algebra $\mathbf{g}$, by virtue of the fact that restricting the index set $I = \{0, 1, 2, \ldots, \ell\}$ of the rows and columns of the GCM of $\mathbf{g}$ to the index set

$I_0 = \{1, 2, \ldots, \ell\}$ gives the GCM of $\mathbf{g}_0$. These natural embeddings $\mathbf{g}_0 \subset \mathbf{g}$ are

$$A_\ell \subset A_\ell^{(1)}; \qquad B_\ell \subset B_\ell^{(1)}, \ A_{2\ell}^{(2)}, \ D_{\ell+1}^{(2)}; \qquad C_\ell \subset C_\ell^{(1)}, \ A_{2\ell-1}^{(2)};$$

$$D_\ell \subset D_\ell^{(1)}; \qquad E_6 \subset E_6^{(1)}; \qquad E_7 \subset E_7^{(1)};$$

$$E_8 \subset E_8^{(1)}; \qquad F_4 \subset F_4^{(1)}, \ E_6^{(2)}; \qquad G_2 \subset G_2^{(1)}, \ D_4^{(3)}.$$

3 Branching Rules for Simple Lie Algebras

Each finite-dimensional irreducible representation (irrep), V^λ, of a simple Lie algebra $\mathbf{g}$ of rank ℓ is specified by its integral dominant highest weight vector $\lambda = \sum_{i=1}^{\ell} a_i(\lambda)\omega_i$. The notation is such that each ω_i is the highest-weight of a fundamental irrep of $\mathbf{g}$ and the Dynkin labels $a_i(\lambda)$ are non-negative integers. On restriction from a simple Lie algebra $\mathbf{g}$ to a semisimple Lie subalgebra $\widetilde{\mathbf{g}}$ each irrep V^λ of $\mathbf{g}$ decomposes into a direct sum of irreps $\widetilde{V}^\mu$ of $\widetilde{g}$ in accordance with a branching rule of the form

$$V^\lambda \, to \sum_\mu b_\mu^\lambda \widetilde{V}^\mu.$$

The coefficients b_μ^λ are known as branching rule multiplicities. They have been tabulated for a vast number of irreps in Ref. [10] in the case of the restriction of all simple Lie algebras of rank $\ell \le 8$ to each of their maximal semisimple Lie subalgebras.

Example 3.1. For $A_\ell \supset A_{\ell-1}$ and $\lambda = \omega_1 + 2\omega_2$ the tables of [10] contain the following data:

$\mathbf{g}$	λ	dim	$\sum_\mu (b_\mu^\lambda)\mu$
A_2	12	15	0+(2)1+(2)2+3
A_3	120	60	12+21+02+30+11+20
A_4	1200	175	120+210+020+300+110+200
A_5	12000	420	1200+2100+0200+3000+1100+2000
A_6	120000	882	12000+21000+02000+30000+11000+20000
A_7	1200000	1680	120000+210000+020000+300000+110000+200000
A_8	12000000	2970	1200000+2100000+0200000+3000000+1100000+2000000

This data suggests both that $\dim V^{\omega_1 + 2\omega_2} = 1/6\ell(\ell+1)^2(\ell+2)(\ell+3)$ for $\ell \ge 2$ and that the branching rule multiplicities are independent of ℓ for $\ell \ge 3$. The former may be verified directly through the use of Weyl's dimension formula. The latter may be derived as shown in the next section through the use of Young diagrams.

Example 3.2. For $A_\ell \supset A_{\ell-1}$ and $\lambda = \omega_1 + \omega_2 + \omega_{\ell-1}$ the tables of [10] contain the following data:

g	λ	dim	$\sum_\mu (b_\mu^\lambda)\mu$
A_3	120	60	30+20+12+02+21+11
A_4	1110	280	120+020+210+110+111+011+201+101
A_5	11010	840	1110+0110+2010+1010+1101+0101+2001+1001
A_6	110010	2016	11010+01010+20010+10010+11001+01001+20001+10001
A_7	1100010	4200	110010+010010+200010+100010+110001+010001+200001+100001

This data suggests both that $\dim V^{\omega_1+\omega_2+\omega_{\ell-1}} = 1/24(\ell-2)\ell(\ell+1)(\ell+2)(\ell+3)$ for $\ell \geq 3$ and that the branching rule multiplicities are independent of ℓ for $\ell \geq 4$. Once again the former may be verified directly through the use of Weyl's dimension formula, while the latter may be derived as shown in the next section through the use, this time, of composite Young diagrams.

4　Young Diagrams and Branching Rules

In the case of A_ℓ each integral dominant highest-weight λ may be expressed in the form $\lambda = \sum_{i=1}^{\ell} a_i(\lambda)\omega_i = \sum_{i=1}^{\ell+1} \lambda_i \varepsilon_i$ with $a_i(\lambda) = \lambda_i - \lambda_{i+1} \in \mathbb{Z}^+$ for $i = 1, 2, \ldots, \ell$, $\lambda_i \in \mathbb{Z}$ for $i = 1, 2, \ldots, \ell+1$ and $\sum_{i=1}^{\ell+1} \varepsilon_i = 0$. If we choose $\lambda_{\ell+1} = 0$, then $\lambda = (\lambda_1, \lambda_2, \ldots, \lambda_\ell)$ is a partition that defines a Young diagram F^λ. For example, if $\ell \geq 3$ and $\lambda = \omega_1 + 2\omega_2 = 3\varepsilon_1 + 2\varepsilon_2 = (3, 2)$, then $F^{32} = $ ⬚

Weyl's branching rule for $A_\ell \supset A_{\ell-1}$ is such that $V^\lambda \to \sum_\mu \tilde{V}^\mu$, where the sum is over those partitions μ such that F^μ is obtained from F^λ by the deletion of no more than one box at the foot of each column. For example, the diagrams

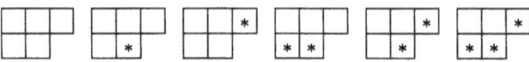

imply that for $A_\ell \supset A_{\ell-1}$ we have the branching rule appropriate to Example 3.1:

$$V^{\{32\}} \to \tilde{V}^{\{32\}} + \tilde{V}^{\{31\}} + \mathbf{g}V^{\{2^2\}} + \tilde{V}^{\{3\}} + \tilde{V}^{\{21\}} + \tilde{V}^{\{2\}},$$

where brackets $\{..\}$ are used to indicate that we are working in the ε-basis. As suggested by the earlier tabulation for Example 3.1, this result is valid for all $\ell \geq 3$.

If on the other hand we choose $\lambda_{\ell+1} \in \mathbb{Z}^-$, then $\lambda = (\lambda_1, \lambda_2, \ldots, \lambda_{\ell+1})$ is not a partition but serves to define a composite Young diagram $F^\lambda = F^{\mu;\bar{\nu}}$.

For example, if $\ell \geq 3$ and $\lambda = \omega_1 + \omega_2 + \omega_{\ell-1} = 2\varepsilon_1 + \varepsilon_2 - \varepsilon_\ell - \varepsilon_{\ell+1}$, then

$$F^{21;\overline{11}} = \qquad \text{or, more succinctly,}$$

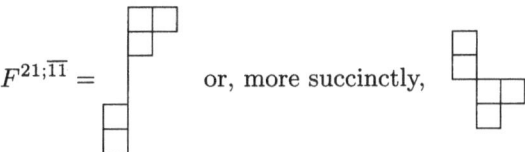

Weyl's branching rule again involves deleting no more than one box from each column so as to leave a standard composite diagram. For example, the following diagrams

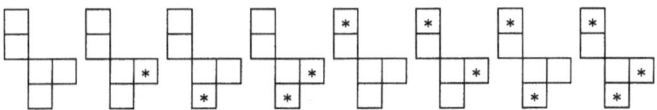

imply that for $A_\ell \supset A_{\ell-1}$ we have the branching rule appropriate to Example 3.2:

$$V^{\{21;\overline{1^2}\}} \to \widetilde{V}^{\{21;\overline{1^2}\}} + \widetilde{V}^{\{1^2;\overline{1^2}\}}$$
$$+ \widetilde{V}^{\{2;\overline{1^2}\}} + \widetilde{V}^{\{1;\overline{1^2}\}} + \widetilde{V}^{\{21;\overline{1}\}} + \widetilde{V}^{\{1^2;\overline{1}\}} + \widetilde{V}^{\{2;\overline{1}\}} + \widetilde{V}^{\{1^2;\overline{1}\}},$$

where brackets $\{..\}$ are again used to indicate that we are working in the ε-basis. This conforms with our previous tabulation for Example 3.2 and is valid for all $\ell \geq 4$. Even for $\ell = 3$ the above set of diagrams enables one to recover the appropriate branching rule. However, to do so it is necessary to invoke the appropriate modification rule, which implies that the contributions of the first two diagrams are null.

5   Weight Multiplicities of Simple Lie Algebras

The multiplicity, m_κ^λ, of the weight κ in the irrep V^λ of highest-weight λ is defined by the expansion of the character of V^λ in terms of formal exponentials:

$$\text{ch}\, V^\lambda = \sum_\kappa m_\kappa^\lambda e^\kappa.$$

The weight multiplicities of each simple Lie algebra $\mathbf{g}$ of rank $\ell \leq 12$ have been very extensively tabulated in [3]. In the case of certain class 1 irreps of A_ℓ these tables include the following data:

$\ell{=}5,\lambda\backslash\kappa$	0	1	2	3	4	5
0	1					
1	4	1				
2	5	1	1			
3	5	2		1		
4	6	3		1	1	
5	15	6	3	2		1

$\ell{=}6,\lambda\backslash\kappa$	0	1	2	3	5	6
0	1					
1	5	1				
2	6	1	1			
3	9	3		1		
5	10	4		1	1	
6	24	8	4	2		1

$\ell{=}7,\lambda\backslash\kappa$	0	1	2	3	4	6
0	1					
1	6	1				
2	7	1	1			
3	14	4		1		
4	15	5		1	1	
6	35	10	5	2		1

$\ell{=}8,\lambda\backslash\kappa$	0	1	2	3	4	5
0	1					
1	7	1				
2	8	1	1			
3	20	5		1		
4	21	6		1	1	
5	48	12	6	2		1

Taking care with the identification of the various highest weights, a polynomial fit to all the above data is provided by the following tabulation:

$\lambda\backslash\kappa$	$100\cdots000$	$010\cdots001$	$200\cdots001$	$001\cdots010$	$001\cdots002$	$110\cdots010$
$100\cdots000$	1					
$010\cdots001$	$\ell-1$	1				
$200\cdots001$	ℓ	1	1			
$001\cdots010$	$\frac{1}{2}(\ell^2-3\ell)$	$\ell-3$		1		
$001\cdots002$	$\frac{1}{2}(\ell^2-3\ell+2)$	$\ell-2$		1	1	
$110\cdots010$	$\ell^2-2\ell$	$2\ell-4$	$\ell-2$	2		1

This provides evidence in support of the fact that the weight multiplicities of A_ℓ may be expressed as polynomials in the rank ℓ. More generally, the data published in [3] suggest that the weight multiplicities of A_ℓ, B_ℓ, C_ℓ and D_ℓ are all polynomial in the rank ℓ, a fact now well established [1, 8].

6   Young Tableaux and Weight Multiplicities

The character $\mathrm{ch}\,V^\lambda$ of an irrep V^λ of highest-weight λ of a simple Lie algebra $\mathbf{g}$ is given by Weyl's character formula:

$$\mathrm{ch}\,V^\lambda = \sum_{w\in W} \varepsilon(w)e^{w(\lambda+\rho)-\rho} \Big/ \sum_{w\in W} \varepsilon(w)e^{w(\rho)-\rho}.$$

In the case of A_ℓ the Weyl group W is just the symmetric group $S_{\ell+1}$ and $\mathrm{ch}\,V^\lambda$ takes the form of a Schur function $s^\lambda(x)$. The Schur function itself may be defined in terms of $(\ell+1)$-standard Young tableaux T^λ. These are the tableaux formed by numbering the boxes of the corresponding Young diagram with entries taken from the set $\{1, 2, \ldots, \ell+1\}$, arranged so as to be weakly increasing across rows from left to right and strictly increasing down columns from top to bottom. To be precise

$$\operatorname{ch} V^\lambda = \sum_{T^\lambda} e^{\operatorname{wgt}(T^\lambda)},$$

where $\operatorname{wgt}(T^\lambda) = \sum_{i=1}^{\ell+1} n_i \varepsilon_i$ with n_i equal to the number of entries i appearing in T^λ. It follows that the weight multiplicity m_κ^λ is the number of $(\ell+1)$-standard Young tableaux T^λ of weight κ.

Example 6.1. For all A_ℓ with $\ell \geq 3$, $\lambda = \omega_1 + 2\omega_2 = 3\varepsilon_1 + 2\varepsilon_2$ and $\kappa = \omega_2 + \omega_3 = 2\varepsilon_1 + 2\varepsilon_2 + \varepsilon_3$ there exist just two standard Young tableaux T^λ of weight κ:

$$\begin{array}{|c|c|c|} \hline 1 & 1 & 2 \\ \hline 2 & 3 \\ \cline{1-2} \end{array} \qquad \begin{array}{|c|c|c|} \hline 1 & 1 & 3 \\ \hline 2 & 2 \\ \cline{1-2} \end{array}$$

Hence, for all A_ℓ with $\ell \geq 3$ we have $m_{\omega_2+\omega_3}^{\omega_1+2\omega_2} = 2$.

All the above supposes that $\lambda_{\ell+1} = 0$, so that the components of λ in the ε-basis define a partition. If instead, $\lambda_{\ell+1} \in \mathbb{Z}^-$, then it is necessary as we have seen earlier to use composite Young diagrams. Weyl's character formula then leads to the result

$$\operatorname{ch} V^\lambda = \sum_\kappa m_\kappa^\lambda e^\kappa = \operatorname{ch} V^{\mu;\bar\nu} = \sum_\kappa m_{\sigma;\bar\tau}^{\mu;\bar\nu}(\ell) e^{\sigma;\bar\tau} = \sum_{T^{\mu;\bar\nu}} e^{\operatorname{wgt}(T^{\mu;\bar\nu})},$$

where the weight multiplicity $m_{\sigma;\bar\tau}^{\mu;\bar\nu}(\ell)$ is the number of $(\ell+1)$-standard composite Young tableaux $T^{\mu;\bar\nu}$ of weight $\sigma;\bar\tau$. The entries in the two parts F^μ and $F^{\bar\nu}$ of the composite diagram are unbarred and barred, respectively, and contribute positively and negatively to weight vectors in the ε-basis. In each part of the composite diagram the numbers are weakly increasing from left to right across rows and strictly increasing from top to bottom down columns. Moreover each composite subdiagram consisting of those boxes containing unbarred entries no greater than k and barred entries no less than $\bar k$ must occupy a total of no more than k rows [7].

Example 6.2. For A_ℓ, $\lambda = \omega_1 + \omega_2 + \omega_{\ell-1} = 2\varepsilon_1 + \varepsilon_2 - \varepsilon_3 - \varepsilon_4 = \mu;\bar\nu$ and $\kappa = \omega_2 + \omega_\ell = \varepsilon_1 + \varepsilon_2 - \varepsilon_{\ell+1} = \sigma;\bar\tau$, in the case $\ell = 3$ there exist just two standard composite Young tableaux:

$$\begin{array}{cc} \begin{array}{|c|} \hline \bar 4 \\ \hline \bar 3 \\ \hline \end{array} & \\ & \begin{array}{|c|c|} \hline 1 & 2 \\ \hline 3 \\ \cline{1-1} \end{array} \end{array} \qquad \begin{array}{cc} \begin{array}{|c|} \hline \bar 4 \\ \hline \bar 3 \\ \hline \end{array} & \\ & \begin{array}{|c|c|} \hline 1 & 3 \\ \hline 2 \\ \cline{1-1} \end{array} \end{array}$$

Hence for A_3 we have as before $m_{\omega_2+\omega_3}^{\omega_1+2\omega_2}(3) = 2$. However, for $\ell \, 3$ we know from our tabulation that the corresponding weight multiplicity depends on

ℓ. In fact, in the case of Example 6.2, we have the following $(\ell+1)$-standard tableaux:

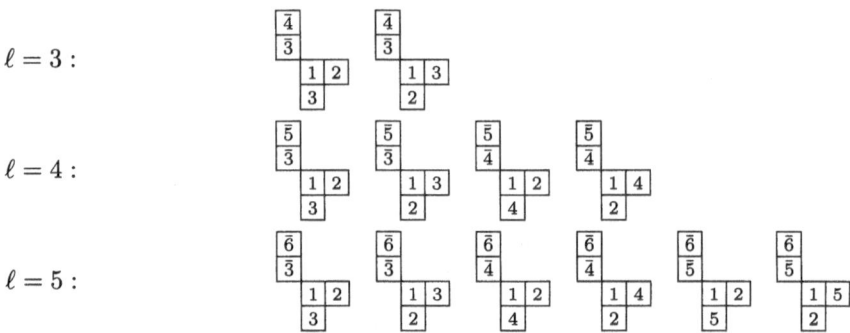

$\ell = 3$:

$\ell = 4$:

$\ell = 5$:

These enumerations imply that $m_{\sigma;\bar{\tau}}^{\mu;\bar{\nu}}(\ell) = m_{\omega_2+\omega_\ell}^{\omega_1+\omega_2+\omega_{\ell-1}} = 2\ell - 4$ for all $\ell \geq 3$. This is in accordance with our earlier polynomial fit to the weight multiplicity data.

More generally the weight multiplicities $m_{\sigma;\bar{\tau}}^{\mu;\bar{\nu}}(\ell)$ of A_ℓ are polynomials in ℓ of degree $|\mu| - |\sigma| = |\nu| - |\tau|$ [8].

7   Branching Rule Multiplicities for the Restriction from Affine to Simple Lie Algebras

The branching rule multiplicities $b_{d,\nu}^\lambda(\ell)$ associated with the restriction of the irrep V^λ of an affine Lie algebra $\mathbf{g}$ of rank ℓ to irreps V_0^ν of its naturally embedded simple Lie algebra $\mathbf{g}_0$ are defined through the identity

$$\text{ch } V^\lambda = e^{L(\lambda)\Lambda_0 - d(\lambda)\delta} \sum_{d,\nu} b_{d,\nu}^\lambda(\ell) e^{-d\delta} \text{ ch } V_0^\nu,$$

where $L(\lambda)$ and $d(\lambda)$ are the level and depth, respectively, of λ, and d specifies the relative depth of the irrep V_0^ν in the restriction of V^λ. Here the fundamental irreps of $\mathbf{g}$ have highest-weights Λ_i for $i = 0, 1, 2, \ldots, \ell$ and δ is the imaginary root.

Example 7.1. In Ref. [6] the extensive tabulation of branching rules for the restriction of affine Lie algebras to simple Lie algebras includes the following data in the case $A_\ell^{(1)} \supset A_\ell$ and $\lambda = \Lambda_1 + \Lambda_\ell$:

$\ell = 2$:

$d\backslash\{\nu\}$	$\{0;\bar{0}\}$	$\{1;\bar{1}\}$	$\{1^2;\bar{1}^2\}$	$\{1^2;\bar{2}\}$	$\{2;\bar{1}^2\}$	$\{2;\bar{2}\}$
0	0	1				
1	1	2	$-$	1	1	
2	2	5	$-$	2	2	2
3	3	12	$-$	5	5	5

$\ell = 3$:

$d\backslash\{\nu\}$	$\{0;\bar{0}\}$	$\{1;\bar{1}\}$	$\{1^2;\overline{1^2}\}$	$\{1^2;\bar{2}\}$	$\{2;\overline{1^2}\}$	$\{2;\bar{2}\}$
0	0	1				
1	1	2	1	1	1	
2	2	6	3	3	3	2
3	4	15	9	9	9	6

$\ell = 4$:

$d\backslash\{\nu\}$	$\{0;\bar{0}\}$	$\{1;\bar{1}\}$	$\{1^2;\overline{1^2}\}$	$\{1^2;\bar{2}\}$	$\{2;\overline{1^2}\}$	$\{2;\bar{2}\}$
0	0	1				
1	1	2	1	1	1	
2	2	6	4	3	3	2
3	4	16	12	10	10	6

$\ell = 5, 6, 7, 8$:

$d\backslash\{\nu\}$	$\{0;\bar{0}\}$	$\{1;\bar{1}\}$	$\{1^2;\overline{1^2}\}$	$\{1^2;\bar{2}\}$	$\{2;\overline{1^2}\}$	$\{2;\bar{2}\}$
0	0	1				
1	1	2	1	1	1	
2	2	6	4	3	3	2
3	4	16	13	10	10	6

This would lead one to conjecture that the branching rule multiplicities $b_{d,\nu}^{\lambda}(\ell)$ for $A_{\ell}^{(1)} \supset A_{\ell}$ are independent of ℓ for sufficiently large ℓ. To establish this it is necessary to have recourse to an explicit character formula.

8 Branching Rules Derived from Characters

The Weyl–Kac character formula for an affine Lie algebra $\mathbf{g}$ takes the form $\mathbf{g}$ [5]

$$\operatorname{ch} V^{\lambda} = \sum_{w \in W} \varepsilon(w) e^{w(\lambda+\rho)-\rho} \Big/ \sum_{w \in W} \varepsilon(w) e^{w(\rho)-\rho}.$$

By a judicious choice [4, 9] of the set W_s of right coset representatives of the Weyl group W of $\mathbf{g}$ with respect to the Weyl group W_0 of $\mathbf{g}_0$ it is possible to rewrite the character formula as the quotient of two depth-dependent infinite series N^{λ} and D in characters of $\mathbf{g}_0$. In fact for each dominant integral highest-weight λ of $\mathbf{g}$ of level $L(\lambda)$ and depth $d(\lambda)$ we can write

$$\operatorname{ch} V^{\lambda} = e^{L(\lambda)\Lambda_0 - d(\lambda)\delta} N^{\lambda}/D \quad \text{where} \quad D = \prod_{\alpha \in \Delta^+ \backslash \Delta_0^+} (1 - e^{-\alpha})^{\operatorname{mult}(\alpha)},$$

where Δ^+ and Δ_0^+ are the sets of positive roots of $\mathbf{g}$ and $\mathbf{g}_0$, respectively, and $\operatorname{mult}(\alpha)$ is the multiplicity of α in $\mathbf{g}$.

In the case $\mathbf{g} = A_{\ell}^{(\ell)}$ and $\mathbf{g}_0 = A_{\ell}$ we have the denominator formula [7]

$$D = \sum_{\zeta} (-1)^{|\zeta|} e^{-|\zeta|\delta} \operatorname{ch} V_0^{\{\zeta;\bar{\zeta}'\}},$$

where the summation is over all partitions ζ, and ζ', the partition conjugate to ζ, is such that $F^{\zeta'}$ is obtained from F^{ζ} by interchanging rows and columns. Setting $q = e^{-\delta}$, this gives

$$D = 1 - q \operatorname{ch} V_0^{\{1;\bar{1}\}} + q^2 \left(\operatorname{ch} V_0^{\{1^2;\bar{2}\}} + \operatorname{ch} V_0^{\{2;\overline{1^2}\}} \right)$$
$$- q^3 \left(\operatorname{ch} V_0^{\{1^3;\bar{3}\}} + \operatorname{ch} V_0^{\{21;\overline{21}\}} + \operatorname{ch} V_0^{\{3;\overline{1^3}\}} \right) - \cdots$$

This may be inverted up to any required depth by using the familiar product rules of A_ℓ to give

$$D^{-1} = 1 + q \operatorname{ch} V_0^{\{1;\bar{1}\}}$$
$$+ q^2 \left(\operatorname{ch} V_0^{\{2;\bar{2}\}} + \operatorname{ch} V_0^{\{1^2;\overline{1^2}\}} + 2 \operatorname{ch} V_0^{\{1;\bar{1}\}} + \operatorname{ch} V_0^{\{0;\bar{0}\}} \right)$$
$$+ q^3 \left(\operatorname{ch} V_0^{\{3;\bar{3}\}} + \operatorname{ch} V_0^{\{21;\overline{21}\}} + \operatorname{ch} V_0^{\{1^3;\overline{1^3}\}} + 2 \operatorname{ch} V_0^{\{2;\bar{2}\}} \right.$$
$$+ 2 \operatorname{ch} V_0^{\{2;\overline{1^2}\}} + 2 \operatorname{ch} V_0^{\{1^2;\bar{2}\}} + 2 \operatorname{ch} V_0^{\{1^2;\overline{1^2}\}}$$
$$\left. + 5 \operatorname{ch} V_0^{\{1;\bar{1}\}} + 2 \operatorname{ch} V_0^{\{0;\bar{0}\}} \right) + \cdots$$

To evaluate the numerator N^λ it is necessary to introduce a new numbering of composite Young diagrams [4]. Let the boxes of $F^{\zeta;\bar{\zeta'}}$ in the ith row and jth column of the part F^{ζ} and the jth row and ith column of the part $F^{\bar{\zeta'}}$ contain the entry $\eta_{ij} = (i - j) \mod (\ell + 1)$. Typically for $\ell = 9$ and $\zeta = (3, 2)$ we have

$$F^{\zeta;\bar{\zeta'}} = F^{32;\overline{2^21}} = \qquad$$

With this notation it may be shown [4] that for all $A_\ell^{(1)}$ and all integral dominant highest-weights λ of level $L(\lambda)$ and depth $d(\lambda)$:

$$N^\lambda = \sum_\zeta (-1)^{|\zeta|} e^{-(|\zeta| + |\zeta(\lambda)|)\delta} \operatorname{ch} V_0^{\bar{\lambda} + \{\zeta;\bar{\zeta'}\} + \{\zeta(\lambda);\overline{\zeta'(\lambda)}\}},$$

with $\bar{\lambda} = \lambda - L(\lambda)\Lambda_0 + d(\lambda)\delta$ and

$$\{\zeta(\lambda); \overline{\zeta'(\lambda)}\} = \sum_{(ij) \in F^\zeta} \lambda_{\eta_{ij}} \varepsilon_i - \sum_{(ji) \in F^{\bar{\zeta'}}} \lambda_{\eta_{ij}} \varepsilon_{\ell-j+2},$$

while the summation is over all partitions ζ.

Example 8.1. For $A_\ell^{(1)}$ in the case $\lambda = \Lambda_0 + \Lambda_1$ we have

$$N^\lambda = \operatorname{ch} V_0^{\{1;\bar{0}\}} - q^2 \operatorname{ch} V_0^{\{3;\bar{2}\}} + q^3 \operatorname{ch} V_0^{\{4;\overline{12}\}} - \cdots$$

By evaluating the product $N^\lambda D^{-1}$ in the case $\lambda = \Lambda_0 + \Lambda_1$ it then follows that

$$
\begin{aligned}
\operatorname{ch} V^\lambda = e^{2\Lambda_0} \Big(&\operatorname{ch} V_0^{\{1;\bar{0}\}} + q(\operatorname{ch} V_0^{\{2;\bar{1}\}} + \operatorname{ch} V_0^{\{1^2;\bar{1}\}} + \operatorname{ch} V_0^{\{1;\bar{0}\}}) \\
&+ q^2(\operatorname{ch} V_0^{\{21;\bar{2}\}} + \operatorname{ch} V_0^{\{21;\overline{1^2}\}} + \operatorname{ch} V_0^{\{1^3;\overline{1^2}\}} + 3\operatorname{ch} V_0^{\{2;\bar{1}\}} \\
&\qquad\qquad + 3\operatorname{ch} V_0^{\{1^2;\bar{1}\}} + 3\operatorname{ch} V_0^{\{1;\bar{0}\}}) \\
&+ q^3(\operatorname{ch} V_0^{\{2^2;\overline{21}\}} + \operatorname{ch} V_0^{\{21^2;\overline{21}\}} + \operatorname{ch} V_0^{\{21^2;\overline{1^3}\}} + \operatorname{ch} V_0^{\{1^4;\overline{1^3}\}} \\
&\qquad + \operatorname{ch} V_0^{\{3;\bar{2}\}} + \operatorname{ch} V_0^{\{3;\overline{1^2}\}} + 4\operatorname{ch} V_0^{\{21;\bar{2}\}} + 5\operatorname{ch} V_0^{\{21;\overline{1^2}\}} \\
&\qquad\qquad + 2\operatorname{ch} V_0^{\{1^3;\bar{2}\}} + 3\operatorname{ch} V_0^{\{1^3;\overline{1^2}\}} + 8\operatorname{ch} V_0^{\{2;\bar{1}\}} \\
&\qquad\qquad\qquad + 9\operatorname{ch} V_0^{\{1^2;\bar{1}\}} + 7\operatorname{ch} V_0^{\{1;\bar{0}\}}) + \dots \Big).
\end{aligned}
$$

The above result is valid for all $\ell \geq 5$ without the necessity of applying A_ℓ modification rules.

More generally, it follows that all the branching rule multiplicities for $A_\ell^{(1)} \supset A_\ell$ down to any prescribed depth are independent of ℓ for ℓ sufficiently large that A_ℓ modification rules are not required. For lower values of ℓ the results remain valid but it may be necessary to invoke A_ℓ modification rules so as to ensure that each irrep $V_0^{\{\sigma;\bar\tau\}}$ is labeled by a dominant A_ℓ weight $\mu = \{\sigma;\bar\tau\}$.

9 Weight Multiplicities of Affine Lie Algebras

The weight multiplicities, $m_\kappa^\lambda(\ell)$, of affine Lie algebras $\mathbf{g}$ appearing in the expansion

$$
\operatorname{ch} V^\lambda = \sum_\kappa m_\kappa^\lambda(\ell) e^\kappa,
$$

have been very extensively tabulated in Ref. [6, Vol. 2]. These results include the following data for $A_\ell^{(1)}$ in the case $\lambda = \Lambda_0 + \Lambda_1$ and $\kappa = \Lambda_0 + \Lambda_1 - d\delta$:

ℓ	$d=0$	$d=1$	$d=2$	$d=3$
1	1	2	4	8
2	1	4	13	36
3	1	6	27	98
4	1	8	46	208
5	1	10	70	380
6	1	12	99	628
7	1	14	133	966
8	1	16	172	1408
.				
ℓ	1	2ℓ	$\frac{1}{2}(5\ell^2 + 3\ell)$	$\frac{1}{3}(7\ell^3 + 9\ell^2 + 8\ell)$

The last row of this table represents the polynomial fit to the preceding data. To see more generally that the weight multiplicities of $A_\ell^{(1)}$ are indeed polynomial in the rank ℓ, it is merely necessary to note that the branching rule multiplicities for $A_\ell^{(1)} \supset A_\ell$ are rank-independent whilst the weight multiplicities of A_ℓ are polynomial in ℓ. This result was first exemplified and then established by Benkart and her colleagues [2].

By way of example one arrives, by means of the branching rules for $A_\ell^{(1)} \supset A_\ell$ and the weight multiplicities for A_ℓ, at the following polynomial expressions for the weight multiplicities $m_\kappa^\lambda(\ell)$ of $A_\ell^{(1)}$ for various irreps λ and weights $\kappa = \gamma - d\delta$:

λ	γ	$d=0$	$d=1$	$d=2$	$d=3$
Λ_i	Λ_i	1	ℓ	$\frac{1}{2}(\ell^2+3\ell)$	$\frac{1}{6}(\ell^3+9\ell^2+8\ell)$
$2\Lambda_0$	$2\Lambda_0$	1	ℓ	$\ell^2+2\ell$	$\frac{1}{6}(5\ell^3+15\ell^2+10\ell)$
$2\Lambda_0$	$\Lambda_1+\Lambda_\ell$		1	2ℓ	$\frac{1}{2}(5\ell^2+\ell+2)$
$2\Lambda_0$	$\Lambda_2+\Lambda_{\ell-1}$		2	$5\ell-4$	
$2\Lambda_0$	$\Lambda_3+\Lambda_{\ell-2}$			5	
$\Lambda_0+\Lambda_1$	$\Lambda_0+\Lambda_1$	1	2ℓ	$\frac{1}{2}(5\ell^2+3\ell)$	$\frac{1}{3}(7\ell^3+9\ell^2+8\ell)$
$\Lambda_0+\Lambda_1$	$\Lambda_2+\Lambda_\ell$		2	$5\ell-3$	$7\ell^2-7\ell+8$
$\Lambda_0+\Lambda_1$	$\Lambda_3+\Lambda_{\ell-1}$		5	$14\ell-20$	
$\Lambda_0+\Lambda_1$	$\Lambda_4+\Lambda_{\ell-2}$			14	
$\Lambda_1+\Lambda_\ell$	$2\Lambda_0$	ℓ	$\frac{1}{2}(3\ell^2+\ell)$	$\frac{1}{2}(3\ell^3+5\ell^2)$	$\cdots$
$\Lambda_1+\Lambda_\ell$	$\Lambda_1+\Lambda_\ell$	1	$3\ell-2$	$\frac{1}{2}(9\ell^2-7\ell+4)$	$\cdots$
$\Lambda_1+\Lambda_\ell$	$\Lambda_2+\Lambda_{\ell-1}$		3	$9\ell-2$	$\cdots$
$\Lambda_1+\Lambda_\ell$	$\Lambda_3+\Lambda_{\ell-2}$		9	$28\ell-60$	
$\Lambda_1+\Lambda_\ell$	$\Lambda_4+\Lambda_{\ell-3}$			28	

If $\lambda = L(\lambda)\Lambda_0 + d(\lambda)\delta + \{\mu; \bar\nu\}$ and $\kappa = L(\lambda)\Lambda_0 + d(\lambda)\delta - d\delta + \{\sigma; \bar\tau\}$ then the degree of the weight multiplicity polynomial is $d+|\mu|-|\sigma| = d+|\nu|-|\tau|$.

It is further conjectured [4] that the branching rule multiplicities of at least six of the seven natural embeddings of $\mathbf{g}_0$ in $\mathbf{g}$ indexed by ℓ are independent of ℓ and that the weight multiplicities of all seven infinite sequences of affine Lie algebras $\mathbf{g}$ indexed by the rank ℓ, save possibly $D_\ell^{(1)}$, are polynomial in ℓ.

REFERENCES

1. G.M. Benkart, D.J. Britten, and F.W Lemire, *Stability in modules for classical Lie algebras—a constructive approach*, Mem. Amer. Math. Soc. **85** (1990), no. 430.

2. G. Benkart, S.-J. Kang, and K.C. Misra, *Indefinite Kac–Moody algebras of classical type*, Adv. Math. **105** (1994), 76–110.

3. M.R. Bremner, R.V. Moody, and J. Patera, *Tables of dominant*

weight multiplicities for representations of simple Lie algebras, Marcel Dekker, New York, 1985.

4. A. Hussin, PhD Thesis, University Southampton, 1995.

5. V.G. Kac, *Infinite-dimensional Lie algebras*, Cambridge University Press, Cambridge, 1990.

6. S. Kass, R.V. Moody, J. Patera, and R. Slansky, *Affine Lie algebras, weight multiplicities and branching rules*, Vols. 1–2, University California Press, Berkeley and Los Angeles, 1990.

7. R.C. King, *S-functions and characters of Lie algebras and superalgebras*, Invariant Theory and Tableaux (Minneapolis, MN, 1988) (D. Stanton, ed.), IMA Vol. Math. Appl., Vol. 19, Springer-Verlag, New York-Berlin, pp. 226–261, 1990.

8. R.C. King and S.P.O. Plunkett, *The evaluation of weight multiplicities using characters and S-functions*, J. Phys. A **9** (1976), 863–887.

9. L. Liu, *Kostant's formula for Kac–Moody Lie algebras*, J. Algebra **149** (1992), 155–178.

10. W.G. McKay and J. Patera, *Tables of dimensions, indices, and branching rules for representations of simple Lie algebras*, Marcel Dekker, New York, 1981.

10

Conditions for the Existence of Higher Symmetries and Nonlinear Evolutionary Equations on the Lattice

D. Levi and R. Yamilov

ABSTRACT In this paper we derive a set of five conditions necessary for the existence of generalized symmetries. We apply them to a class of dynamical equations on the lattice, depending on nearest neighboring interaction, which has a four-dimensional Lie group of continuous point symmetries.

1 Introduction

When considering differential equations we have many effective tools to integrate them, for example, Lie Group techniques [2, 12, 16]. This is not the case for difference equations, where few results are known [5–8, 13]. Thus it is very important to introduce new tools to be able to treat difference equations, as often discrete equations present "discrete" features that are lost in the continuum limit approximation. This is especially the case for nonlinear differential-difference equations, which are important in applications.

As in the case of partial differential equations, there are a certain number of difference equations that are integrable [1, 4], as, for example, the Toda lattice, which has a Lax pair, Bäcklund transformations, an infinity of conserved quantities and symmetries, and an infinity of explicit exact solutions. However, in order to improve the techniques for studying difference equations we first need us to find new integrable difference equations. There are various possible approaches to this problem. One possibility is to construct integrable difference equations of a given form by starting from the integrability conditions (i.e., Lax pair). A second possibility is to start from a class of equations of the desired form and deduce the integrable cases by applying some integrability conditions.

This second approach will be considered in the following. This relies on the so-called formal symmetry approach introduced by A.B. Shabat and collaborators in Ufa (see, e.g., review articles [10, 11, 15]) by which the authors classified all equations of a certain class that possess a few generalized symmetries of a certain kind. Such an approach has been introduced at first to study partial differential equations but then the procedure has been extended to the case of differential-difference equations [14, 17, 18]. In such an approach, one introduces conditions under which one can prove the existence of at least one (or more) generalized symmetries.

The class of nonlinear differential-difference equations we will consider in the following is given by

$$u_{n,t}(t) = f_n\big(u_{n-1}(t), u_n(t), u_{n+1}(t)\big), \tag{1.1}$$

where $u_n(t)$ is a dependent field expressed in terms of its independent variables, t varying over the reals and n varying over the integers. Equation (1.1) is a differential functional relation that correlates the time evolution of a function calculated at the point n to its values in its nearest neighboring points $(n+1, n-1)$. A peculiarity of the choice of Eq. (1.1) is the fact that its right-hand side is not a function—i.e., it is not the same for all points in the lattice—but for each point of the lattice one has an a priori different right-hand side. In fact, we can think of Eq. (1.1) as an infinite system of different differential equations for the infinite number of functions u_n. By proper choices of the functions f_n, Eq. (1.1) can be reduced to a system of k coupled differential-difference equations for the k unknown u_m^k or to a system of dynamical equations on the lattice. Let us assume that f_n and u_n are periodic functions of n of period k, i.e.,

$$f_n\big(u_{n-1}(t), u_n(t), u_{n+1}(t)\big) = \sum_{j=0}^{k-1} P_{n-j}^k f^j\big(u_{m-1}(t), u_m(t), u_{m+1}(t)\big),$$

$$u_n = \sum_{j=0}^{k-1} P_{n-j}^k u_m^j,$$

where P_n^k is a projection operator such that for any integer m, such that $n = km + j$ with $0 \le j \le k - 1$. The following relations are true:

$$P_{km}^k = 1, \quad P_{km+j}^k = 0, \quad (j = 1, 2, \ldots, k - 1);$$

then Eq. (1.1) becomes, for example, in the case $k = 2$, the system

$$\begin{aligned} u_{m,t}^0 &= f^0(u_{m-1}^1, u_m^0, u_m^1), \\ u_{m,t}^1 &= f^1(u_m^0, u_m^1, u_{m+1}^0). \end{aligned} \tag{1.2}$$

A subclass of Eq. (1.2) of particular relevance for its physical applications is given by dynamical systems on the lattice, i.e., equations of the type

$$\chi_{n,tt} = g(\chi_{n+1} - \chi_n, \chi_n - \chi_{n-1}). \tag{1.3}$$

So part of the results presented in Ref. [8] can be further analyzedd for the existence of generalized symmetries and of their integrability using the conditions presented here as these equations can be written in the form (1.3).

Section 2 is devoted to the construction of a certain number of conditions (the simpler ones) necessary to prove that an equation of the class (1.1) has generalized symmetries and higher-order conservation laws. The obtained conditions thus obtained are applied in Section 3 to the case

$$u_{n,t} = P_{n+1}^2 e^{u_{n+1}} g_n(u_{n+1} - u_{n-1}) + P_n^2 \lambda_n(u_{n+1} - u_{n-1}), \qquad (1.4)$$

where $g_n(\zeta_n)$ is a set of functions and λ_n is a set of constants. This equation describes a class of dynamical equations

$$v_{ktt} = \exp\left(\frac{v_{k+1} - v_k}{\varepsilon_{k+1}}\right) G_k\left(\frac{v_{k+1} - v_k}{\varepsilon_{k+1}} - v_k - v_{k-1}\varepsilon_k\right),$$

$$u_{2k} = \frac{v_{k+1} - v_k}{\varepsilon_{k+1}}, \quad u_{2k-1} = v_{kt}, \quad g_{2k-1} = G_k, \quad \lambda_{2k} = \varepsilon_{k+1}^{-1}, \qquad (1.5)$$

having a four-dimensional group of point symmetries and including the Toda lattice as one of its members [8]. A more complete set of results on the conditions associated with the class of Eqs. (1.1), together with other examples of equations that pass the test, can be found in Ref. [9].

2 Construction of the Classifying Conditions

Before considering in detail the problem of constructing generalized symmetries to Eq. (1.1) we will introduce a few definitions necessary for future calculations.

A function g_n depending on the set of fields u_n, for n varying on the lattice, will be called a *restricted function* and will be denoted by the symbol RF if it is defined on a compact support; i.e., if

$$g_n = g_n(u_{n+i}, u_{n+i-1}, \dots, u_{n+j+1}, u_{n+j}), \quad i \geq j, \qquad (2.1)$$

and i and j are finite integer numbers. If there exist, in the range of possible values of n, values k and m such that

$$\frac{\partial g_k}{\partial u_{k+i}} \neq 0, \quad \frac{\partial g_m}{\partial u_{m+j}} \neq 0, \qquad (2.2)$$

then we say that the function g_n has a *length* $i - j + 1$.

Let us define the shift operator D by

$$D g_n(u_{n+i}, \dots, u_{n+j}) = g_{n+1}(u_{n+i+1}, \dots, u_{n+j+1}).$$

Then we can split the RF into equivalence classes.

Definition. Two RF $a_n(u_{n+i_a}, \ldots, u_{n+j_a})$ and $b_n(u_{n+i_b}, \ldots, u_{n+j_b})$ are said to be *equivalent*, denoted by $a_n \sim b_n$, if

$$a_n - b_n = (D - 1)c_n, \tag{2.3}$$

where c_n is an RF.

Let us notice that any function that is equal to a total difference is equivalent to zero; i.e., $a_n = (D - 1)c_n \sim 0$. If an RF a_n of length $i - j + 1$ $(i > j)$ is equivalent to zero, then there will exist, by necessity, a RF c_n of length $i - j$ such that $a_n = (D - 1)c_n$ and consequently

$$\frac{\partial^2 a_n}{\partial u_{n+i} \partial u_{n+j}} = 0. \tag{2.4}$$

In the case $i = j$, $(da_n)/(du_{n+i}) = 0$, i.e., a_n is an *invariant* function, that is a function which depends only on n.

We can moreover define the "*formal*" *variational derivative* of an RF a_n of length $i - j + 1$ as

$$\frac{\delta a_n}{\delta u_n} = \sum_{k=n-i}^{n-j} \frac{\partial a_k}{\partial u_n}. \tag{2.5}$$

If a_n is linear in u_n, then $\delta a_n / \delta u_n$ is an invariant function. However, if it is nonlinear, then $\delta a_n / \delta u_n = \tilde{g}_n(u_{n+N}, \ldots, u_{n-N})$, where $\partial \tilde{g}_k / \partial u_{k+N} \neq 0$ for some k and $\partial \tilde{g}_m / \partial u_{m-N} \neq 0$ for some m. It is easy to prove that if a_n is an RF equivalent to zero, then the formal variational derivative of a_n is zero. The inverse is also true; i.e., if $\delta a_n / \delta u_n = 0$, then a_n is equivalent to zero.

Given a nonlinear chain (1.1), we will say that the RF $g_n(u_{n+i}, \ldots, u_{n+j})$ is a generalized (or higher) local symmetry of *order* i (more precisely, of left order i) of our equation if

$$u_{n,\tau} = g_n(u_{n+i}, \ldots, u_{n+j}), \tag{2.6}$$

is compatible with (1.1), i.e., if

$$\partial_t \partial_\tau(u_n) = \partial_\tau \partial_t(u_n). \tag{2.7}$$

Making explicit condition (2.7), we get

$$(\partial_t - f_n^*)g_n = 0, \tag{2.8}$$

where by f_n^* we mean the Frechet derivative of the function f_n, given by

$$f_n^* = \frac{\partial f_n}{\partial u_{n+1}} D + \frac{\partial f_n}{\partial u_n} + \frac{\partial f_n}{\partial u_{n-1}} D^{-1} = f_n^{(1)} D + f_n^{(0)} + f_n^{(-1)} D^{-1}. \tag{2.9}$$

Equation (2.8) is an equation for the symmetries g_n once the function f_n is given.

Given a symmetry we can construct a new symmetry by applying a recursive operator, i.e., an operator that transforms symmetries into symmetries. Given a symmetry g_n of Eq. (1.1), an operator

$$L_n = \sum_{j=-\infty}^{m} l_n^{(j)}(t) D^j \tag{2.10}$$

is a recursive operator for Eq. (1.1) if $\tilde{g}_n$ given by

$$\tilde{g}_n = L_n g_n \tag{2.11}$$

is a new generalized symmetry associated with (1.1). Equations (2.8) and (2.11) imply that

$$A(L_n) = L_{n,t} - [f_n^*, L_n] = 0. \tag{2.12}$$

Moreover from (2.8) it follows that

$$A(g_n^*) = \partial_\tau(f_n^*). \tag{2.13}$$

Equation (2.13) implies that, as its right-hand side is an operator of order 1 (see (2.9)), the highest terms in the left-hand side must be zero.

Thus we can define as *approximate* symmetry of *order* i and *length* m, the operator $G_n = \sum_{k=i-m+1}^{i} g_n^{(k)} D^k$, such that the m terms of highest order of the operator $A(G_n) = \sum_{k=i-m}^{i+1} a_n^{(k)} D^k$ are zeros. Taking into account Eq. (2.13), we find that we must have $i - m + 2 > 1$ if the equation

$$A(G_n) = 0, \tag{2.14}$$

is to be satisfied. From this result we can derive the first integrability condition, which can be stated in the following theorem:

Theorem 2.1. *If Eq. (1.1) has a local generalized symmetry of order $i \geq 2$, then it must have a conservation law given by*

$$\partial_t \log f_n^{(1)} = (D - 1) q_n^{(1)}, \tag{2.15}$$

where $q_n^{(1)}$ is an RF.

The next canonical conservation laws could be obtained in the same way, namely, by assuming the existence of a higher symmetry so that we may consider an approximate symmetry of higher length. This procedure can be carried out, and leads to the following theorem:

Theorem 2.2. *If Eq. (1.1) has two generalized local symmetries of order i and $i + 1$, with $i \geq 4$, then the following conservation laws must be true:*

$$\partial_t p_n^{(k)} = (D - 1) q_n^{(k)} \quad (k = 1, 2, 3),$$

$$p_n^{(1)} = \log \frac{\partial f_n}{\partial u_{n+1}}, \quad p_n^{(2)} = q_n^{(1)} + \frac{\partial f_n}{\partial u_n}, \tag{2.16}$$

$$p_n^{(3)} = q_n^{(2)} + \frac{1}{2}(p_n^{(2)})^2 + \frac{\partial f_n}{\partial u_{n+1}} \frac{\partial f_{n+1}}{\partial u_n},$$

where $q_n^{(k)}$ $(k = 1, 2, 3)$ are some RFs.

Thus, if Eq. (1.1) has local generalized symmetries of high enough order, we can construct a few conservation laws depending on the function on the right-hand side of Eq. (1.1).

One can divide the conservation laws into conjugacy classes under an equivalence condition. Two conservation laws $p_{n,t} = (D - 1)q_n$ and $r_{n,t} = (D - 1)s_n$ are *equivalent* if

$$p_n \sim r_n. \tag{2.17}$$

A local conservation law is *trivial* if $p_n \sim 0$. If $p_n \sim r_n(u_n)$, with $r'_n \neq 0$ for at least some n, then we have a conservation law of *zeroth order*, while if $p_n \sim r_n(u_{n+N}, \ldots, u_n)$, $N > 0$, and $(\partial^2 r_n)/(\partial u_n \partial u_{n+N}) \neq 0$ for at least some n, the conservation law is of *order N*.

An alternative way to define equivalence classes of local conservation laws is via the formal variational derivative. Let us denote by $\tilde{p}_n$ the formal variational derivative of the density p_n of a local conservation law, i.e.,

$$\tilde{p}_n = \frac{\delta p_n}{\delta u_n}. \tag{2.18}$$

If the local conservation law is trivial, then $\tilde{p}_n = 0$. If it is of order zero, then $\tilde{p}_n = \tilde{p}_n(u_n) \neq 0$ for at least some n. And if it is of order N, then $\tilde{p}_n = \tilde{p}_n(u_{n+N}, \ldots, u_n, \ldots, u_{n-N})$, where $\partial \tilde{p}_n / \partial u_{n+N} \neq 0$, $(\partial \tilde{p}_n)/(\partial u_{n-N}) \neq 0$ for at least some n. Then, for any conserved density p_n, by direct calculation, we derive the following relation:

$$p_{n,t} \sim \tilde{p}_n f_n \sim 0. \tag{2.19}$$

By carrying out the formal variational derivative of Eq. 2.19 we get that the formal variational derivative $\tilde{p}_n$ of a conserved density p_n satisfies the following equation:

$$(\partial_t + f_n^{*T})\tilde{p}_n = 0, \tag{2.20}$$

where the transposed Frechet derivative of f_n is given by

$$f_n^{*T} = \frac{\partial f_{n+1}}{\partial u_n}D + \frac{\partial f_n}{\partial u_n} + \frac{\partial f_{n-1}}{\partial u_n}D^{-1} = f_{n+1}^{(-1)}D + f_n^{(0)} + f_{n-1}^{(1)}D^{-1}. \tag{2.21}$$

Let us consider the Frechet derivative of $\tilde{p}_n$ for a local conservation law of order N. In this case, we have

$$\tilde{p}_n^* = \sum_{k=-N}^{N} \tilde{p}_n^{(k)} D^k, \quad \tilde{p}_n^{(k)} = \frac{\partial \tilde{p}_n}{\partial u_{n+k}}. \tag{2.22}$$

We can construct the following operator:

$$B(S_n) = S_{n,t} + S_n f_n^* + f_n^{*T} S_n, \tag{2.23}$$

where

$$S_n = \sum_{j=-\infty}^{l} s_n^{(j)}(t) D^j. \tag{2.24}$$

Taking into account (2.8) and (2.20), one can easily prove that

$$\tilde{p}_n = S_n g_n. \tag{2.25}$$

Let us construct

$$B(\tilde{p}_n^*) = \sum_k b_n^{(k)}(t) D^k, \tag{2.26}$$

where $b_n^{(k)}(t)$ are some RFs. It then follows from Eq. (2.23) that $b_n^{(k)}$ are different from zero only for $-2 \leq k \leq 2$.

In this way, for a sufficiently high order conserved density p_n, we can require that

$$B(\tilde{p}_n^*) = 0 \tag{2.27}$$

is approximately solved. If the first $m < N - 1$ terms of the Frechet derivative of $\tilde{p}_n$ satisfy Eq. (2.27), then we say that we have an *approximate conserved density* of *order* N and *length* m. Taking into account all the results up to now, we can state the following theorem:

Theorem 2.3. *If the chain* (1.1) *has a conservation law of order* $N \geq 3$, *and condition* (2.16) *is satisfied, then the following conditions must take place:*

$$r_n^{(k)} = (D-1)s_n^{(k)} \quad (k=1,2), \tag{2.28a}$$

with

$$r_n^{(1)} = \log[-f_n^{(1)}/f_n^{(-1)}], \quad r_n^{(2)} = s_{n,t}^{(1)} + 2f_n^{(0)}, \tag{2.28b}$$

where $s_n^{(k)}$ *are RFs.*

Conditions (2.16) require only that $f_n^{(1)} \neq 0$. An analogous set of conditions could be derived if we required that just $f_n^{(-1)} \neq 0$ for all n. They can be derived in a straightforward way by considering expansions in negative powers of D instead of positive, as we have done up to now. This set of conditions also will have the form of canonical conservation laws,

$$\partial_t \hat{p}_n^{(k)} = (D-1)\hat{q}_n^{(k)}, \tag{2.29}$$

and conserved densities will be symmetric to those of (2.16). For example,

$$\hat{p}_n^{(1)} = \log\left(-\frac{\partial f_n}{\partial u_{n-1}}\right). \tag{2.30}$$

Let us notice moreover that if $H_n^{(1)}$ and $H_n^{(2)}$ are two solutions of (2.27) of different order, the operator

$$K_n = (H_n^{(1)})^{-1} H_n^{(2)}, \tag{2.31}$$

satisfies (2.12) and thus it is a recursive operator. Consequently, if we start from two approximate solutions of (2.27), i.e., two Frechet derivatives of formal variational derivatives of conserved densities, we can, using (2.27), get an approximate symmetry.

If we compare conditions (2.16), (2.28), and (2.29), we can see, for example, that

$$r_n^{(1)} = p_n^{(1)} - \hat{p}_n^{(1)}; \tag{2.32}$$

i.e., the first of conditions (2.28) implies that $p_n^{(1)} \sim \hat{p}_n^{(1)}$. Thus the first canonical conservation laws of (2.16) and (2.29) are equivalent.

The solutions of the conditions, be they those obtained by requiring existence of the generalized symmetries or those of local conservation laws, provide the highest-order coefficients of the Frechet derivative of a symmetry or of the formal variational derivative of a conserved density. Those coefficients are the building blocks for the reconstruction of the symmetries or of the formal variational derivatives of the conserved densities. In fact, the knowledge of $g_n^{(k)} = (\partial g_n)/(\partial u_{n+k})$ with $k = i, i-1, \ldots$ for a few values of k, gives a set of partial right-hand equations for g_n with respect to its variables, whose solution provides the needed symmetry. In the same way we can reconstruct variational derivatives of conserved densities. There is, however, a more direct way to obtain conserved densities. In fact, if we know the highest coefficients of L_n, the solution of Eq. (2.12), we can obtain several conserved densities by computing $p_n^{(j)} = \mathrm{res}(L_n^j)$ for $j = 1, 2, \ldots$, where $\mathrm{res}(L_n^j)$ denotes the coefficient of D^0 in L_n^j.

It is worthwhile to show here how all five conditions (2.16), (2.28) can be rewritten in explicit form. Such explicit conditions can be easily verified using a computer and thus they can be the starting point for the construction of a program like DELIA [3] to check the integrability of differential-difference equations of the form (1.1).

A condition is *explicit* if it has the form $A_n = 0 \ \forall n$, where A_n is a function depending only on f_n and its partial derivatives with respect to all u_{n+i}. Let us define the functions $P_n^{(k)} = \delta/\delta u_n \partial_t p_n^{(k)}$, $R_n^{(k)} = \delta/\delta u_n r_n^{(k)}$ for $k = 1, 2$ and $P_n^{(3)} = \delta/\delta u_n \partial_t p_n^{(3)} - q_n^{(1)} P_n^{(2)}$. The five explicit conditions are given by

$$P_n^{(k)} = 0, \quad R_n^{(l)} = 0 \quad \forall n, \quad k = 1, 2, 3; \quad l = 1, 2. \tag{2.33}$$

The functions $p_n^{(1)}$, $r_n^{(1)}$ are already explicit, and from them one can derive all partial derivatives $\partial q_n^{(1)}/\partial u_{n+i}$, $\partial s_n^{(1)}/\partial u_{n+i}$ and then express $\partial_t p_n^{(2)}$, $r_n^{(2)}$ in an explicit form. For example, from (2.28) we have $r_n^{(2)} = \partial r_{n-1}^{(1)}/\partial u_n f_n - \partial r_n^{(1)}/\partial u_{n-1} f_{n-1} + 2 \partial f_n/\partial u_n$. Let us now consider $P_n^{(3)}$. We have

$$\partial_t p_n^{(3)} = \partial_t \left(q_n^{(2)} + \frac{\partial f_n}{\partial u_{n+1}} \frac{\partial f_{n+1}}{\partial u_n} \right) + \left(q_n^{(1)} + \frac{\partial f_n}{\partial u_n} \right) \partial_t p_n^{(2)}.$$

Using (2.16) with $k = 2$, one can find all the partial derivatives of $q_n^{(2)}$ and consequently get an explicit expression for $\partial_t q_n^{(2)}$. Using (2.16) with $k = 1$, we can obtain not only the functions $\partial q_{n+i}^{(1)}/\partial u_n$ but also all differences $q_{n+i}^{(1)} - q_n^{(1)}$. Consequently, as

$$\frac{\delta}{\delta u_n}(q_n^{(1)} \partial_t p_n^{(2)}) - q_n^{(1)} P_n^{(2)}$$

$$= \sum_i \frac{\partial q_{n+i}^{(1)}}{\partial u_n} \partial_t p_{n+i}^{(2)} + \sum_i (q_{n+i}^{(1)} - q_n^{(1)}) \frac{\partial}{\partial u_n} \partial_t p_{n+i}^{(2)}, \quad (2.34)$$

we can easily write down the explicit form of $P_n^{(3)}$.

3 The Toda Lattice Class

In the following we learn about integrability of differential-difference equations of the form (1.4). The following conditions must be imposed to ensure that (1.1) represents an evolutionary difference equation:

$$\frac{\partial f_n}{\partial u_{n+1}} = P_{n+1}^2 e^{u_{n+1}}(g_n + g_n') + P_n^2 \lambda_n \neq 0, \quad (3.1)$$

$$-\frac{\partial f_n}{\partial u_{n-1}} = P_{n+1}^2 e^{u_{n+1}} g_n' + P_n^2 \lambda_n \neq 0. \quad (3.2)$$

This means, in particular, that $\lambda_n \neq 0$ for n even. As the λ_n do not exist in our equation for n odd, we can make them arbitrary for n odd and then assume that $\lambda_n \neq 0$ for all n. Analogously, we have to require that $g_n' \neq 0$, $g_n + g_n' \neq 0$ for all n. We can than formulate the following theorem:

Theorem. *A chain of the form* (1.5) *satisfies the classifying conditions* (2.16), (2.28) *iff it is related by a point transformation of the form* $\tilde{u}_n = \alpha_n u_n + \beta_n$ *to one of two following chains:*

$$u_{n,t} = P_{n+1}^2(\exp u_{n+1} - \exp u_{n-1}) + P_n^2(u_{n+1} - u_{n-1}), \quad (3.3a)$$

$$u_{n,t} = P_{n+1}^2 \exp\left(\frac{u_{n+1} - u_{n-1}}{a_n}\right) + P_n^2 a_{n+1} a_{n-1}(u_{n+1} - u_{n-1}), \quad (3.3b)$$

with $a_n = \alpha n + \beta$ *where* α *and* β *are arbitrary constants.*

To prove this theorem we first consider the conditions (2.28). As we have (2.32),

$$p_n^{(1)} = \log \frac{\partial f_n}{\partial u_{n+1}} = P_{n+1}^2(u_{n+1} + \log(g_n + g_n')) + P_n^2 \log \lambda_n, \quad (3.4)$$

$$\tilde{p}_n^{(1)} = \log\left(-\frac{\partial f_n}{\partial u_{n-1}}\right) = P_{n+1}^2(u_{n+1} + \log g_n') + P_n^2 \log \lambda_n, \quad (3.5)$$

and consequently $r_n^{(1)} = P_{n+1}^2 H_n$, where $H_n = \log(g_n + g_n') - \log g_n'$. Hence

$$P_{n+1}^2 H_n'' = 0, \quad \text{i.e.,} \quad P_{n+1}^2 H_n = a_n v_n + b_n, \tag{3.6}$$

where $v_n = u_{n+1} - u_{n-1}$, and a_n and b_n are constants that depend on n. Consequently we must have $a_{n-1} = a_{n+1}$ for all n. Thus from (3.6) we get that

$$a_n = P_{n+1}^2 a, \tag{3.7}$$

where a is a pure constant. So, the condition $r_n^{(1)} \sim 0$ implies (3.6) and (3.7). We find moreover that $r_n^{(2)} = \partial_t s_n^{(1)}$. Consequently

$$a P_{n+1}^2 (\lambda_{n-1} - \lambda_{n+1}) = 0. \tag{3.8}$$

Let us consider the first canonical conservation law. It follows from (2.32) that this condition is equivalent to $\partial_t \widetilde{p}_n^{(1)} \sim 0$. Since $\widetilde{p}_n^{(1)} \sim P_n^2 u_n + P_{n+1}^2 \log g_n'$ (see (3.5)). Then we must have

$$P_{n+1}^2 (\log g_n')'' = 0, \quad \text{i.e.,} \quad P_{n+1}^2 \log g_n' = c_n v_n + d_n, \tag{3.9}$$

where c_n and d_n are constants depending on n.

By comparing (3.6) and (3.9) we obtain an explicit formula for g_n. Since for any function θ_n the formula $P_{n+1}^2 \exp(P_{n+1}^2 \theta_n) = P_{n+1}^2 \exp \theta_n$ is valid, we get from Eq. (3.9)

$$P_{n+1}^2 g_n' = P_{n+1}^2 \exp(c_n v_n + d_n). \tag{3.10}$$

Using (3.6) we are led to the following formula for g_n:

$$P_{n+1}^2 g_n = P_{n+1}^2 \exp\big((a + c_n) v_n + b_n + d_n\big) - P_{n+1}^2 \exp(c_n v_n + d_n). \tag{3.11}$$

Moreover the consistency between (3.10) and (3.11) implies that we must have

$$a(a - 1) = 0, \tag{3.12}$$
$$P_{n+1}^2 (1 + c_n) a = 0. \tag{3.13}$$

Let us first consider the case $a = 1$. Condition (3.13) gives $P_{n+1}^2 c_n = -P_{n+1}^2$ and it then follows from (3.8) that $\lambda_{2k} = \lambda_{2k-2} = \lambda$, where λ is a constant different from zero. Taking into account formula (3.11), we get that the chain thus obtained is of the form

$$u_{n,t} = P_{n+1}^2 \big(\exp(u_{n+1} + \alpha_n) - \exp(u_{n-1} + \beta_n)\big) + (u_{n+1} - u_{n-1}) P_n^2 \lambda,$$

where α_n, β_n are some constants depending on n. This chain can be further simplified, using simple point transformations. If we apply first the transformation $\tilde{u}_n = (P_{n+1}^2 \lambda + P_n^2) u_n$ and then $\tilde{u}_n = u_n + P_n^2 \alpha_{n-1}$, we can reduce it to the form

$$u_{n,t} = P_{n+1}^2 \big(\exp u_{n+1} - \exp(u_{n-1} + \gamma_n) \big) + (u_{n+1} - u_{n-1}) P_n^2,$$

where γ_n are some n-dependent constants. Moreover, we have $\partial_t p_n^{(2)} \sim 2 P_n^2 \exp(u_n)(1 - \exp \gamma_{n+1})$. Since $P_{n+1}^2 \exp \gamma_n = P_{n+1}^2$ for all n, the chain takes the form (3.3a).

Let us consider now the case $a = 0$. It follows from (3.11) that $P_{n+1}^2 g_n = P_{n+1}^2 \exp(c_n v_n + d_n)(\exp b_n - 1)$ so that the function $\exp b_n - 1$ cannot be zero for odd n, as $g_n \neq 0$. This means we can redefine d_n so that (1.4) takes the form

$u_{n,t}$
$$= P_{n+1}^2 \exp\big[(1 + c_n) u_{n+1} - c_n u_{n-1} + d_n\big] + (u_{n+1} - u_{n-1}) P_n^2 \lambda_n. \quad (3.14)$$

As in the previous case, the chain (3.14) can be simplified, using point transformations of the form $\tilde{u}_n = \alpha_n u_n + \beta_n$, and we get the following chain:

$$u_{n,t} = f_n = P_{n+1}^2 \exp \frac{v_n}{a_n} + P_n^2 b_n v_n, \quad (3.15)$$

where $v_n = u_{n+1} - u_{n-1}$ and a_n and b_n are some nonzero constants depending on n. To fix a_n and b_n we use two of the conditions (2.16) that will give us two other constraints. It is easy to see that $p_n^{(1)} = P_{n+1}^2 (v_n)/(a_n) + \delta_n$, and that

$$\partial_t p_n^{(2)} \sim 2 P_{n+1}^2 \frac{b_{n-1}}{a_{n-2}} - \frac{b_{n+1}}{a_{n+2}} \exp \frac{v_n}{a_n}.$$

Thus we get the constraint

$$P_{n+1}^2 A_n = 0. \quad (3.16)$$

Introducing $\tilde{b}_n$ such that $b_n = \tilde{b}_n a_{n+1} a_{n-1}$, we obtain from (3.16) that $P_n^2 \tilde{b}_n = P_n^2 b$, where b is a constant different from zero. Therefore

$$P_n^2 b_n = b P_n^2 a_{n+1} a_{n-1}, \quad (3.17)$$

and we obtain

$$P_{n+1}^2 (a_{n+2} - 2a_n + a_{n-2}) = 0.$$

Taking into account the form of (3.15), it follows that we can set $a_n = cn + d$ for all n. The chain (3.15) with b_n satisfying (3.17) has the form

$$u_{n,t} = P_{n+1}^2 \exp \frac{v_n}{a_n} + b P_n^2 a_{n+1} a_{n-1} v_n,$$

i.e., coincides with (3.3b) up to the constant b. This constant, however, can be easily removed, using an obvious point transformation.

If we go over to the class (1.5), we see that, in case (3.3a),

$$\varepsilon_k = 1, \quad G_k = 1 - \exp\big((v_k - v_{k-1}) - (v_{k+1} - v_k)\big),$$

and this is the Toda model for the function v_k. The chain (3.3b) is a new example of an integrable (and n-dependent) equation. In this case, the chain equation can be rewritten, setting, for simplicity, $c_k = a_{2k-1}$ as

$$v_{k,tt} = \exp\big[c_{k+1}(v_{k+1} - v_k) - c_{k-1}(v_k - v_{k-1})\big]. \tag{3.18}$$

It belongs to the class (1.5), as

$$G_k(\zeta_k) = \exp(\delta_k \zeta_k), \quad \delta_k = c_{k-1}\varepsilon_k, \quad c_{k+1}\varepsilon_{k+1} - c_{k-1}\varepsilon_k = 1.$$

As c_k is linear in k, Eq. (3.18) can be written as

$$v_{k,tt} = \exp(c_{k+1}v_{k+1} - 2c_k v_k + c_{k-1}v_{k-1})$$

and by an obvious point transformation, we can remove the c_k and obtain the potential Toda equation:

$$v_{k,tt} = \exp(v_{k+1} - 2v_k + v_{k-1}). \tag{3.19}$$

This reduces to the Toda equation by the transformation

$$\widetilde{v}_k = v_{k+1} - v_k.$$

This implies that Eq. (3.18) is completely integrable.

In conclusion we state that the only equation in the class (1.4)–(1.5) that possesses generalized symmetries characterized by RF is the Toda lattice. Point symmetries by themselves are not sufficient to discriminate between integrable and nonintegrable equations. At least in the case of dynamical systems on the lattice depending on nearest neighbor interaction, the existence of a large group of continuous Lie point symmetries is not an indication that an equation is more integrable.

References

1. M.J. Ablowitz and J. Ladik, *Nonlinear differential-difference equations*, J. Math. Phys. **16** (1975), 598–603; *Nonlinear differential-difference equations and Fourier transform*, J. Math. Phys. **17** (1976), 1011–1018; *A nonlinear difference scheme and inverse scattering*, Stud. Appl. Math. **55** (1976) 213–229; *On the solution of a class of nonlinear partial difference equations*, Stud. Appl. Math. **57** (1976/77), No. 1, 1–12.

2. G.W. Bluman and S. Kumei, *Symmetries and differential equations*, Springer, New York, 1989.

3. A.V. Bocharov, *DEliA: A system of exact analysis of differential equations using S. Lie approach*, Report by Joint Venture OWIMEX Program Systems Institute of the USSR, Academy of Sciences, Pereslavl-Zalessky, USSR, 1989; *DEliA: project presentation*, SIGSAM Bulletin **24** (1990), 37–38; *Will DEliA grow into an expert system?*, Design and Implementation of Symbolic Computation Systems (A. Miola, ed.) (Capri, 1990), Lecture Notes in Comput. Sci., Vol. 429, Springer Verlag, Berlin, 1990, pp. 266–267.

4. D. Levi and O. Ragnisco, *Nonlinear differential-difference equations with n-dependent coefficients.* I, J. Phys. A **12** (1979), L157–L162; *Nonlinear differential-difference equations with n-dependent coefficients.* II, J. Phys. A **12** (1979), L163–L167; *The inhomogeneous Toda Lattice: Its hierarchy and Darboux–Bäcklund transformations*, J. Phys. A **24** (1991), 1729–1739; R. Benguria and D. Levi, *Bäcklund transformations and nonlinear differential-difference equations*, Proc. Nat. Acad. Sci. USA **77** (1980), No. 9, part 1, 5025–5027; D. Levi, L. Pilloni, and P.M. Santini, *Integrable three-dimensional lattices*, J. Phys. A **14** (1981), 1567–1575; M. Bruschi, D. Levi, and O.Ragnisco, *Evolution equations associated with the discrete analogue of the matrix Schrödinger spectral problem solvable by the inverse spectral transform*, J. Math. Phys. **22** (1981), 2463–2471; *Toda lattice and generalized Wronskian technique*, J. Phys. A **13** (1980), 2531–2533; *The discrete chiral field hierarchy*, Lett. Nuovo Cimento **33** (1982), 284–288; *Continuous and discrete matrix Burgers hierarchies*, Nuovo Cimento B (11) **74** (1983), 33–51.

5. D. Levi, L. Vinet, and P. Winternitz, *Lie group formalism for difference equations*, J. Phys. A **30** (1997), 633–649.

6. D. Levi and P. Winternitz, *Continuous symmetries of discrete equations*, Phys. Lett. A **152** (1991), 335–338.

7. D. Levi and P. Winternitz, *Symmetries and conditional symmetries of differential-difference equations*, J. Math. Phys. **34** (1993), 3713–3730.

8. D. Levi and P. Winternitz, *Symmetries of discrete dynamical systems*, J. Math. Phys. **37** (1996), 5551–5576.

9. D. Levi and R. Yamilov, *Conditions for the existence of higher symmetries of evolutionary equations on the lattice*, J. Math. Phys. **38** (1997), 6648–6674.

10. A.V. Mikhailov, A.B. Shabat, and V.V. Sokolov, *The symmetry approach to classification of integrable equations*, What is Integrability?, Springer, Berlin, 1991, pp. 115–184.

11. A.V. Mikhailov, A.B. Shabat, and R.I. Yamilov, *The symmetry approach to the classification of nonlinear equations. Complete lists of integrable systems*, Uspekhi Mat. Nauk **42** (1987), No. 4, 3–53 (Russian).

12. P.J. Olver, *Applications of Lie group to differential equations*, Springer, New York, 1986.

13. G.R. Quispel, H.W. Capel, and R. Sahavedan, *Continuous symmetries of difference equations; the Kac-van Moerbeke equation and Painlevé reduction*, Phys. Lett. A **170** (1992), 379–383.

14. A.B. Shabat and R.I. Yamilov, *Symmetries of nonlinear chains*, Algebra i Analiz **2** (1990), No. 2, 183–208 (Russian); English transl. Leningrad Math. J. **2** (1991), No. 2, 377–400.

15. V.V. Sokolov and A.B. Shabat, *Classification of integrable evolution equations*, Mathematical Physics Reviews, Vol. 4, Soviet Sci. Rev., Sect. C, Vol. 4, 1984, pp. 221–280.

16. P. Winternitz, *Lie groups and solutions of nonlinear partial differential equations*, Integrable Systems, Quantum groups and Quantum Field theories (L.A. Ibort and M.A. Rodriguez, eds.) Kluwer, Dordrecht, 1993, pp. 429–495.

17. R.I. Yamilov, *Classification of discrete evolution equations*, Uspekhi Mat. Nauk **38** (1983), No. 6, 155–156 (Russian).

18. R.I. Yamilov, *Generalizations of the Toda model and conservation laws*, Preprint, Inst. of Mathematics, Ufa, 1989 (Russian); *Classification of Toda type scalar lattices*, Proc. Internat. Conf. NEEDS'92, World Scientific, 1993, pp. 423-431.

11

Complete Description of the Voronoï Cell of the Lie Algebra A_n Weight Lattice. On the Bounds for the Number of d-Faces of the n-Dimensional Voronoï Cells

Louis Michel[1]

À mes amis Jiři et Pavel
pour leur 60ᵉ anniversaire

ABSTRACT Denoting these bounds by $N_d(n)$, $0 \leq d \leq n$ we prove that $N_d(n)/(n+1)!$ is a polynomial $p_d(n)$ of degree d with rational coefficients. We give the polynomials for $d \leq 5$ explicitly. The proof uses the fact that these bounds $N_d(n)$ are also the number of d-faces of the Voronoï cell of the weight lattice of the Lie algebra A_n (it is also the Cayley diagram of the symmetric group $\mathcal{S}_{n+1}$, which is isomorphic to the Weyl group of A_n). Each d-face of this cell is a zonotope that can be defined by a symmetry group $\sim G_d(\alpha)$, d-dimensional reflection subgroup of the A_n Weyl group. We show that for a given d and n large enough, all such subgroups of A_n are represented, and we compute explicitly $N\big(G_d(\alpha), n\big)$ the number of d-faces of type $G_d(\alpha)$ in the Voronoï cell of $L = A_n^w$. The final result is obtained by summing over α. That also yields the simple expression $N_d(n) = (n+1-d)!\mathbf{S}_{n+1}^{(n+1-d)}$ where the last symbol is the Stirling number of second kind.

1 Introduction

The proximity cell of a lattice of points was defined and studied by (Lejeune-) Dirichlet for two-dimensional lattices (= two variable quadratic forms) and also by Hermite. The three-dimensional case was thoroughly treated

---

[1]This paper was received in August 1997.

in the book [11] of Fedorov: *An introduction to the theory of figures*.[2] The last Voronoï memoir is the fundamental study for arbitrary dimension; it appeared in two parts [24, 25], the second has been printed after Voronoï's death at the age of 40. The proximity cells are usually now called[3] Voronoï cells. We shall give their definition in Eqs. (1.1a), (1.1b), and (1.1c) after introducing the necessary notations.

Let E_n be a n-dimensional real orthogonal vector space whose scalar product is denoted by $(\vec{x}, \vec{y})$; we define $N(\vec{x}) = (\vec{x}, \vec{x})$. A lattice $L \subset E_n$ is the set of vectors generated from an E_n basis by the addition of vectors. It is a group $L \sim Z^n$, free Abelian group of rank n, closed subgroup of R^n. Three equivalent definitions of the Voronoï cell $D(L)$ of L are

$$D(L) = \{\vec{x} \in E_n; \forall \ell \in L, N(\vec{x}) \le N(\vec{x} - \vec{\ell})\}; \tag{1.1a}$$

$$D(L) = \{\vec{x} \in E_n; \forall \ell \in L, 2(\vec{\ell}, \vec{x}) \le N(\vec{\ell})\}; \tag{1.1b}$$

$$D(L) = \{\vec{x} \in E_n; \vec{x} \text{ is a shortest vector in the } \vec{x} + L \text{ coset of } R^n\}. \tag{1.1c}$$

If $\{\vec{b}_j\}$, $1 \le j \le n$, is a basis of L, any other basis is of the form $\sum_j m_{ij}\vec{b}_j$, $m \in GL_n(Z)$. So $|\det \vec{b}_j|$ is an invariant of L that we denote simply $\mathrm{vol}(L)$. By definition, two lattices are isomorphic if they can be transformed into each other by an orthogonal transformation. So a class of isomorphism depends only on the Gram matrices $q_{ij} = (\vec{b}_i, \vec{b}_j)$; each one is a symmetric positive $n \times n$ matrix that defines a positive quadratic form $\sum_{ij} q_{ij}x_i x_j$. Since the sum of two positive quadratic forms is a positive quadratic form, the set of positive quadratic forms form a convex cone that we denote by $\mathcal{C}_+(\mathcal{Q}_n)$ in the $n(n+1)/2$-dimensional vector space of n variable quadratic forms. To the change of basis of a lattice $\vec{b}_i \mapsto \sum_j m_{ij}\vec{b}_j$ by the matrix $m \in GL_n(Z)$ corresponds the transformation $q \mapsto mqm^\top$ for the corresponding quadratic forms. So there is a natural bijective map between the set of isomorphic classes of lattices and the orbit space $\mathcal{C}_+(\mathcal{Q}_n) \mid GL_n(Z)$. The generic lattices (represented by an open dense set in the $n(n+1)/2$ dimensional manifold $\mathcal{C}_+(\mathcal{Q}_n)$) and their Voronoï cells are studied in Refs. [24, 25] where they are called *primitive*. These cells are combinatorially equivalent for $n = 2$, 3. Voronoï established that they form three combinatorially distinct classes for $n = 4$. In Ref. [25] he also gave an expression for the bounds $N_d(n)$. The aim of this paper is to give a different and more explicit expression.

---

[2] E.S. Fedorov wrote this book between the ages of 16 and 26, while serving in the army, or studying medicine, chemistry, and physics. Then he became a mineralogist and six years later his book was accepted for publication in a crystallography series. No translation in a Western language is known. There exists a detailed analysis of it in [23].

[3] They are often called Wigner and Seitz cells for crystals; that of the dual of the crystal lattice is called Brillouin zone. Indeed these scientists introduced their use in physics at the beginning of the 1930s.

2   The Expression of the Bounds $N_d(n)$ Obtained by Voronoï

That expression is given in Eq. (2.10), at the end of the section. Before, we recall the fundamental concepts and objects introduced by Voronoï. Instead of giving a summary of his papers, we shall introduce a more modern (and faster) presentation of the basic facts on lattice and their Voronoï cells. This can be found for instance in Refs. [4, 7, 8]. Here we shall follow [16], a monograph in preparation by M. Senechal.

The symmetry point group $P \subset O_n$ of the lattice is the symmetry group of $D(L)$; hence the origin o is its symmetry center. Definition (1.1b) shows that $D(L)$ is convex since it is the intersection of half spaces bounded by hyperplanes. Let $\mathcal{E}_n$ be the Euclidean space built from E_n. Its points are $x = o + \vec{x}$, the translate of o by $\vec{x} \in E_n$. Conversely, any pair of points x, $y \in \mathcal{E}_n$ defines a vector $y - x \equiv \vec{xy} \in E_n$. The set of translates of the Voronoï cell by all the vectors $\vec{\ell} \in L$, i.e., $\{D(L) + \vec{\ell} = D_{o+\vec{\ell}}(L); \ell \in L\}$, form a face to face paving of the space $\mathcal{E}$. We notice that $D(L)$ is a fundamental domain of L, so $\mathrm{vol}(D(L)) = \mathrm{vol}(L)$.

Following [16], we say that the cells that have contact with $D_o(L)$ form its *corona* and the centers of these cells define the *corona vectors*. Their set is

$$C = \{\vec{c} \in L; D_o(L) \cap D_c(L) \neq \emptyset\} = L \cap \partial D_o(2L). \qquad (2.1)$$

It follows from Eq. (1.1c) that the corona vectors are shortest in their $L/2L$ cosets, i.e.,

$$\vec{c} \in C: \forall \vec{\ell} \in L, \quad N(\vec{c} + 2\vec{\ell}) - N(\vec{c}) \geq 0 \iff (\vec{c}, \vec{\ell}) + N(\vec{\ell}) \geq 0. \qquad (2.2)$$

The converse also holds. Note that $\vec{c} \in C \implies -\vec{c} \in C$. Let us replace 2 by $m > 2$ in (2.2):

$$\vec{c} \in C, \; \forall \vec{\ell} \neq 0, \; \vec{\ell} \in L,$$
$$m^{-1}\left(N(\vec{c} + m\vec{\ell}) - N(\vec{c})\right) = 2\left((\vec{c}, \vec{\ell}) + N(\vec{\ell})\right) + (m - 2)N(\vec{\ell}) > 0. \qquad (2.3)$$

This proves [16] that *a corona vector is* **the** *shortest vector in its coset* L/mL when $m > 2$. So we obtain

$$|C| \leq 3^n - 1. \qquad (2.4)$$

This result was first obtained by Minkowski [19] for the more general cases of lattice packings of any convex domain (instead of the Voronoï cell). Equation (2.4) implies that the number of supporting hyperplanes of the faces of $D_0(L)$ is finite. Since $\mathrm{vol}(D_0(L))$ is finite, they cannot all be parallel to a direction, so $D(L)$ *is a polytope*. To simplify notation when there is no ambiguity, we will use D for $D(L)$ from now on. Since $\forall \vec{\ell} \in L$, $\frac{1}{2}\ell = o + \frac{1}{2}\vec{\ell}$

is a symmetry center of L, then $\frac{1}{2}c$ is a symmetry center of the face $\mathcal{D}_o \cap \mathcal{D}_c$ (that face is convex and that symmetry exchanges the points o and c, so it transforms this face into itself). From now on we will call the $(n-1)$-dimensional faces of $D(L)$, *facets*. The set of *facet vectors* $F \subset C$ is

$$F = \{\vec{f} \in C; \dim(D_o \cap D_f) = n - 1\}. \tag{2.5}$$

One easily proves the

Proposition 2.1. $\vec{f} \in L$ *is a facet vector if and only if* $\pm \vec{f}$ *are strictly shorter than the other vectors of their* $L/2L$ *coset.*

That proposition was first proven in Ref. [25, §55, pp. 67–69]. As a corollary, combined with (2.4), we obtain

$$2n \leq |F| \leq 2(2^n - 1) \leq |C| \leq 3^n - 1. \tag{2.6}$$

The second inequality is in Ref. [25, p. 70]; it was also proven before in Ref. [18].

At least n facets meet at each vertex of D. Each vertex v belongs to n_v Voronoï cells D_{o_α}, $1 \leq \alpha \leq n_v > n$, and the o_α are on a sphere of center v. Voronoï defined ([25, §60, p. 74]) and used the polytope that is the convex hull of the points o_α; we denote it Δ_v because it was thoroughly studied by Delone and his school. In dimension n a sphere is determined by $n+1$ points in general position, i.e., they are the vertices of a simplex. So for the generic lattices (forming an open dense set in $\mathcal{C}_+(\mathcal{Q}_n)$ see Ref. [24]) $n_v = n + 1$ for each vertex. Each Voronoï cell containing a vertex v meets at v the n others, along n facets. Voronoï called such cells *primitive*; we call here the corresponding lattices *primitive*. It is easy to prove [24, §17, pp. 228–229].

Proposition 2.2. $D(L)$ *primitive* $\Longleftrightarrow$ *each of its d-faces is the intersection of* $n + 1 - d$ *cells, for* $d = 0, 1, 2, \ldots, n$.

For $d = n - 1$, we have as a corollary

$$L \text{ is a primitive lattice} \iff \forall v, \Delta_v \text{ is a simplex.} \tag{2.7}$$

From Propositions 2.1 and 2.2 we obtain:

$$\text{primitive lattice} \implies |F| = |C| = 2(2^n - 1). \tag{2.8}$$

The converse of (2.8) is not true for $n \geq 4$. All four-dimensional lattices that are tensor product of two primitive two-dimensional lattices have the same combinatorial type of Voronoï cells with $|F| = |C| = 30$ but those cells, which are studied thoroughly in Ref. [16], are not primitive: among their 102 vertices, 12 of them are the intersection of 5 facets.[4]

---

[4]In dimension 5 [10] has found 225 combinatorial types of nonprimitive Voronoï cells with the greatest possible number of facets, i.e., 62.

Let $\vec{f_i}(v)$, $1 \le i \le n$ be the facet vectors of the n facets meeting at the vertex v of the Voronoï cell D of a primitive n-dimensional lattice L; these vectors form a basis of the space. The lattice generated by these n linearly independent $\vec{f_i}$ may be only a sublattice of L of index w_v; so

$$w_v \text{vol}(L) = \det(\vec{f_i}) \mid = n! \text{vol}(\Delta_v), \tag{2.9}$$

where the last equality is obtained from the known formula giving the volume of a simplex. Let us study the sets $V(D)$ of vertices of the Voronoï cell D. If $v \in V(D)$, the n points $v_i = v - \vec{f_i}$ are also vertices of D and one verifies that the set $V(D)$ is the disjoint union of subsets, each one containing exactly $n + 1$ vertices of D that are obtained from each other by translations of L ([25, p. 71]). Since the set of all Delone cells of L pave the Euclidean space $\mathcal{E}_n$, by choosing a representative v_ρ for each of these subsets, $\cup_\rho \Delta(v_\rho)$ is a fundamental domain of L; so

$$\sum_{v \in V(D)} \text{vol}(\Delta_v) = (n+1)\text{vol}(L). \tag{2.9'}$$

From this equation and the sum of (2.9) over the vertices, we obtain

$$|V(D)| \le \sum_{v \in V(D)} w_v = (n+1)!; \tag{2.9''}$$

the equality holds if and only if $w_v = 1$ at each vertex. As we have seen, that means that at *at each vertex* v of a primitive Voronoï cell, the corresponding n facet vectors $\vec{f_i}$ generate the full lattice; we call these lattices and their cells, *principal primitive*.

The proof of (2.9'') is only a particular case of the proof Voronoï gave in Ref. [25, §63–66, pp. 78–83] of the following theorem:

Theorem (Voronoï). *The number of d-faces of a principal primitive n-dimensional Voronoï cell is*

$$N_d(n) = (n+1-d)\sum_{\ell=0}^{n-d}(-1)^{n-d-\ell}\binom{n-d}{\ell}(1+\ell)^n,$$
$$0 \le d < nx. \tag{2.10}$$

Voronoï proof uses a discretization of the space and finite differences[5]. In Ref. [25, §101 p. 136] Voronoï proved that the $N_d(n)$ are upper bounds for the number of d-faces of **any** n-dimension Voronoï cell. That was a remarkable achievement.

---

[5] At the top of p. 82, there is a misprint in the equation that defines the finite difference operator (replace the last symbol μ by k). At the bottom of the page the Voronoï equation (2.10) (the first of §66) is equivalent to (2.10) above.

Finally Voronoï gave for all dimensions an open set in $\mathcal{C}_+(\mathcal{Q}_n)$ of primitive principal lattice, and he called *type* I the combinatorial type of their Voronoï. cells.[6] These lattices are described by the quadratic forms

$$Q_{\lambda_{ij}}(x_i) = \sum_i \lambda_{ii} x_i^2 + \sum_{i,j,i<j} \lambda_{ij}(x_i - x_j)^2 \equiv \sum_{ij} q(\lambda)_{ij} x_i x_j; \qquad \lambda_{ij} > 0. \quad (2.11)$$

Voronoï showed that each one of these quadratic forms has a formal symmetry $\mathcal{S}_{n+1}$, the permutation group of $n+1$ objects; it was a generalization to dimension n of the Selling's study [22] of the two and three variable quadratic forms. In the particular case where all the λ's are equal, $\mathcal{S}_{n+1}$ becomes a geometric symmetry.[7] In that case, Voronoï computed the coordinates of the $(n+1)!$ vertices (see (3.15)) and showed that they form a principal orbit of $\mathcal{S}_{n+1}$. At that time it was not known that, up to a dilation of scale, these lattices were A_n^w, the weight lattices of the simple Lie algebra A_n. Moreover, the Voronoï cells $\mathcal{D}(A_n^w)$ are the *Cayley graphs* of the permutation groups $\mathcal{S}_{n+1}$ (see Ref. [6, pp. 65–66]). That is one more incentive to study them in detail in the next section.

3   Detailed Description of the Voronoï Cells of the A_n^w Lattices

A representative quadratic form q of A_n^w is obtained from (2.11) with $\lambda_{ij} = (n+1)^{-1}$; so

$$q = I - \frac{1}{(n+1)} J, \text{ with } I_{ij} = \delta_{ij}, \ J_{ij} = 1. \quad (3.1)$$

From $J^2 = nJ$ one obtains easily q^{-1}, the quadratic form of the dual lattice $A_n^r = (A_n^w)^*$, the root lattice of the simple Lie algebra A_n:

$$q^{-1} = I + J. \quad (3.2)$$

To write explicitly the corresponding bases of this pair of dual lattices we introduce the following notation. Let $\mathcal{N}_n$ and $\mathcal{N}_n^+$ be the sets of integers m satisfying respectively $0 \le m \le n$ and $1 \le m \le n$. To exploit easily the

---

[6] As we said at the end of the introduction, Voronoï [25] showed that there are three types of primitive lattices (all principal) in dimension 4 and, in a side remark (p. 84), he stated that he found some nonprincipal primitive lattices in dimension 5. In that dimension, with the correction by Engel [10] of the Baranovskii and Ryshkov result [2], it is now known that there are 222 types of primitive lattices, including 21 nonprincipal ones. Voronoï did not introduce a word for the concept of principal primitive lattices.

[7] As shown between (3.3) and (3.5); see also (3.17). In fact these lattices are the primitive ones with the largest symmetry; see Ref. [16].

action of the symmetry group[8] $A_n \sim S_{n+1}$, it is usual to consider the $n+1$ dimensional space E_{n+1} with the orthonormal basis and the vector $\vec{e}$:

$$\alpha, \beta \in \mathcal{N}_n, \quad (\vec{e}_\alpha, \vec{e}_\beta) = \delta_{\alpha\beta},$$

$$\vec{e} = \frac{1}{(n+1)} \sum_\alpha \vec{e}_\alpha, \quad \text{so } (\vec{e}, \vec{e}) = (\vec{e}, \vec{e}_\alpha) = \frac{1}{(n+1)}. \quad (3.3)$$

The group S_{n+1} is the group of permutations of the $n+1$ vectors $\vec{e}_\alpha$; it leaves $\vec{e}$ fixed. We denote by $\mathcal{H}_e$ the hyperplane (= vector subspace) orthogonal to $\vec{e}$. It contains the vectors

$$(\vec{e})^\perp \equiv \mathcal{H}_e \ni \vec{u}_\alpha; \quad \vec{u}_\alpha = \vec{e}_\alpha - \vec{e}, \quad (\vec{u}_\alpha, \vec{u}_\beta) = \delta_{\alpha\beta} - \frac{1}{(n+1)},$$

$$\sum_\alpha \vec{u}_\alpha = 0. \quad (3.4)$$

The vectors of $\mathcal{H}_e$ are those of E_{n+1} with a vanishing sum of coordinates. S_{n+1} acts linearly (and irreducibly) on $\mathcal{H}_e$ and is the group of permutations of the $n+1$ vectors $\vec{u}_\alpha$. Those vectors generate the weight lattice A_n^w. From the last equality of (3.4) one can take as a basis of A_n^w the n vectors

$$i \in \mathcal{N}_n^+, \quad \{\vec{u}_i = \vec{e}_i - \vec{e}\} \text{ basis of } A_n^w; \quad (\vec{u}_i, \vec{u}_j) = \delta_{ij} - \frac{1}{(n+1)}. \quad (3.5)$$

The last equality shows that its Gram matrix is q of Eq. (3.1).

We denote by $\mathcal{A}$ a nonempty proper subset of $\mathcal{N}_n$ and its complement by $\bar{\mathcal{A}}$, in order to define the set W of lattice vectors of A_n^w:

$$\emptyset \neq \mathcal{A} \subset \mathcal{N}_n,$$

$$W = \left\{ \vec{w}_\mathcal{A} = \sum_{\alpha \in \mathcal{A}} \vec{u}_\alpha \right\}, \quad \vec{w}_\mathcal{A} + \vec{w}_{\bar{\mathcal{A}}} = 0; \quad |W| = 2(2^n - 1). \quad (3.6)$$

From the defining quadratic form (3.1) Voronoï proved ([25, §102, pp. 137–139]) that for each $L/2L$ coset of A_n^w, the shortest vectors form a pair: $\vec{w}_\mathcal{A}$, $\vec{w}_{\bar{\mathcal{A}}} = -\vec{w}_\mathcal{A}$; by Proposition 2.1 we see that W is the set of the facet vectors of A_n^w:

$$W = F(A_n^w). \quad (3.7)$$

Another proof is given in Ref. [16].

We now use well-known properties of the A_n Lie algebra that were not all discovered at Voronoï's time. The roots of the Lie algebra A_n form the set R of $n(n+1)$ vectors:

$$R = \{\vec{r}_{\alpha\beta} = \vec{e}_\alpha - \vec{e}_\beta = \vec{u}_\alpha - \vec{u}_\beta\}, \quad N(\vec{r}_{\alpha\beta}) = 2. \quad (3.8)$$

---

[8]We use the same notation A_n for the simple Lie algebra and its Weyl group, but to avoid any ambiguity we shall always note whether if A_n denotes the group or the Lie algebra.

These vectors generate the the root lattice A_n^r. One can choose for its basis

$$\{\vec{r}_{i0}\} \text{ basis of } A_n^r; \quad \text{its vectors satisfy: } (\vec{r}_{i0}, \vec{u}_j) = \delta_{ij}. \quad (3.9)$$

This proves the duality $A_n^w = (A_n^r)^*$. The weights of the Lie algebra A_n are the nonzero vectors whose scalar products with every root are ± 1 or 0; they form the set $W = F(A_n^w)$ of facet vectors of the weight lattice.

For the study of Lie algebras the traditional basis of the root lattice A_n^r is

$$\{\vec{r}_i = \vec{r}_{i-1,i} = \vec{e}_{i-1} - \vec{e}_i = \vec{u}_{i-1} - \vec{u}_i\} \text{ basis of } A_n^r. \quad (3.10)$$

Notice that

$$(\vec{r}_i, \vec{r}_j) = \begin{cases} 2 & \text{when } i = j, \\ -1 & \text{when } |i - j| = 1, \\ 0 & \text{when } |i - j| > 1. \end{cases} \quad (3.11)$$

The corresponding quadratic form is called the Cartan matrix of A_n^r. The dual basis of (3.10) has to be made of weights:

$$\left\{ \vec{w}_k = \sum_{\alpha=0}^{k-1} \vec{u}_\alpha \right\} \text{ is the basis of } A_n^w \text{ satisfying } (\vec{r}_i, \vec{w}_k) = \delta_{ik}. \quad (3.12)$$

Notice also that

$$k \le \ell, \quad (\vec{w}_k, \vec{w}_\ell) = \frac{k(n + 1 - \ell)}{n + 1} > 0. \quad (3.13)$$

The inequality is a necessary condition for the n facet vectors $\vec{w}_k$ to be those of the facets meeting at a vertex v. Here this condition is sufficient since there is a unique vector v satisfying the system of linear equations:

$$(\vec{w}_k, \vec{v}) = \frac{1}{2} N(\vec{w}_k). \quad (3.14)$$

The solution

$$\vec{v} = \frac{1}{(n + 1)} \left(\frac{n}{2} \vec{e} - \sum_{\alpha=0}^{n} \alpha \vec{e}_\alpha \right); \quad N(\vec{v}) = \frac{n(n + 2)}{12(n + 1)}. \quad (3.15)$$

was obtained by Voronoï [25, §103, pp. 140–143)], who remarked that all coordinates of v are distinct, so the vertices of $\mathcal{D}(A_n^w)$ form at least a $\mathcal{S}_{n+1}$ orbit of $(n+1)!$ vertices. Since this value attains the bound (2.10), there are no other vertices and (2.9'') proves that A_n^w is a principal primitive lattice. No other symmetry arguments were used by Voronoï; it is time to produce more.

In an orthogonal vector space, we denote by $\sigma(\vec{r})$ the reflection through the hyperplane $\mathcal{H}_r$ orthogonal to $\vec{r}$. Its action on the vectors is given by

$$\sigma(\vec{r}).\vec{x} = \vec{x} - 2(\vec{x}, \vec{r})N(\vec{r})^{-1}\vec{r}. \quad (3.16)$$

The Weyl group A_n is the group generated by the reflections $\sigma(\vec{r}_{\alpha\beta})$ with $\vec{r}_{\alpha\beta} \in R$. Since the roots $\vec{r}_{\alpha\beta}$ generate A_n^r, the Weyl group is a symmetry group of this lattice. That group leaves the vector $\vec{e}$ of (3.3) fixed and acts irreducibly on the n-dimensional subspace $\mathcal{H}_e$. Applying the reflection $\sigma(\vec{r}_{\alpha\beta})$ to the basis vectors $\vec{e}_\gamma$ of E_{n+1}, one verifies that the $\vec{e}_\gamma$'s, $\alpha \neq \gamma \neq \beta$, stay fixed while the vectors $\vec{e}_\alpha$ and $\vec{e}_\beta$ are exchanged. This verifies the isomorphism of the Weyl group $A_n \sim S_{n+1}$, the group of permutations of the $n+1$ coordinates of the vectors $\in E_{n+1}$ in the basis (3.3). With the use of the classical notation of cycles for the permutations, this isomorphism corresponds to

$$\sigma(\vec{r}_{\alpha\beta}) \sim (\alpha\beta). \tag{3.17}$$

It is well known that the permutation group S_{n+1} is generated by the n involutive permutations $(i-1, i)$. They are represented by the n reflections $\sigma(\vec{r}_i)$ and from (3.11) and (3.16) one obtains the relations between these generators for the group A_n:

$$1 \leq i \leq n; \quad \sigma(\vec{r}_i)^2 = 1;$$
$$|i - j| = 1, \quad \left(\sigma(\vec{r}_i)\sigma(\vec{r}_j)\right)^3 = 1; \tag{3.18}$$
$$|i - j| > 1, \quad \left(\sigma(\vec{r}_i)\sigma(\vec{r}_j)\right)^2 = 1.$$

Note that the first equality implies that the last one is equivalent to

$$|i - j| > 1; \quad \sigma(\vec{r}_i)\sigma(\vec{r}_j) = \sigma(\vec{r}_j)\sigma(\vec{r}_i). \tag{30'}$$

The Coxeter diagram for the group A_n has n vertices labeled by i, $1 \leq i \leq n$, each representing the reflection $\sigma(\vec{r}_i)$, and the $n - 1$ edges joining the pair of vertices i, $i + 1$.

Coxeter diagram of A_{10} :

A pair of vertices not joined by an edge represents commuting reflections. As we have seen, this orthogonal n-dimensional representation of A_n is also the group of permutations of the vectors $\vec{u}_\alpha$, generators of A_n^w. So the Weyl group is a symmetry group of the lattices A_n^w, A_n^r. It is not their full symmetry group because, for $n > 1$, it does not contain $-I_n$, the symmetry through the origin on the n-dimensional vector space $\mathcal{H}_e$. It will be more convenient here to use only the symmetry A_n.

We have proven after (3.15) that the group $A_n \sim S_{n+1}$ acts transitively on the set V of vertices. We leave as an exercise the computation of the action of some of its elements on the vertex v (given in Eq. (3.15)):

$$\sigma(\vec{r}_{\alpha\beta}).\vec{v} \equiv (\alpha\beta).\vec{v} = \vec{v} + \frac{1}{n+1}(\alpha - \beta)\vec{r}_{\alpha\beta}. \tag{3.19}$$

$$\left(\sigma(\vec{r}_1)\sigma(\vec{r}_2)\sigma(\vec{r}_3)\ldots\sigma(\vec{r}_n)\right)^k.\vec{v} \equiv (1, 2, \ldots, n, n+1)^k.\vec{v} = \vec{v} - \vec{w}_k. \tag{3.20}$$

For $1 \le k \le n$ those are the n other vertices of $D_o(A_n^w)$ obtained from $\vec{v}$ by translations of the lattice; it could have been obtained geometrically that these n translations are the opposite of the n facet vectors $\{\vec{w}_k\}$ (3.12) meeting at v. Any edge from the vertex v is the intersection of $n-1$ facets meeting at v, so it is orthogonal to $n-1$ of the facet vectors $\vec{w}_k$. This shows that edges meeting at v are parallel to the roots of the basis $\{\vec{r}_i\}$ defined in (3.10) which is the dual basis of $\{\vec{w}_k\}$. From (3.10) and (3.15) we obtain $(\vec{v}, \vec{r}_i) = 1/(n+1)$. With the symmetry $A_n \sim S_{n+1}$, this shows that any vertex is equidistant from the n walls of the Weyl chamber that contains it. As a particular case of (3.19) we obtain

$$\sigma(\vec{r}_i).\vec{v} \equiv (i, i+1).\vec{v} = \vec{v} - \frac{1}{(n+1)}\vec{r}_i. \qquad (3.21)$$

This shows that the neighboring vertices of v are obtained by reflection through the Weyl chamber walls. Given another vertex v', there exists a unique $g \in A_n \sim S_{n+1}$ transforming v into $v' = g.v$; it transforms roots into roots and $\sigma(g.\vec{r}_i) = g\sigma(\vec{r}_i g^{-1})$. Hence—

Proposition 3.1. *All edges of the Voronoï cell of A_n^w are parallel to the roots ($\in R$ defined in Eq. (3.8)) and have the same length $\sqrt{2}/(n+1)$.*

In Ref. [6, at the end of §6.2 pp. 65–66], it is explained that this cell is combinatorially equivalent to the Cayley diagram of S_{n+1}. The proof is easily obtained from the isomorphism based on (3.17) and from (3.18). The vertex v of (3.15) can be chosen to represent $1 \in S_{n+1}$ so the n vertices of (3.21) represent the n group generators $\sigma(\vec{r}_i)$. Repeating the action of the group generators on the newly obtained vertices one defines the correspondence between the $(n+1)!$ elements of S_{n+1} and the $(n+1)!$ vertices of $\mathcal{D}(A_n^w)$.

Proposition 3.2. *The Voronoï cell $\mathcal{D}(A_n^w)$ is a zonotope.*

We must recall first the definition of the *vector sum*[9] $\dotplus$ of two convex polytopes P', P'' in the Euclidean space $\mathcal{E}_n$. We follow here [13] the beginning of Chapter 15 (written by G.C. Shephard). The following two equivalent definitions are given:

$$P' \dotplus P'' = \{x' + x'' \mid x' \in P', x'' \in P''\} \qquad (3.22)$$

with $\{v'_i\}$ and $\{v''_j\}$ the sets $V(P')$, $V(P'')$ of vertices of P', P''. The second definition is

$$P' \dotplus P'' = \text{ convex hull } (\{v'_i + v''_j\}),$$
$$1 \le i \le |V(P')|, \ 1 \le j \le |V(P'')|. \qquad (3.23)$$

---

[9]It is also called the Minkowski sum.

A change of origin translates the vector sum; so it should be considered as a binary operator on the classes of translated polytopes. It is straightforward to prove that this sum is commutative and associative. A trivial case of such a sum is the parallelepiped built on a basis of vectors (identified with one-dimensional polytopes = line segment). In the general case of eventually nonlinearly independent vectors we have—

Definition. A zonotope is the vector sum of a finite number of segments.

Proof of Prop. 3.2. For $\mathcal{D}(A_n^w)$, these segments are defined by the roots. Following the literature on semisimple Lie algebras, we choose as the set R_+ of *positive roots* the $n(n+1)/2$ roots $r_{\alpha\beta}$ satisfying $\alpha < \beta$. To conclude the proof one verifies that by adding to v, for each subset of R_+, the sum of its elements, one obtains the set of vertices

$$V(A_n^w) = \left\{ v + \frac{1}{n+1} \sum_{0 \leq \alpha < \beta \leq n} \eta_{\alpha\beta} \vec{r}_{\alpha\beta} \mid \eta_{\alpha\beta} = 0 \text{ or } 1 \right\}. \qquad (3.24)$$

Notice that most vertices can be expressed by different sums.

Let us recall well known properties of zonotopes. A zonotope has a symmetry center; for the Voronoï cell A_n^w it is

$$v + \frac{1}{n+1} \vec{\rho}, \text{ with } \vec{\rho} = \frac{1}{2} \sum_{0 \leq \alpha < \beta \leq n} \vec{r}_{\alpha\beta}; \qquad (3.25)$$

(this half sum of all positive roots is often used in the study of semisimple Lie algebras). The faces of zonotopes are zonotopes; so they have a symmetry center. Notice that it is generally not true for the d-faces, $1 < d < n-1$, of Voronoï cells. That is the case for the two other types of primitive Voronoï cells in dimension 4; they have pentagons among their 2-cells, e.g., [9].

It is well known that the number of edges of a primitive principal Voronoï cell is $N_1(n) = (n+1)!(n/2)$ (indeed n edges meet at any vertex and each edge has two vertices). Let us apply to this elementary case $d = 1$ the method we shall develop for computing the $N_d(n)$'s. Since the stabilizer of any vertex v is trivial, the stabilizer of an edge is the A_n subgroup $A_1 \sim S_2$, which permutes the two vertices of an edge. Note that this group leaves fixed the middle of the edge, so it is a stabilizer of the Weyl group acting on E_n. There are $|A_n|/|A_1| = (n+1)!/2!$ edges in each A_n orbit. From the fact that the stabilizer of a vertex is trivial, no element of the group A_n can be a permutation of the edges meeting at the vertex. Hence the n edges meeting at v (or at any vertex) belong to the n different orbits of edges. Moreover, each orbit can be associated with a $\vec{r}_i$, that is the position of A_1 at one of the n vertex of the Coxeter diagram of the reflection group A_n. In (2.9″) and now, we have determined two polynomials defined in the abstract:

$$p_0(n) = 1, \quad p_1(n) = \frac{n}{2}. \quad \square \qquad (3.26)$$

We now study the general case of a d-face Φ_d^α of the Voronoï cell $\mathcal{D}$ of A_n^w. It is defined by its set of vertices $V(\Phi_d^\alpha) \subset V$. This d-face defines its supporting d-plane $\mathcal{H}(\Phi_d^\alpha)$ (the affine span of $V(\Phi_d^\alpha)$) and $V(\Phi_d^\alpha) = V(\mathcal{D}) \cap \mathcal{H}(\Phi_d^\alpha)$. So the edges of Φ_d^α are parallel to the roots parallel to the d-plane $\mathcal{H}(\Phi_d^\alpha)$. Since this d-plane does not contain the origin (center of $\mathcal{D}$), its normalizer in the group A_n (i.e., the subgroup of A_n made of all elements that transform $\mathcal{H}(\Phi_d^\alpha)$ into itself) is a reflection group generated by the reflections $\sigma(\vec{r}_{\alpha\beta})$ whose roots $\vec{r}_{\alpha\beta}$ are parallel to the d-plane. From Proposition 3.1 and (3.19), this group is also the normalizer in A_n of the d-face Φ_d^α and it acts transitively on its vertices. We will denote this reflection group (or anyone of its conjugacy class) simply by $G_d^{(\alpha)}$. All the d-dimensional reflection subgroups of A_n are known. Their Coxeter graph is obtained from that of A_n by removing $n - d$ vertices and the edges issued from them; in general it is not connected but it is the disjoint union of diagrams A_{m_i} with $d = \sum_i m_i$. From (30'), these distinct subgroups A_{m_i} commute between each other. Hence the group $G_d^{(\alpha)}$ is a direct product:

$$G_d^{(\alpha)} = \times_i A_{m_i}, \quad \sum_i m_i = d. \tag{3.27}$$

It is convenient to write the direct product of s isomorphic group A_m as A_m^s. So the general preceding equation can be written

$$G_d^{(\alpha)} = \times_i A_{m_i}^{s_i}, \quad \sum_i m_i s_i = d; \tag{3.28}$$

when $s_i = 1$, we write simply A_{m_i}. Conversely, given a d-dimensional reflection group of type (3.27) (i.e., direct product of irreducible reflection groups A_{m_i}), it defines a zonotope with all edges equal that, for n large enough, is a d-face of a Voronoï cell $\mathcal{D}(A_n^w)$. Each factor A_{m_i} defines the Voronoï cell $D(A_{m_i}^w)$. The direct product of factors defines a particular case of the vector sum of the cells; indeed these polytopes are in linearly independent (in our case, orthogonal direct sum of) vector subspaces[10]. We will use a different notation, $P' * P''$, for this particular case of the polytope vector sum defined in Eqs. (3.22), (3.23) because it has richer properties. For example, with the notation $N_d(P_n)$ for the number of d-faces of the n-dimensional convex polytope P_n and the conventions $N_d(P_n) = 1$ for $d = n$ and 0 for $d > n$, we have

$$N_{d'+d''}(P_{n'}' * P_{n''}'') = N_{d'}(P_{n'}')N_{d''}(P_{n''}'') \iff \tag{3.29}$$

$$\iff N_d(P_{n'}' * P_{n''}'' = \sum_{d=d'+d''} N_{d'}(P_{n'}')N_{d''}(P_{n''}''). \tag{3.30}$$

---

[10]This particular vector sum of polytopes is studied in Ref. [5, pp. 123–124]; it was called "rectangular product" by Pólya.

In this paper we replace the symbol P_n by its symmetry group: for instance, A_1 is a line segment, $N_0(A_1) = 2$; then A_1^s represents the s-dimensional hypercube and we obtain immediately the well-known formula

$$0 \leq d \leq s, \quad N_d(A_1^s) = \binom{s}{d} 2^{s-d}. \tag{3.31}$$

Similarly we obtain immediately the number of vertices of the d-faces labeled by the general reflection group (3.28); it is exactly its number of elements:

$$N_0(\times_i A_{m_i}^{s_i}) = | \times_i A_{m_i}^{s_i}| = \prod_i ((m_i + 1)!)^{s_i}. \tag{3.32}$$

It is important to emphasize that, as A_n was not the full symmetry point group of the lattices A_n and of their Voronoï cell (see end of paragraph containing (30')), $\times_i A_{m_i}^{s_i}$ is not the full symmetry group of the d-face type it labels. For instance, the symmetry of the hypercube is the wreath product[11] $A_1 \uparrow s = O_s(Z) = B_s$; it is again a group generated by reflection, hence the notation B_s of the Weyl group of the simple Lie algebra $B_s \sim O_{2s+1}$. Note that for $d > 1$, $s > 1$, $A_d \uparrow s$ is not a group generated by reflection; note also that it is not a subgroup of A_{sd} but it is a subgroup of $GL_{sd}(Z)$.

We shall denote by $Q_d(\times_i A_{m_i}^{s_i}, n)$, $\sum_i s_i m_i = d$, the number of such d-faces of $\mathcal{D}(A_n^w)$ per orbit of the group A_n:

$$Q_d(\times_i A_{m_i}^{s_i}, n) = \frac{(n+1)!}{\prod_i ((m_i+1)!)^{s_i}}. \tag{3.33}$$

Then the number of d-faces of $\mathcal{D}(A_n^w)$ for a given symmetry group of (3.28) is

$$N_d(\times_i A_{m_i}^{s_i}, n) = Q_d(\times_i A_{m_i}^{s_i}, n) K_d(\times_i A_{m_i}^{s_i}, n), \tag{3.34}$$

where $K_d(\times_i A_{m_i}^{s_i}, n)$ is the number of d-face of symmetry $\times_i A_{m_i}^{s_i}$ at the vertex v (or at any vertex); that is also the number of different Coxeter subdiagrams (of the diagram of A_n) corresponding to the group isomorphism class $\times_i A_{m_i}^{s_i}$. For instance there are $n + 1 - d$ possible positions of the Coxeter diagram of A_d in that of A_n. Hence the number of $\mathcal{D}(A_n^w)$ d-faces that are $D(A_d^w)$ zonotopes is

$$N_d(A_d, n) = (n+1)! \frac{n+1-d}{(d+1)!}; \tag{3.35}$$

for example,

$$N_2(A_2, n) = (n+1)! \frac{n-1}{6},$$

$$N_1(A_1, n) = (n+1)! \frac{n}{2}.$$

---

[11]The group $G \uparrow s = G^s \rtimes S_s$ is generated by the direct product G^s and the group of permutations of the s factors of this direct product. So $|G \uparrow s| = |G|^s \, s!$

The last expression is the maximum number of edges of an n-dimensional Voronoï cell; we have already computed it before (3.26). The 2-faces of the Voronoï cell $\mathcal{D}(A_n^w)$ are either regular hexagons, with symmetry A_2, or squares, with symmetry A_1^2; so $N_2(A_2, n)$ is the number of hexagonal 2-faces.

By recursion with all m_i different we find

$$\neq m_i, \ 1 \leq i \leq k, \ d = \sum_i m_i,$$
$$K_d(\times_i A_{m_i}) = \prod_{\ell=0}^{k-1}(n+1-d-\ell) = k!\binom{n+1-d}{k}. \quad (3.36)$$

The vanishing of this expression when $d + k \leq n$ corresponds to the necessity, for obtaining the diagram of a direct product of k factors as a subdiagram of A_n, that one has to remove at least $k-1$ vertices in order to separate the subdiagrams of the A_{m_i}'s. When, in the direct product A_m^s, the factors are identical, one cannot distinguish between their diagrams in that of A_n, so one must divide the expression of K by the factorial of the number of factors:

$$d = ms, \quad K_d(A_m^s) = (s!)^{-1}\prod_{\ell=0}^{s-1}(n+1-d-\ell) = \binom{n+1-d}{s}. \quad (3.37)$$

Using this expression for the particular case of the d-dimensional hypercube, $m = 1$, $d = s$, and (3.33)–(3.34), we obtain for the number of d-hypercubes in $\mathcal{D}(A_n^w)$

$$N_d(A_1^d) = (n+1)! \, 2^{-d}\binom{n+1-d}{d}; \quad (3.38)$$

for example:

$$N_2(A_1^2, n) = (n+1)! \, (n-1)(n-2)/8.$$

Adding this expression with the middle expression of (3.35), we obtain

$$N_2(n) = (n+1)! \, p_2(n) \text{ with } p_2(n) = (n-1)(3n-2)/24. \quad (3.39)$$

The general expression of K_d for a general reflection group of (3.28) is

$$d = \sum_i m_i s_i, \quad k = \sum_i s_i,$$
$$K_d(\times_i A_{m_i}^{s_i}, n) = \frac{\prod_{\ell=0}^{k-1}(n+1-d-\ell)}{\prod_i s_i!} = \frac{k!}{\prod_i s_i!}\binom{n+1-d}{k}. \quad (3.40)$$

With (3.33)–(3.34) we finally obtain for $N_d(\times_i A_{m_i}^{s_i}, n)$ with $d = \sum_i m_i s_i$, $k = \sum_i s_i$:

$$N_d(\times_i A_{m_i}^{s_i}, n) = (n+1)! \frac{\prod_{\ell=0}^{k-1}(n+1-d-\ell)}{\prod_i s_i!(m_i+1)!)^{s_i}}$$

$$= \frac{(n+1)!\,k!}{\prod_i s_i!(m_i+1)!)^{s_i}} \binom{n+1-d}{k}. \qquad (3.41)$$

Note that d is the dimension of the face of symmetry type $G_d^{(\alpha)}$ and k is the number of factors of the direct product group $G_d^{(\alpha)}$ (we always denote by n the dimension of the space spanned by the lattice).

To summarize this section, we have obtained detailed knowledge of the combinatorial structure of the Voronoï cell of the A_n^w lattice by giving—

   (i) The list of the d-dimensional reflection subgroups of A_n; it describes the different types (α) of the d-faces. Each one is either the Voronoï cell of dimension d or the $*$-sum of such cells of smaller dimension.

  (ii) The number of (α)-cells at each vertex; it is given by $K_d(G_d^{(\alpha)})$ in Eq. (3.40).

 (iii) The total number of (α)-cells; it is given by $N_d(G_d^{(\alpha)})$ in Eq. (3.41).

As a last example, let us study the symmetry types and the number of facets of the Voronoï cell A_n^w. Their symmetry type are A_{n-1}, $A_{n-2} \times A_1$, $A_{n-3} \times A_2$, $A_{n-4} \times A_3, \ldots$. (This structure of the facets was given by theorem of Ref. [21]; the theorem also gives information on the structure of the d-faces). Then, from (3.40), it is easy to complete the proof of—

Proposition 3.3. *At each vertex two facets of each of the $[n/2]$ symmetry types $A_m \times A_{n-1-m}$ with $m \in Z$, $0 \leq m < (n-1)/2$ meet; and, when n is odd, one facet of type $A_{(n-1)/2}^2$.*

Hence, the total numbers of facets of the different symmetry types are (with m integer and A_0 the trivial group)

$$0 \leq m < \frac{n-1}{2}, \quad N_{n-1}(A_m \times A_{n-1-m}, n) = 2\binom{n+1}{m+1}, \qquad (3.42)$$

$$\text{and for odd } n, \quad m = \frac{n-1}{2}, \quad N_{n-1}(A_m^2, n) = \binom{n+1}{m+1}. \qquad (3.42')$$

Adding these numbers for all values of m yields $N_{n-1}(n) = 2(2^n - 1)$, already given in Eq. (2.8).

Table 11.1 summarizes the nature of d-faces for $0 \leq d \leq 5$ and for $d = n - 1$ and n (i.e., the cell itself); and for each symmetry type, their number per vertex and their total number. Some results on the structure of the Voronoï cell of the lattice A_n^w have been given in Refs. [3, 4]. A general algorithm for studying the structure of the Voronoï cells of root and weight lattices is established in Ref. [20].

TABLE 11.1. The d-faces of the Voronoï cell $\mathcal{D}(A_n^w)$. Column 2: k–(Vcell)=A_k^w Voronoï cell. $G_k^{(\alpha)} \times A_1$ is a prism of basis $G_k^{(\alpha)}$. Column 3: number of $G_k^{(\alpha)}$ faces per vertex. Column 4: total number of $G_k^{(\alpha)}$ faces. The last line is a natural convention.

d	$G_d^{(\alpha)}$	$K_d(G_d^{(\alpha)})$	$N_d(G_d^{(\alpha)})$
0	$1 = $ vertex	1	$(n+1)!$
1	$A_1 = $ edge	n	$(n+1)!\frac{n}{2}$
2	$A_2 = $ hexagon	$n-1$	$\frac{(n+1)!}{6}(n-1)$
	$A_1^2 = $ square	$\frac{(n-1)(n-2)}{2}$	$\frac{(n+1)!}{8}\prod_{k=1}^{2}(n-k)$
3	$A_3 = 3$-(Vcell)	$n-2$	$\frac{(n+1)!}{24}(n-2)$
	$A_2 \times A_1$ prism	$(n-2)(n-3)$	$\frac{(n+1)!}{12}\prod_{k=2}^{3}(n-k)$
	$A_1^3 = 3$-cube	$(\prod_{k=2}^{4}(n-k))/6$	$\frac{(n+1)!}{48}\prod_{k=2}^{4}(n-k)$
4	$A_4 = 4$-(Vcell)	$n-3$	$\frac{(n+1)!}{120}(n-3)$
	$A_4 \times A_1$ prism	$(n-3)(n-4)$	$\frac{(n+1)!}{48}\prod_{k=3}^{4}(n-k)$
	A_2^2	$(\prod_{k=3}^{4}(n-k))/2$	$\frac{(n+1)!}{72}\prod_{k=3}^{4}(n-k)$
	$A_2 \times A_1^2$	$(\prod_{k=3}^{5}(n-k))/2$	$\frac{(n+1)!}{48}\prod_{k=3}^{5}(n-k)$
	$A_1^4 = 4$-cube	$(\prod_{k=3}^{6}(n-k))/24$	$\frac{(n+1)!}{384}\prod_{k=3}^{6}(n-k)$
5	$A_5 = 5$-(Vcell)	$n-4$	$\frac{(n+1)!}{720}(n-4)$
	$A_4 \times A_1$ prism	$(n-4)(n-5)$	$\frac{(n+1)!}{240}\prod_{k=4}^{5}(n-k)$
	$A_3 \times A_2$	$(n-4)(n-5)$	$\frac{(n+1)!}{144}\prod_{k=4}^{5}(n-k)$
	$A_3 \times A_1^2$	$(\prod_{k=4}^{6})/2$	$\frac{(n+1)!}{192}\prod_{k=4}^{6}(n-k)$
	$A_2^2 \times A_1$	$(\prod_{k=4}^{6})/2$	$\frac{(n+1)!}{144}\prod_{k=4}^{6}(n-k)$
	$A_2 \times A_1^3$	$(\prod_{k=4}^{7}(n-k))/6$	$\frac{(n+1)!}{288}\prod_{k=4}^{7}(n-k)$
	$A_1^5 = 5$-cube	$(\prod_{k=4}^{8}(n-k))/120$	$\frac{(n+1)}{3840}\prod_{k=4}^{8}(n-k)$
$n-1$	$A_{n-1} = (n-1)$-(Vcell)	2	$2(n+1)$
	$A_m \times A_{n-m-1}$	$2, 1 \le m < \frac{n-1}{2}$	$2\binom{n+1}{m+1}$
	A_m^2	$1, n = 2m+1$	$\binom{n+1}{m+1}$
n	$A_n = n$-(Vcell)	1	1

4   The New Explicit Expression of Bounds $N_d(n)$

When d is given, (3.41) shows that $N_d(G_d^{(\alpha)})/(n+1)!$ is a polynomial of n of degree $k \leq d$; the maximum value of the degree is reached for $G_d^{(\alpha)} = A_1^d$ (the d-dimensional hypercube). To find the expression of $N_d(n)$, which is both the total number of d-faces of the Voronoï cell $\mathcal{D}(A_n^w)$ and the upper bound of the number of d-faces of any n-dimensional Voronoï cell, we just add the polynomials $N_d(G_d^{(\alpha)}, n)$ for all d-dimensional reflection subgroups $G_d^{(\alpha)}$ of A_n. That proves the result announced in the abstract: for a given d, $N_d(n)/(n+1)!$ is a polynomial in n of degree d. Moreover, the coefficients of these polynomials are rational numbers.

We can predict an upper bound of the smallest common multiple of the denominators of these coefficients (in the reduce form). These denominators contain two types of terms. One of them appears in the number of elements of an orbit: $Q_d(\times_i A_{m_i}^{s_i})$ in Eq. (3.33); indeed it is the quotient $|A_n|/|\times_i A_{m_i}^{s_i}|$. Since we factorize $(n+1)!$ in the final expression of $N_d(n)$, the number of elements of all reflection subgroups of A_n are in the denominators and their smallest common multiple is $(n+1)! = |A_n|$. For the direct product of identical reflection groups, as in A_m^s, another type of terms appears in denominator: $s!$ in the middle of (3.37). This number of permutations was introduced because one cannot distinguish between the different factors A_m of A_m^s in the Coxeter diagram of the later group. As we pointed it (after (3.31)) for the hypercube (case $m = 1$), that was another way to state that the symmetry of the face A_m^s is the wreath product $A_m \uparrow s$ with

$$|A_m \uparrow s| = ((m+1)!)^s (s!). \qquad (4.1)$$

For $m > 1$, A_m^s is no longer a group generated by reflection but, as for A_1^s, it is also a subgroup of $GL_d(Z)$ with $n \geq ms$.

To summarize: the denominators of the rational coefficients of the polynomial $p_d(n)$ are the orders of different finite subgroups of $GL_n(Z)$. Minkowski [17] has computed the *smallest common multiple of all finite subgroups of* $GL_d(Z)$ and has denoted it by $\overline{d|}$:

$$\overline{d|} = \prod_{q \text{ is prime}} q^{\sum_{k=0}^{\infty} [d/(q^k(q-1))]}; \qquad \Longrightarrow \quad \overline{2d+1|} = 2.\overline{2d|}. \qquad (4.2)$$

He also proved (B_d is the dth Bernouilli number):

$$b_d = \text{ denominator of } (B_d/d),$$

$$\overline{2d|} = 2b_d.\overline{2d-1|} \iff \overline{d|} = 2^d \prod_{k=1}^{[d/2]} b_k. \qquad (4.3)$$

From $\overline{1|} = 2$ with this equation and the last equality of (4.2), it is easy to compute

$$\leq d \leq 7, \quad \overline{|d|} = 2,\ 24,\ 48,\ 5\,760,\ 11\,520,\ 2\,903\,040,\ 5\,806\,080. \tag{4.4}$$

Then

$$d > 0, \quad N_d(n) = (n+1)!(n+1-d)(\overline{|d|})^{-1} P_d(n), \tag{4.5}$$

where $P_d(n)$ is a polynomial in n of degree $d-1$ and integer coefficients. For $1 \leq d \leq 5$ the polynomials $P_d(n)$ can be computed by adding the expressions of the last column of Table 11.1 corresponding to a given d. These $P_d(n)$ are listed in Table 11.2.

5 Expression of $N_d(n)$ as Multiple of a Stirling Number of Second Kind

In this section we first give a reinterpretation of the general formula (3.41). We consider first the cases where all $s_i \doteq 1$. We have studied in Eq. (3.42) the simplest of these cases with $k = n + 1 - d = 2$; we can generalize it to other values of k small enough compared to n (e.g., $k^2 < (n+1-k)$); then $G_d^{(\alpha)} = \times_{i=1}^{k} A_{m_i}$ with $1 \leq m_i$, the m_i's are all different. With these conditions, (3.41) becomes

$$k = n + 1 - d, \quad 1 \leq i \leq d, \quad 1 \leq m_i \in Z, \quad m_i \text{ all different,}$$

$$N_d(n) = k! \sum_{\Sigma m_i = d} \frac{(n+1)!}{\prod_{i=1}^{k}(m_i + 1)!}. \tag{5.1}$$

Changing the notation $N_d(n) = N_k'(n)$, $n_i = m_i + 1$ and ordering the integers n_i's by decreasing values, we can also write this last equation:

$$k = n + 1 - d, \quad 1 \leq i \leq k, \quad n_i > n_{i+1}, \quad n_k \geq 2,$$

$$N_d(n) \equiv N_k'(n) = k! \sum_{\Sigma n_i = n+1} \frac{(n+1)!}{\prod_{i=1}^{k} n_i!}. \tag{5.1'}$$

TABLE 11.2. The polynomials $P_d(n)$ for $1 \leq d \leq 5$.

$P_1(n) = 1$.

$P_2(n) = 3n - 2$.

$P_3(n) = (n-1)(n-2)$.

$P_4(n) = 15n^3 - 105n^2 + 230n - 152$.

$P_5(n) = 3n^4 - 38n^3 + 173n^2 - 330n + 216$.

To obtain the equation equivalent to the general form of (3.41), we have to relax the condition $n_i > n_{i+1}$ to $n_i \geq n_{i+1}$ and count the contiguous numbers of $=$:

$$s_1, s_2, \ldots, s_\alpha, \ldots \text{ in the sequence } n_i \geq n_{i+1}. \tag{5.2}$$

Then (3.41) becomes, with $k = n + 1 - d$,

$$1 \leq i \leq k, \quad n_i > n_{i+1}, \quad n_k \geq 2,$$

$$N_d(n) \equiv N'_k(n) = k! \sum_{\Sigma n_i = n+1} \frac{(n+1)!}{\prod s_\alpha \prod_{i=1}^k n_i!}. \tag{5.3}$$

Finally we have to relax the condition k small compared to n and to consider the cases where $k < n + 1 - d$ (as is the case for instance for the facet of symmetry A_{n-1}). For that we complete the sequence $n_1 \geq \cdots \geq n_k \geq 2$ by adding to it $(n + 1 - d - k)$ "1"'s; this adds a new s_α whose value is $n + 1 - d - k$. Then (5.3), which is equivalent to (3.41), is transformed into

$$ell = n + 1 - d, \quad 1 \leq i \leq \ell, \quad n_i > n_{i+1}, \quad n_\ell \geq 1,$$

$$N_d(n) \equiv N'_\ell(n) = \ell! \sum_{\Sigma n_i = n+1} \frac{(n+1)!}{\prod s_\alpha \prod_{i=1}^\ell n_i!}. \tag{5.4}$$

This is a new general expression for $N_d(n)$. It is easy to verify that for $1 \leq d \leq 5$ and for the different partitions of $n + 1$ into $\ell = n + 1 - d$ terms, the different terms in the the sum over the partitions in Eq. (5.4) have exactly the form given in the last column of Table 11.1.

The expression

$$1 \leq i \leq \ell, \quad n_i > n_{i+1}, \quad n_\ell \geq 1, \quad \sum_{\Sigma n_i = n+1} \frac{(n+1)!}{\prod s_\alpha \prod_{i=1}^\ell n_i!}. \tag{5.5}$$

is the number of ways of partitioning a set of $n+1$ elements into ℓ nonempty subsets. By definition this number is the Stirling number of second kind; it is denoted $\mathbf{S}_{n+1}^{(\ell)}$ in Ref. [1]. Thus we have established that (it is also valid for $d = 0$ and $d = n$):

$$0 \leq d \leq n, \quad N_d(n) = (n + 1 - d)! \mathbf{S}_{n+1}^{(n+1-d)}. \tag{5.6}$$

The right-hand side corresponds to the ordered partitions: it includes the permutations of the different nonempty subsets that define the partition. Through the use of (5.6) and (4.5), any relation between Stirling numbers of second kind yields a relation between the polynomials $P_d(n)$ (defined in Eq. (4.5)). For instance,

$$\mathbf{S}_{n+1}^{n+1-d} = (n + 1 - d)\mathbf{S}_n^{n+1-d} + \mathbf{S}_n^{n-d}, \tag{5.7}$$

yields

$$(n+1)P_d(n) - (n-d)P_d(n-1) = (n+1-d)\frac{\overline{d|}}{d-1|}P_{d-1}(n-1). \quad (5.8)$$

This gives another possibility for computing the $P_d(n)$ by recursion.

One closed expression given in Ref. [1] for the Stirling numbers of the second kind is

$$S_n^{(m)} = (m!)^{-1}\sum_{k=1}^{m}(-1)^{m-k}\binom{m}{k}k^n. \quad (5.9)$$

That yields

$$N_d(n) = \sum_{k=1}^{n+1-d}(-1)^{n+1-d-k}\binom{n+1-d}{k}k^{n+1}. \quad (5.10)$$

With the change of dumb index $k \mapsto \ell+1$, one obtains the expression (2.10) given by Voronoï.

So we can either consider that we have given another proof of the Voronoï formula (2.10) or, more modestly, that (5.6) is a simpler form of the Voronoï expression. Equations (5.6) and (4.5) show that the Stirling numbers of the second kind can be expressed as a family of polynomials; this was known long ago, e.g., Ref. [14, §58]. However, the only reference I know[12] to a form similar to

$$S_{n+1}^{(n+1-d)} = (\overline{d|})^{-1}P_d(n)\prod_{\ell=0}^{d}(n+1-\ell) \quad (5.11)$$

is in the book [12, Eqs. (6.45), (6.5)]. These authors defined the "Stirling polynomials" $\sigma_d(n+1)$ from a formula identical to (5.11) but written for the Stirling numbers of the first kind[13]. The relation between the two families of polynomials is

$$P_d(n) = (-1)^{d+1}\overline{d|}\sigma_d(-(n+1-d)). \quad (5.12)$$

When Eq. (5.12) is used, the recursion (5.8) yields that to be proven in Ref. [12, Exercise 6.18] .

6   Final Remarks

As we noted, we have also studied the Cayley graph of the symmetric group S_{n+1}. The use of its natural realization (3.17) as an n-dimensional reflec-

---

[12]I am very grateful to Dr Thomas Scharf (in Bayreuth) who pointed this reference out to me.

[13]Their Stirling numbers of first kind do not change sign, so they are those of Ref. [1] multiplied by $(-1)^d$. In their book, the denominators of the σ_d polynomials are not recognized as Minkowski numbers.

tion group imbeds this graph as a polytope of the orthogonal vector space E_n. That introduces the roots, their orthogonal hyperplanes, and the Weyl chambers. Then we can define the set of vertices of the Cayley diagram, not as an abstract principal orbit of S_{n+1}, but as an orbit of a point in the interior of a Weyl chamber and equidistant to its walls. The Coxeter elements of this reflection group are the permutations corresponding to a cycle of length $n + 1$. They play an important role (see Eq. (3.20) that involves the weights = facet vectors of the Cayley graph). Then the Theorem 1.7.7 of Ref. [15], which in some cases leads to set ordered partitions, is probably the shortest way to obtain (5.6). So one could have started the study of Cayley graphs of symmetric groups without any reference to their identification (given, e.g., in Ref. [5]) with the Voronoï cells of the lattices A_n^w. But in this paper, we wanted first to describe the structure of the Voronoï cell of A_n^w.

REFERENCES

1. M. Abramowitz and I.A., *Handbook of mathematical functions*, National Bureau of Standards (1964). Since, many corrected editions and reprintings have been published by Dover Publications.

2. E.P. Baranovskii and S.S. Ryškov, *Primitive five-dimensional parallelohedra*, Soviet Math. Dokl. **14** (1973), 1391–1395.

3. J.H. Conway and N.J.A. Sloane, *The cell structures of certain lattices*, Miscellaneous Mathematics (P. Hilton, F. Hirzebruch, R. Remmert, eds.), Springer, Berlin, 1991, pp. 71–109.

4. J.H. Conway and N.J.A. Sloane , *Sphere packings, lattices and groups*, 3rd ed., Grundlehren Math. Wiss., Vol. 290, Springer-Verlag, New York, 1988.

5. H.S.M. Coxeter, *Regular polytopes*, Mac Millan, 1947; 2nd ed., 1963, Dover reprint, 1973.

6. H.S.M. Coxeter and W.O.J. Coxeter, *Generators and relations for discrete groups*, Springer-Verlag, New York, 1957; 4th ed., 1980.

7. B.N. Delone, R.V. Galiulin, and M.I. Shtogrin, *Bravais theory and its generalization to n-dimensional lattices*, Appendix, Auguste Bravais: Collected scientific works, Nauka, Leningrad, 1974, pp. 309–415 (in Russian).

8. P. Engel, *Geometric crystallography: An axiomatic introduction to crystallography*, D. Reidel Publ. Co., Dordrecht, 1986.

9. P. Engel, *On the symmetry classification of the four-dimensional parallelohedra*, Z. Krist. **200** (1992), 199–213.

10. P. Engel, *Investigations on lattices and parallelohedra in R^d*, Voronoï Impact on Modern Science, Proc. Inst. Math. Acad. Sci. Ukraine, Vol. 21, 1998, pp. 22–60.

11. E.S. Fedorov, *An introduction to the theory of figures*, Verhandlungen der russisch-kaiserlichen mineralogischen Gesellschaft zu St Petersbourg **21** (1885), 1–279 (in Russian).

12. R.L. Graham, D.E. Knuth, and O. Patashnik, *Concrete mathematics*, Addison-Wesley Publishing Co., Reading, MA, 1989.

13. B. Grünbaum, *Convex polytopes*, Pure Appl. Math., Vol. 16, John Wiley & Sons, Inc., New York, 1967.

14. C. Jordan, *Calculus of finite differences*; reprinted by Chelsea, New York, 1960.

15. A. Kerber, *Algebraic combinatorics via finite group actions*, Preprint, Bayreuth University, 1996.

16. L. Michel and M. Senechal, *Geometry of Euclidean lattices* (to appear).

17. H. Minkowski, *Zur Theorie der positiven quadratischen Formen*, J. Reine Angew. Math. **106** (1890), 5–26; Collected works, Vol. 1, pp. 212–218.

18. H. Minkowski, Allgemeine, *Lehrsätze über konvexe Polyeder*, Nachr. Akad. Wiss. Göttingen, Math.-Phys. Kl. (1897); Ges. Abhand. **2**, 103–121.

19. H. Minkowski, *Diophantische Approximationen*, Teubner, Leipzig (1907); reprinted, Chelsea Publishing Co., New York, 1957.

20. R.V. Moody and J. Patera, *Voronoï domains and dual cells in the generalized kaleidoscope with applications to root and weight lattices*, Canad. J. Math. **47** (1995), 573–605.

21. S.S. Ryškov, *The structure of a n-dimensional parallelohedron of the first kind*, Soviet Math. Dokl. **3** (1962), 1451–1454.

22. E. Selling, *Ueber die binären und ternären quadratischen Formen*, J. Reine Angew. Math. **77** (1874), 143–229.

23. M. Senechal and R.V. Galiulin, *An introduction to the theory of figures: The geometry of E.S. Fedorov*, Structural Topology, No. 10, (1984), 5–22.

24. G. Voronoï, *Recherches sur les paralléloèdres primitifs. I. Propriétés générales des paralléloèdres*, J. Reine Angew. Math. **133** (1908), 198–287.

25. G. Voronoï, *Recherches sur les paralléloèdres primitifs. II. Domaines de formes quadratiques correspondant aux différents types de paralléloèdres primitifs*, J. Reine Angew. Math. **136** (1909), 67–181.

12

The Relativistic Oscillator and the Mass Spectra of Baryons

M. Moshinsky

ABSTRACT An analysis of the baryon spectra using harmonic oscillator states has been carried out by many authors, usually in a nonrelativistic approximation, and employing a great number of parameters. In the present paper we perform this analysis using a single relativistic equation that has only one parameter, the frequency ω of the oscillator. By eliminating the center of mass motion, the problem can be formulated in terms of the generators of an SU(2) group, and Racah algebra allows us then to obtain a reasonable energy spectra for nonstrange baryons, by appropriately choosing our single variational parameter ω.

I had the pleasure of meeting both Patera and Winternitz at a mathematical physics conference almost 30 years ago. Through all the following years I kept in close contact with them at my Institution in México; at the Centre de Recherches Mathématiques at the Université de Montréal; and, in collaboration also with Robert T. Sharp, at the University of McGill. I have greatly benefitted from these contacts, which have resulted in several joint papers, and wish on the occasion of their 60th birthdays to send to them my bests wishes. I expect their future work will be as outstanding as their past accomplishments.

1 Introduction

Starting from the well-known single equation for a system of n relativistic noninteracting particles, we generalize it to include an oscillator interaction. Eliminating then the center of mass motion we have a Hamiltonian expressible in terms of the generators of a U($n-1$) group whose energy or, equivalently, mass spectra, can be obtained from group theoretical considerations.

As our final objective is to apply this approach to a system of three quarks, we will consider in detail only the case $n = 3$, which will be associated with the very well known group U(2).

In Section 2 we shall discuss the scalar relativistic three-particle problem, while in Section 3 we apply the results, essentially equivalent to the above, to the spinorial three-particle problem. Finally with the help of the single parameter of our formalism, the frequency ω of the oscillator, we proceed to derive the mass spectra of the nonstrange baryons.

2   The System of Three Relativistic Scalar Particles with Oscillator Interactions

Rather than discuss the system of n relativistic particles, we shall restrict ourselves to $n = 3$, as we will see that the case is general enough, with only the algebraic steps becoming more complicated as n increases.

In our units $\hbar = m = c = 1$ the total energy for a system of three free relativistic particles can be written as

$$E = \pm \Pi_1 \pm \Pi_2 \pm \Pi_3, \tag{2.1}$$

where Π_s, $s = 1$, 2, 3 is defined as

$$\Pi_s \equiv (p_s^2 + 1)^{1/2}. \tag{2.2}$$

with the $\mathbf{p}_s$ being the momentum of the sth particle.

It is very important to note that, in our units, Einstein relation is $E^2 = p^2 + 1$, and when reduced to the E itself gives both the square root in Eq. (2.2) and the $\pm$ signs in Eq. (2.1).

Obviously we can not get a Schroedinger equation from the relation (2.1), but we can take $\pm \Pi_3$ to the right-hand side and square both sides. Then we can square again and again appropriately, and we easily arrive at the fact that (2.1) becomes an eight-degree equation in E (actually of fourth-degree in E^2) of the form

$$\Phi(E^2, \Pi_s^2) \equiv E^8 - 4AE^6 + (4A^2 + 2B)E^4$$
$$- (4C^2 + 4AB)E^2 + B^2 = 0, \quad (2.3)$$

where A, B, C are functions of Π_s^2, $s = 1$, 2, 3 given by

$$A \equiv \Pi_1^2 + \Pi_2^2 + \Pi_3^2, \tag{2.4a}$$

$$B \equiv \Pi_1^4 + \Pi_2^4 + \Pi_3^4 - \Pi_1^2\Pi_2^2 - \Pi_2^2 P i_3^2 - \Pi_1^2\Pi_3^2 - \Pi_2^2\Pi_1^2$$
$$- \Pi_3^2\Pi_2^2 - \Pi_3^2\Pi_1^2 \tag{2.4b}$$

$$C^2 \equiv \frac{8}{3}(\Pi_1^2\Pi_2^2\Pi_3^2 + \Pi_2^2\Pi_1^2\Pi_3^2 + \Pi_3^2\Pi_1^2\Pi_2^2 + \Pi_1^2\Pi_3^2\Pi_2^2$$
$$+ \Pi_3^2\Pi_2^2\Pi_1^2 + \Pi_2^2\Pi_3^2\Pi_1^2). \tag{2.4c}$$

Note that as we are working with classical observables that commute we could express A, B, C^2 of (2.4) in a much more compact form. We prefer though to write them in an explicitly invariant way under permutation of the indices $s = 1, 2, 3$, as later Π_s^2 will be replaced by operators that do not commute.

Now we can write an equation that does not have E as an eigenvalue, but in which it appears as a parameter, if we replace $\mathbf{p}_s$ by $-i\partial/\partial\mathbf{x}_s$, so that Π_s^2 become the operators

$$\widehat{\Pi}_s^2 = (-\nabla_s^2 + 1), \tag{2.5}$$

and we get

$$\Phi(E^2, \widehat{\Pi}_s^2)\psi = 0. \tag{2.6}$$

Thus far we have obtained nothing useful because $\mathbf{p}_1$, $\mathbf{p}_2$, $\mathbf{p}_3$, considered as operators, commute with the operator Φ and are integrals of motion, so that ψ can be written as

$$\psi = \exp\big[i(\mathbf{p}_1 \cdot \mathbf{x}_1 + \mathbf{p}_2 \cdot \mathbf{x}_2 + \mathbf{p}_3 \cdot \mathbf{x}_3)\big], \tag{2.7}$$

where now $\mathbf{p}_1$, $\mathbf{p}_2$, $\mathbf{p}_3$ are ordinary numbers and we are returned to Eq. (2.3) whose eight roots for the energy E are obviously given by (2.1) with all the possible combination of the signs $\pm$.

Before proceeding further, we remark that we would like to work in the center of mass frame, as our interest is restricted to the internal energy of the system. Thus we go to the Jacobi momenta $\dot{\mathbf{p}}_s$, $s = 1, 2, 3$, which are given now by the matrix relation

$$\begin{pmatrix} \dot{\mathbf{p}}_1 \\ \dot{\mathbf{p}}_2 \\ \dot{\mathbf{p}}_3 \end{pmatrix} = \begin{pmatrix} \frac{1}{\sqrt{2}} & -\frac{1}{\sqrt{2}} & 0 \\ \frac{1}{\sqrt{6}} & \frac{1}{\sqrt{6}} & -\sqrt{\frac{2}{3}} \\ \frac{1}{\sqrt{3}} & \frac{1}{\sqrt{3}} & \frac{1}{\sqrt{3}} \end{pmatrix} \begin{pmatrix} \mathbf{p}_1 \\ \mathbf{p}_2 \\ \mathbf{p}_3 \end{pmatrix}. \tag{2.8}$$

As the matrix is orthogonal, transposing it we get $\mathbf{p}_s$ in terms of $\dot{\mathbf{p}}_s$, and as we want to be in the center of mass frame $\dot{\mathbf{p}}_3 = 0$, so we get finally

$$\mathbf{p}_1 = \frac{1}{\sqrt{2}}\dot{\mathbf{p}}_1 + \frac{1}{\sqrt{6}}\dot{\mathbf{p}}_2,$$

$$\mathbf{p}_2 = -\frac{1}{\sqrt{2}}\dot{\mathbf{p}}_1 + \frac{1}{\sqrt{6}}\dot{\mathbf{p}}_2, \tag{2.9}$$

$$\mathbf{p}_3 = -\sqrt{\frac{2}{3}}\dot{\mathbf{p}}_2.$$

As Eq. (2.6) contains only powers of $\widehat{\Pi}_s^2$, $s = 1, 2, 3$, we can write the latter using the hermitian property of $\dot{\mathbf{p}}_s$, now considered as operators, as

$$\widehat{\Pi}_1^2 = \frac{1}{2}\dot{\mathbf{p}}_1^\dagger \cdot \dot{\mathbf{p}}_1 + \frac{1}{6}\dot{\mathbf{p}}_2^\dagger \cdot \dot{\mathbf{p}}_2 + \frac{1}{2\sqrt{3}}(\dot{\mathbf{p}}_1^\dagger \cdot \dot{\mathbf{p}}_2 + \dot{\mathbf{p}}_2^\dagger \cdot \dot{\mathbf{p}}_1) + 1, \qquad (2.10a)$$

$$\widehat{\Pi}_2^2 = \frac{1}{2}\dot{\mathbf{p}}_1^\dagger \cdot \dot{\mathbf{p}}_1 + \frac{1}{6}\dot{\mathbf{p}}_2^\dagger \cdot \dot{\mathbf{p}}_2 - \frac{1}{2\sqrt{3}}(\dot{\mathbf{p}}_1^\dagger \cdot \dot{\mathbf{p}}_2 + \dot{\mathbf{p}}_2^\dagger \cdot \dot{\mathbf{p}}_1) + 1, \qquad (2.10b)$$

$$\widehat{\Pi}_3^2 = \frac{2}{3}\dot{\mathbf{p}}_2^\dagger \cdot \dot{\mathbf{p}}_2 + 1. \qquad (2.10c)$$

The interesting point is to introduce the oscillator interaction, by the replacement $\dot{\mathbf{p}}_s \to \dot{\mathbf{p}}_s - iw\dot{\mathbf{x}}_s,\ \dot{\mathbf{p}}_s^\dagger \to \dot{\mathbf{p}}_s + iw\dot{\mathbf{x}}_s$ as in Ref. [3, formula (67.7), p. 368]. For notational purposes we introduce the creation and annihilation operators

$$\dot{\boldsymbol{\eta}}_s \equiv \frac{1}{\sqrt{2}}(\omega^{1/2}\dot{\mathbf{x}}_s - i\omega^{-1/2}\dot{\mathbf{p}}_s), \qquad s = 1, 2, \qquad (2.11a)$$

$$\dot{\boldsymbol{\xi}}_s \equiv \frac{1}{\sqrt{2}}(\omega^{1/2}\dot{\mathbf{x}}_s + i\omega^{-1/2}\dot{\mathbf{p}}_s), \qquad s = 1, 2, \qquad (2.11b)$$

so that the above relations can be written as

$$\dot{\mathbf{p}}_s \to -i\omega^{1/2}\sqrt{2}\dot{\boldsymbol{\xi}}_s, \qquad (2.12a)$$

$$\dot{\mathbf{p}}_s^\dagger \to i\omega^{1/2}\sqrt{2}\dot{\boldsymbol{\eta}}_s. \qquad (2.12b)$$

Under this replacement the $\widehat{\Pi}_s^2$ operators then become

$$\widehat{\Pi}_1^2 = \omega\left[\dot{C}_{11} + \frac{1}{3}\dot{C}_{22} + \frac{1}{\sqrt{3}}(\dot{C}_{12} + \dot{C}_{21})\right] + 1, \qquad (2.13a)$$

$$\widehat{\Pi}_2^2 = \omega\left[\dot{C}_{11} + \frac{1}{3}\dot{C}_{22} - \frac{1}{\sqrt{3}}(\dot{C}_{12} + \dot{C}_{21})\right] + 1, \qquad (2.13b)$$

$$\widehat{\Pi}_3^2 = \frac{4}{3}\omega\dot{C}_{22} + 1, \qquad (2.13c)$$

where the operator $\dot{C}_{st},\ s, t = 1,\ 2$ is defined by

$$\dot{C}_{st} = \dot{\boldsymbol{\eta}}_s \cdot \dot{\boldsymbol{\xi}}_t. \qquad (2.14)$$

From the fact that

$$[\dot{\xi}_{it}, \dot{\eta}_{js}] = \delta_{ij}\delta_{st}; \qquad i, j = 1, 2, 3; \quad s = 1, 2, \qquad (2.15)$$

we have the commutation relations

$$[\dot{C}_{st}, \dot{C}_{s't'}] = \dot{C}_{st'}\delta_{s't} - \dot{C}_{s't}\delta_{st'}, \qquad (2.16)$$

and thus they are generators [4] of a U(2) group. Therefore the operators $\widehat{\Pi}_1^2, \widehat{\Pi}_2^2, \widehat{\Pi}_3^2$, appearing in the Eq. (2.6), are linear functions of the generators of this group.

To obtain from Eqs. (2.6), (2.13) the eigenvalues of the energy for this relativistic oscillator problem we can proceed as follows: First we note that the first order Casimir operator of U(2) group is

$$\widehat{N} = \dot{C}_{11} + \dot{C}_{22}, \tag{2.17}$$

and that it has an SU(2) subgroup whose generators are

$$\widehat{F}_+ \equiv \dot{C}_{12}, \tag{2.18a}$$
$$\widehat{F}_o \equiv (\tfrac{1}{2})(\dot{C}_{11} - \dot{C}_{22}), \tag{2.18b}$$
$$\widehat{F}_- \equiv \dot{C}_{21}, \tag{2.18c}$$

with a corresponding Casimir operator of the form

$$\widehat{F}^2 \equiv \widehat{F}_-\widehat{F}_+ + \widehat{F}_o(\widehat{F}_o + 1). \tag{2.19}$$

The $\widehat{N}$, $\widehat{F}^2$ by definition commute with all C_{st} and among themselves, so from Eq. (2.13), they will be integrals of motion of the operator $\Phi(E^2, \widehat{\Pi}_s^2)$. Thus the eigenstates of the Eq. (2.6) can be characterized by the eigenvalues of $\widehat{N}$, $\widehat{F}^2$, which we denote, respectively, by

$$N, \quad \mathcal{F}(\mathcal{F}+1), \tag{2.20}$$

with $\mathcal{F}$ taking the values $(N/2)$, $(N/2) - 1$, down to $1/2$ or 0 depending on whether N is odd or even.

Another operator that commutes with $\widehat{N}$, $\widehat{F}^2$ is obviously $\widehat{F}_o$ and we shall designate its eigenvalue by

$$\nu = \mathcal{F}, \quad \mathcal{F} - 1, \dots, -\mathcal{F}, \tag{2.21}$$

so the eigenstates associated with $\widehat{N}$, $\widehat{F}^2$, $\widehat{F}_o$ could be represented by the ket

$$|N\mathcal{F}\nu\rangle, \tag{2.22}$$

and the solution ψ of Eq. (2.6) is necessarily a linear combination of these kets, i.e.,

$$\psi = \sum_{\nu=-\mathcal{F}}^{\mathcal{F}} a_\nu |N\mathcal{F}\nu\rangle, \tag{2.23}$$

as $\widehat{N}$, $\widehat{F}^2$ are integrals of motion.

To obtain the eigenvalues of the internal energy E as function of N, $\mathcal{F}$ we need first to consider the matrix elements of the operator Φ of Eq. (2.6) in the basis (2.22), i.e.,

$$\langle N\mathcal{F}\nu'|\Phi(E^2, \widehat{\Pi}_s^2)|N\mathcal{F}\nu\rangle. \tag{2.24}$$

These elements can be obtained straightforwardly, as from Eqs. (2.13), (2.17), and (2.18) we have

$$\widehat{\Pi}_1^2 = 1 + \omega\left[\left(\frac{2}{3}\right)(\widehat{N} + \widehat{F}_o) + \left(\frac{1}{\sqrt{3}}\right)(\widehat{F}_+ + \widehat{F}_-)\right], \tag{2.25a}$$

$$\widehat{\Pi}_2^2 = 1 + \omega\left[\left(\frac{2}{3}\right)(\widehat{N} + \widehat{F}_o) - \left(\frac{1}{\sqrt{3}}\right)(\widehat{F}_+ + \widehat{F}_-)\right], \tag{2.25b}$$

$$\widehat{\Pi}_3^2 = 1 + \omega\left[\left(\frac{2}{3}\right)(\widehat{N} - 2\widehat{F}_o)\right], \tag{2.25c}$$

and besides we have the well-known angular momentum relations that, in the language used here for the SU(2) group, are

$$\widehat{F}_\pm|N\mathcal{F}\nu\rangle = [(\mathcal{F} \mp \nu)(\mathcal{F} \pm \nu + 1)]^{1/2}|N\mathcal{F}\nu \pm 1\rangle, \tag{2.26a}$$

$$\widehat{F}_o|N\mathcal{F}\nu\rangle = \nu|N\mathcal{F}\nu\rangle. \tag{2.26b}$$

To get the internal energy

$$E(N, \mathcal{F}, \alpha), \tag{2.27}$$

with α indicating the rest of the indices, we need to evaluate the determinant of the $(2\mathcal{F} + 1) \times (2\mathcal{F} + 1)$ matrix whose elements are (2.24) and equate it to zero. This gives us a numerical equation of degree $4(2\mathcal{F} + 1)$ in the variable E^2 and its solution provides us with the values indicated symbolically in Eq. (2.27).

We shall discuss the results for $N = 0, 1, 2$ and different values of ω. We start with

$$N = \mathcal{F} = \nu = 0, \tag{2.28}$$

which implies that

$$\langle 000|\widehat{\Pi}_s^2|000\rangle = 1, \quad s = 1, 2, 3, \tag{2.29}$$

and so A, B, C^2 of Eq. (2.4) become, respectively,

$$A = 3, \quad B = -3, \quad C^2 = 16, \tag{2.30}$$

and Eq. (2.3)) for the energy is given by

$$E^8 - 12E^6 + 30E^4 - 28E^2 + 9 = 0, \tag{2.31}$$

whose four roots for E^2 are $E^2 = 1$, repeated three times, and $E^2 = 9$, which leads to the values $E = \pm 3$ and $E = \pm 1$ as expected from Eq. (2.1). In this case, of course, the value of the energy is independent of ω.

When $N = 1$ we have $\mathcal{F} = 1/2$, $\nu = \pm 1/2$ and if we calculate the matrix elements (2.24) using Eqs. (2.25), (2.26) we see that the 2×2 matrix is a multiple of the unit matrix, whose coefficient, when equated to 0, gives the fourth-degree algebraic equation in E^2 of the form

$$E^8 - 4(2\omega + 3)E^6 + \left[\left(\frac{56}{3}\right)\omega^2 + 40\omega + 30\right]E^4$$
$$- \left[\left(\frac{32}{27}\right)\omega^3 + \left(\frac{80}{3}\right)\omega^2 + 56\omega + 28\right]E^2$$
$$+ \left[\left(\frac{4}{3}\right)\omega^2 - 4\omega - 3\right]^2 = 0. \quad (2.32)$$

The roots of this equation will be the energy levels for the problem and now they depend on the frequency ω.

We start with ω small compared to 1, which in ordinary units means $\hbar\omega \ll mc^2$ and take, for example, $\omega = 0.1$, thus getting an algebraic equation whose roots E^2 are

$$E^2 = 0.9095,\ 1.0635,\ 1.2363,\ 9.5906. \quad (2.33)$$

As we expect, these values are close to 1 and 9, which were the values obtained in Eq. (2.31)) to which (2.32) reduces when $\omega = 0$. We note though that if we wish only to consider the root corresponding to positive energy close to $E = 3$, then the value of interest to us is $E = 3.096$, which is the positive square root of the largest E^2 appearing above.

When $\omega = 1$, we get an algebraic equation whose roots E^2 are

$$E^2 = 0.4061,\ 1.4565,\ 3.7821,\ 14.3554. \quad (2.34)$$

We now are already well separated from the values 1 and 9 but the roots are still real and positive. On the other hand, when, for example, $\omega = 10$, the roots E^2 of Eq. (2.32) take complex values

$$E^2 = 0.9659 \pm i1.7083,\ 45.034 \pm i9.5265. \quad (2.35)$$

This result seems at first sight nonsensical, but we must remember that Dirac, when discussing the exact solution of his equation for a Coulomb potential in its time like component gets, in our units, the formula [6]

$$E_{nj} = \left[1 + \gamma^2(s + n)^{-2}\right]^{-1/2}, \quad (2.36)$$

where $n = 0, 1, 2, 3, \dots$, $s = \left[(j + \frac{1}{2})^2 - \gamma^2\right]^{1/2}$ and γ is the fine structure constant which for a nucleus of charge Z is given by $\gamma = (Ze^2/\hbar c) = (Z/137)$. Thus, for example, if $j = 1/2$, $n = 0$ we have the energy levels

$$E_{o1/2} = \left[1 + \gamma^2(1 - \gamma^2)^{-1}\right]^{-1/2}, \quad (2.37)$$

and if $Z > 137$, we have that $E_{o1/2}$ becomes complex.

Turning to our problem, the fact that after a certain value of ω, (which in our case turns out to be $\omega \simeq 6.7$), our energy levels become complex, implies that we can no longer think that the part of positive energies makes sense.

In fact Pauli and Weisskopf [2, 5], and later others [2, 5], showed that for very strong interactions, i.e., in our case $\omega > 1$, we can no longer speak of a fixed number of particles, as pairs can be created even if particles obey Bose statistics.

Turning our attention now to the case when $N = 2$, we have, from the argument following Eq. (2.20), that $\mathcal{F} = 1$ or $\mathcal{F} = 0$. In the first case $\nu = 1, 0, -1$ and we have a 3×3 matrix, while in the second there is only a 1×1 matrix. The former, i.e., $\mathcal{F} = 1$, in fact breaks into 2×2 and 1×1 matrices related, respectively, to $\nu = 1, -1$, and $\nu = 0$. The 2×2 matrix can be diagonalized in a trivial way [3], so that finally we get for $\mathcal{F} = 1$ the following algebraic equations:

$$
\begin{aligned}
& E^8 - 4(4\omega + 3)E^6 + (64\omega^2 + 80\omega + 30)E^4 \\
& \quad - 4\left[7 + 28\omega + 32\omega^2 + \left(\frac{256}{27}\right)\omega^3\right]E^2 + (8\omega + 3)^2 = 0, \qquad (2.38a)
\end{aligned}
$$

$$
\begin{aligned}
& E^8 - 4(4\omega + 3)E^6 + \left[\left(\frac{224}{3}\right)\omega^2 + 80\omega + 30\right]E^4 \\
& \quad - 4\left[7 + 28\omega + \left(\frac{80}{3}\right)\omega^2 + \left(\frac{64}{27}\right)\omega^3\right]E^2 \\
& \qquad\qquad + \left[8\omega + 3 - \left(\frac{16}{3}\right)\omega^2\right]^2 = 0. \qquad (2.38b)
\end{aligned}
$$

Equation (2.38a) corresponds to $\mathcal{F} = 1$, $\nu = 0$ and to one of the solutions of $\mathcal{F} = 1$, $\nu \pm 1$ while Eq. (2.38b) corresponds to the other solution of $\mathcal{F} = 1$, $\nu = \pm 1$. For the case $\mathcal{F} = 0$, $\nu = 0$ we have the equation

$$
\begin{aligned}
& E^8 - 4(4\omega + 3)E^6 + \left[\left(\frac{160}{3}\right)\omega^2 + 80\omega + 30\right]E^4 \\
& \quad - 4\left[\left(\frac{448}{27}\right)\omega^3 + \left(\frac{112}{3}\right)\omega^2 + 28\omega + 7\right]E^2 \\
& \qquad\qquad + \left[8\omega + 3 + \left(\frac{16}{3}\right)\omega^2\right]^2 = 0, \qquad (2.39)
\end{aligned}
$$

which happens to have the exact analytic roots

$$
E^2 = \left[1 + \left(\frac{4}{3}\right)\omega\right] \quad \text{triply degenerate and} \quad E^2 = (9 + 12\omega). \qquad (2.40)
$$

For $\omega = 0.1$ we get, respectively, the following roots for E^2 from Eqs. (2.38a),

(2.38b), and (2.39):

$$0.9139, \quad 1.1274, \quad 1.3764, \quad 10.1823, \qquad (2.41a)$$
$$0.8295, \quad 1.1215, \quad 1.4844, \quad 10.1645, \qquad (2.41b)$$
$$1.1333, \quad 1.1333, \quad 1.1333, \quad 10.2. \qquad (2.41c)$$

As we should expect all the roots are either close to 1 (triply degenerate) for $\omega = 0$, or close to 9. As we said before it is the root close to $E = 3$, i.e., $E^2 = 9$, that has a physical interest.

For $\omega = 1$ we get, respectively, the following roots for E^2 from Eqs. (2.38a), (2.38b), and (2.39):

$$0.5558, \quad 2.0328, \quad 5.3338, \quad 20.0776, \qquad (2.42a)$$
$$0.1390, \quad 1.6923, \quad 7.1943, \quad 18.9743, \qquad (2.42b)$$
$$2.3333, \quad 2.3333, \quad 2.3333, \quad 21.0000. \qquad (2.42c)$$

They are now quite separated from the values of E^2 when $\omega = 0$, i.e., 1 and 9, but the ones that are probably correlated with the latter are the last ones in each row.

Finally for $\omega = 10$ we get, respectively, the following roots for E^2 from Eqs. (2.38a), (2.38b), and (2.39):

$$0.1353, \quad 8.8242, \quad 51.8973, \quad 111.143, \qquad (2.43a)$$
$$1.0712 \pm i4.9498, \quad 84.9288 \pm i26.348, \qquad (2.43b)$$
$$14.333, \quad 14.333, \quad 14.333, \quad 129. \qquad (2.43c)$$

At this value of ω all the roots of Eqs. (2.38b) are complex, while those of Eqs. (2.38a) and (2.39) remain real. As the algebraic equations for E^2 have real coefficients, the complex roots are associated with their conjugate, as shown by the $\pm$ signs that appear before the imaginary values in Eq. (2.43).

In Fig. 12.1 we shall plot the largest real levels for the square of the energy as functions of ω, and will also indicate the possible angular momenta associated with these levels.

The eigenstates of our problem can be expressed in terms of the two Jacobi coordinates $\dot{\mathbf{x}}_1$, $\dot{\mathbf{x}}_2$ by the ket

$$(\dot{\mathbf{x}}_1, \dot{\mathbf{x}}_2 \mid n_1 l_1, n_2 l_2, LM) \qquad (2.44)$$

where L is total orbital angular momentum, while the total number of quanta N is given by

$$N = 2n_1 + l_1 + 2n_2 + l_2. \qquad (2.45)$$

As in Fig. 12.1 we go only up to $N = 2$, the number of states is limited; and in Table 12.1 we give them in a shorthand notation where only the ket appears and the M, which is irrelevant, is suppressed.

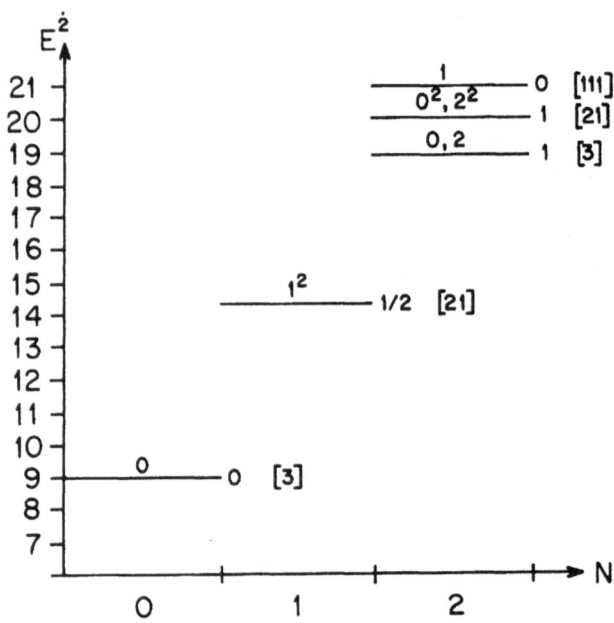

FIGURE 12.1. The square of the energy E^2, as function of the number of quanta N as given by Eqs. (2.31), (2.32), (2.38), (2.39), and (2.40)) when $\omega = 1$. The values of $\mathcal{F}$ are given on the right of each level and the possible orbital angular momenta, with their degeneracies, are given above each level. As the function $\phi(E^2, \Pi_s^2)$ is invariant under permutation of the three-particles the irreps of $S(3)$, i.e., $\{3\}$, $\{21\}$, $\{111\}$, characterize the levels as discussed in Table 12.2, and are also indicated in Fig. 12.1. In the figure the irreps of $S(3)$ were indicated by square brackets as it was difficult to draw the curly ones.

TABLE 12.1. Compact notation $|L_\beta\rangle$ for orbital states of the three-particle system.

$$|00, 00, 0\rangle = |0\rangle, \quad |01, 00, 1\rangle = |1_1\rangle, \quad |00, 01, 1\rangle = |1_2\rangle,$$

$$|02, 00, 2\rangle = |2_1\rangle, \quad |00, 02, 2\rangle = |2_2\rangle, \quad |01, 01, 2\rangle = |2_3\rangle,$$

$$|10, 00, 0\rangle = |0_1\rangle, \quad |00, 10, 0\rangle = |0_2\rangle, \quad |01, 01, 0\rangle = |0_3\rangle,$$

$$|01, 01, 1\rangle = |1\rangle.$$

Once we have the shorthand notation $|L_\beta\rangle$ for the states mentioned in the previous paragraph, we can follow the procedure of Section 17, of reference [3], to characterize the states by irreducible representations of $S(3)$ and, when necessary, by the Yamanouchi symbol r.

3   An Approach to the Spinorial Relativistic Three-Body System

Can the procedure developed in Section 2 be extended to spinorial particles and allow us to discuss the three-body relativistic spinorial system? The answer is affirmative and actually allows us to use many of the results of Section 2, so that we shall proceed to discuss it here.

To begin with we consider three noninteracting spinorial particles whose Hamiltonian will then be the sum of their corresponding Dirac equations. Thus the equation corresponding to Eq. (2.1) for the three-particle system becomes

$$\left[(\alpha_1 \cdot \mathbf{p}_1 + \beta_1) + (\alpha_2 \cdot \mathbf{p}_2 + \beta_2) + (\alpha_3 \cdot \mathbf{p}_3 + \beta_3)\right]\psi = E\psi, \qquad (3.1)$$

or

$$\left[(\alpha_1 \cdot \mathbf{p}_1 + \beta_1) + (\alpha_2 \cdot \mathbf{p}_2 + \beta_2)\right]\psi = \left[E - (\alpha_3 \cdot \mathbf{p}_3 + \beta_3)\right]\psi. \qquad (3.2)$$

Using $\widehat{\Pi}_s^2 \equiv p_s^2 + 1$, $s = 1, 2, 3$ and squaring both sides we get

$$\left[\widehat{\Pi}_1^2 + \widehat{\Pi}_2^2 + 2(\alpha_1 \cdot \mathbf{p}_1 + \beta_1)(\alpha_2 \cdot \mathbf{p}_2 + \beta_2)\right]\psi$$
$$= \left[E^2 + \widehat{\Pi}_3^2 - 2(\alpha_3 \cdot \mathbf{p}_3 + \beta_3)E\right]\psi, \qquad (3.3)$$

or

$$2\left[(\alpha_1 \cdot \mathbf{p}_1 + \beta_1)(\alpha_2 \cdot \mathbf{p}_2 + \beta_2) + E(\alpha_3 \cdot \mathbf{p}_3 + \beta_3)\right]\psi$$
$$= \left[\widehat{\Pi}_1^2 + \widehat{\Pi}_2^2 - \widehat{\Pi}_3^2 - E^2\right]\psi. \qquad (3.4)$$

We square again and get

$$4\left[\widehat{\Pi}_1^2\widehat{\Pi}_2^2 + E^2\widehat{\Pi}_3^2 + 2E(\alpha_1 \cdot \mathbf{p}_1 + \beta_1)(\alpha_2 \cdot \mathbf{p}_2 + \beta_2)(\alpha_3 \cdot \mathbf{p}_3 + \beta_3)\right]\psi$$
$$= (\widehat{\Pi}_1^2 + \widehat{\Pi}_2^2 - \widehat{\Pi}_3^2 - E^2)^2\psi. \qquad (3.5)$$

Finally, passing $4(\widehat{\Pi}_1^2\widehat{\Pi}_2^2 + E^2\widehat{\Pi}_3^2)$ to the right and squaring we obtain

$$64E^2\widehat{\Pi}_1^2\widehat{\Pi}_2^2\widehat{\Pi}_3^2\psi = \left[(\widehat{\Pi}_1^2 + \widehat{\Pi}_2^2 - \widehat{\Pi}_3^2 - E^2)^2 - 4\widehat{\Pi}_1^2\widehat{\Pi}_2^2 - 4E^2\widehat{\Pi}_3^2\right]^2\psi, \qquad (3.6)$$

and developing this we get back Eq. (2.3),

$$\left\{E^8 - 4AE^6 + (4A^2 + 2B)E^4 - (4C^2 + 4AB)E^2 + B^2\right\}\psi = 0, \qquad (3.7)$$

TABLE 12.2. Orbital states of the three-particle system adapted to the permutational symmetry $S(3)$ as function of $|L_\beta\rangle$.

Symmetry adapted orbital states $|NLfr\rangle$	Explicit expression of the states in terms of the $|L_\beta\rangle$ in Table 5.1
$|00\{3\}\rangle$	$|0\rangle$
$|11\{21\}1\rangle$	$|1_2\rangle$
$|11\{21\}2\rangle$	$|1_1\rangle$
$|22\{3\}\rangle$	$\left(\frac{1}{\sqrt{2}}\right)\left[|2_2\rangle + |2_1\rangle\right]$
$|22\{21\}1\rangle$	$\left(\frac{1}{\sqrt{2}}\right)\left[|2_2\rangle - |2_1\rangle\right]$
$|22\{21\}2\rangle$	$|2_3\rangle$
$|20\{3\}\rangle$	$\left(\frac{1}{\sqrt{2}}\right)\left[|0_1\rangle + |0_2\rangle\right]$
$|20\{21\}1\rangle$	$\left(\frac{1}{\sqrt{2}}\right)\left[|0_2\rangle - |0_1\rangle\right]$
$|20\{21\}2\rangle$	$-|0_3\rangle$
$|21\{111\}\rangle$	$|1\rangle$

which we had before, with A, B, C^2 having the same meaning as in Eq. (2.4). We have thus arrived, as we expect, at exactly the same equation (2.6) whch we had in the scalar problem for the case of three free particles.

From there on our analysis for introducing the oscillator interaction is exactly the same as the one carried out in Section 2 and the procedure for finding the eigenvalues of the square of the energy is also identical.

Orbital wave functions for the scalar case are derived in Table 12.2 and, in fact, they can be easily shown to be also eigenfunctions of $\widehat{F}^2$ of Eq. (2.19) by applying to them this operator. Thus the wave function of $|NLfr\rangle$ of Table 12.2 carries a $\mathcal{F}$ value according to the following correspondence,

$$|00\{3\}\rangle \to 0, \quad |11\{21\}r\rangle \to \frac{1}{2}, \quad |2L\{3\}\rangle \to 1,$$
$$|2L\{21\}r\rangle \to 1, \quad |21\{111\}\rangle \to 0, \quad (3.8)$$

where when L appears it has the values 0 or 2, and the states are given in order of increasing value of E^2, as shown in Fig. 12.1.

We shall be concerned with the first three states in Eq. (3.8) whose separation becomes larger as ω increases. The corresponding levels apply also to the spinorial eigenstates of Table 12.3, where the final irrep of $S(3)$

can take the values $\{3\}$, $\{21\}$, $\{111\}$, associated with the decuplet, octet, and singlet baryons as discussed in reference [1].

Starting then with the octet, we consider the states $|NLf', SJ\{21\}fr\rangle$ indicated on the right-hand side of Table 12.3, and look only for those whose orbital part corresponds to the first three states in the sequence (3.8). Thus, disregarding the Yamanouchi symbol r, there is only one state with $N = 0$,

$$|00\{3\}, \tfrac{1}{2}\tfrac{1}{2}\{21\}\rangle; \tag{3.9}$$

five states with $N = 1$,

$$|11\{21\}, SJ\{21\}\rangle, \tag{3.10}$$

where

$$(S, J) = \left(\frac{3}{2}, \frac{1}{2}\right), \left(\frac{1}{2}, \frac{1}{2}\right), \left(\frac{3}{2}, \frac{3}{2}\right), \left(\frac{1}{2}, \frac{3}{2}\right), \left(\frac{3}{2}, \frac{5}{2}\right); \tag{3.11}$$

while for $N = 2$ we have three states,

$$|2L\{3\}, SJ\{21\}\rangle, \tag{3.12}$$

with

$$(L, S, J) = \left(0, \frac{1}{2}, \frac{1}{2}\right), \left(2, \frac{1}{2}, \frac{3}{2}\right), \left(2, \frac{1}{2}, \frac{5}{2}\right). \tag{3.13}$$

TABLE 12.3: Orbital-spin states for the three quark system adapted to the permutational symmetry $S(3)$.

In the ket N stands for a number of quanta, L for the orbital angular momentum, f' for the irrep of $S(3)$ for the orbital part, S for the total spin and $\mathcal{S}$ for the spin of the first two-particles, J for the total angular momentum, while f is the irrep of $S(3)$ for the whole state, and r the Yamanouchi symbol given explicitly only for $f = \{21\}$ in a contracted notation. The orbital states are given in the shorthand notation of Table 12.1. The round kets on the right-hand side of the table are a shorthand notation for the kets given below.

$$|n_1\ell_1, n_2\ell_2(L); \tfrac{1}{2}\tfrac{1}{2}(T)\tfrac{1}{2}(S); JM\rangle$$
$$= [[(\dot{x}_1 \mid n_1\ell_1) \times (\dot{x}_2 \mid n_2\ell_2)]_L \times [[(1 \mid \tfrac{1}{2}) \times (2 \mid \tfrac{1}{2})]_S \times (3 \mid \tfrac{1}{2})]_S]_{JM},$$

N	J	f	r	$|NLf', S, J, fr\rangle = |L_\beta, \mathcal{S}S, J)$
0	$\frac{3}{2}$	$\{3\}$		$\|00\{3\}, \frac{3}{2}, \frac{3}{2}\{3\}\rangle = \|0, 1\frac{3}{2}, \frac{3}{2})$

N	J	f	r	
0	$\frac{1}{2}$	$\{21\}$	1	$\|00\{3\}, \frac{1}{2}, \frac{1}{2}\{21\}1\rangle = \|0, 1\frac{1}{2}, \frac{1}{2})$
0	$\frac{1}{2}$	$\{21\}$	2	$\|00\{3\}, \frac{1}{2}, \frac{1}{2}\{21\}2\rangle = \|0, 0\frac{1}{2}, \frac{1}{2})$

TABLE 12.3. (continued)

N	J	f	r	$\|NLf', S, J, fr\rangle = \|L_\beta, SS, J)$
1	$\frac{5}{2}$	$\{21\}$	1	$\|11\{21\}, \frac{3}{2}, \frac{5}{2}\{21\}1\rangle = \|1_2, 1\frac{3}{2}, \frac{5}{2})$
1	$\frac{5}{2}$	$\{21\}$	2	$\|11\{21\}, \frac{3}{2}, \frac{5}{2}\{21\}2\rangle = \|1_1, 1\frac{3}{2}, \frac{5}{2})$

N	J	f	r	
1	$\frac{3}{2}$	$\{3\}$		$\|11\{21\}, \frac{1}{2}, \frac{3}{2}\{3\}\rangle = \frac{1}{\sqrt{2}}\big[\|1_2, 1\frac{1}{2}, \frac{1}{2}) + \|1_1, 0\frac{1}{2}, \frac{3}{2})\big]$
1	$\frac{3}{2}$	$\{21\}$	1	$\|11\{21\}, \frac{3}{2}, \frac{3}{2}\{21\}1\rangle = \|1_2, 1\frac{3}{2}, \frac{3}{2})$
1	$\frac{3}{2}$	$\{21\}$	1	$\|11\{21\}, \frac{1}{2}, \frac{3}{2}\{21\}1\rangle = \frac{1}{\sqrt{2}}\big[\|1_2, 1\frac{1}{2}, \frac{3}{2}) - \|1_1, 0\frac{1}{2}, \frac{3}{2})\big]$
1	$\frac{3}{2}$	$\{21\}$	2	$\|11\{21\}, \frac{3}{2}, \frac{3}{2}\{21\}2\rangle = \|1_1, 1\frac{3}{2}, \frac{3}{2})$
1	$\frac{3}{2}$	$\{21\}$	2	$\|11\{21\}, \frac{1}{2}, \frac{3}{2}\{21\}2\rangle = \frac{1}{\sqrt{2}}\big[\|1_2, 0\frac{1}{2}, \frac{3}{2}) + \|1_1, 1\frac{1}{2}, \frac{3}{2})\big]$
1	$\frac{3}{2}$	$\{111\}$		$\|11\{21\}, \frac{1}{2}, \frac{3}{2}\{111\}\rangle = \frac{1}{\sqrt{2}}\big[\|1_2, 0\frac{1}{2}, \frac{3}{2}) - \|1_1, 1\frac{1}{2}, \frac{3}{2})\big]$

N	J	f	r	
1	$\frac{1}{2}$	$\{3\}$		$\|11\{21\}, \frac{1}{2}, \frac{1}{2}\{3\}\rangle = \frac{1}{\sqrt{2}}\big[\|1_2, 1\frac{1}{2}, \frac{1}{2}) + \|1_1, 0\frac{1}{2}, \frac{1}{2})\big]$
1	$\frac{1}{2}$	$\{21\}$	1	$\|11\{21\}, \frac{3}{2}, \frac{1}{2}\{21\}1\rangle = \|1_2, 1\frac{3}{2}, \frac{1}{2})$
1	$\frac{1}{2}$	$\{21\}$	1	$\|11\{21\}, \frac{1}{2}, \frac{1}{2}\{21\}1\rangle = \frac{1}{\sqrt{2}}\big[\|1_2, 1\frac{1}{2}, \frac{1}{2}) - \|1_1, 0\frac{1}{2}, \frac{1}{2})\big]$
1	$\frac{1}{2}$	$\{21\}$	2	$\|11\{21\}, \frac{3}{2}, \frac{1}{2}\{21\}2\rangle = \|1_1, 1\frac{3}{2}, \frac{1}{2})$
1	$\frac{1}{2}$	$\{21\}$	2	$\|11\{21\}, \frac{1}{2}, \frac{1}{2}\{21\}2\rangle = -\frac{1}{\sqrt{2}}\big[\|1_2, 0\frac{1}{2}, \frac{1}{2}) + \|1_1, 1\frac{1}{2}, \frac{1}{2})\big]$
1	$\frac{1}{2}$	$\{111\}$		$\|11\{21\}, \frac{1}{2}, \frac{1}{2}\{111\}\rangle = \frac{1}{\sqrt{2}}\big[\|1_2, 0\frac{1}{2}, \frac{1}{2}) - \|1_1, 1\frac{1}{2}, \frac{1}{2})\big]$

N	J	f	r	
2	$\frac{7}{2}$	$\{3\}$		$\|22\{3\}, \frac{3}{2}, \frac{7}{2}\{3\}\rangle = \frac{1}{\sqrt{2}}\big[\|2_2, 1\frac{3}{2}, \frac{7}{2}) + \|2_1, 1\frac{3}{2}, \frac{7}{2})\big]$
2	$\frac{7}{2}$	$\{21\}$	1	$\|22\{21\}, \frac{3}{2}, \frac{7}{2}\{21\}1\rangle = \frac{1}{\sqrt{2}}\big[\|2_2, 1\frac{3}{2}, \frac{7}{2}) - \|2_1, 1\frac{3}{2}, \frac{7}{2})\big]$
2	$\frac{7}{2}$	$\{21\}$	2	$\|22\{21\}, \frac{3}{2}, \frac{7}{2}\{21\}2\rangle = -\|2_3, 1\frac{3}{2}, \frac{7}{2})$

N	J	f	r	
2	$\frac{5}{2}$	$\{3\}$		$\|22\{3\}, \frac{3}{2}, \frac{5}{2}\{3\}\rangle = \frac{1}{\sqrt{2}}\big[\|2_2, 1\frac{3}{2}, \frac{5}{2}) + \|2_1, 1\frac{3}{2}, \frac{5}{2})\big]$
2	$\frac{5}{2}$	$\{3\}$		$\|22\{21\}, \frac{1}{2}, \frac{5}{2}\{3\}\rangle$ $= \frac{1}{2}\big[\|2_2, 1\frac{1}{2}, \frac{5}{2}) - \|2_1, 1\frac{1}{2}, \frac{5}{2})\big] - \frac{1}{\sqrt{2}}\|2_3, 0\frac{1}{2}, \frac{5}{2})$
2	$\frac{5}{2}$	$\{21\}$	1	$\|22\{21\}, \frac{3}{2}, \frac{5}{2}\{21\}1\rangle = \frac{1}{\sqrt{2}}\big[\|2_2, 1\frac{3}{2}, \frac{5}{2}) - \|2_1, 1\frac{3}{2}, \frac{5}{2})\big]$
2	$\frac{5}{2}$	$\{21\}$	1	$\|22\{3\}, \frac{1}{2}, \frac{5}{2}\{21\}1\rangle = \frac{1}{\sqrt{2}}\big[\|2_2, 1\frac{1}{2}, \frac{5}{2}) + \|2_1, 1\frac{1}{2}, \frac{5}{2})\big]$
2	$\frac{5}{2}$	$\{21\}$	1	$\|22\{21\}, \frac{1}{2}, \frac{5}{2}\{21\}1\rangle = \frac{1}{2}\big[\|2_2, 1\frac{1}{2}, \frac{5}{2}) - \|2_1, 1\frac{1}{2}, \frac{5}{2})\big] + \frac{1}{\sqrt{2}}\|2_3, 0\frac{1}{2}, \frac{5}{2})$
2	$\frac{5}{2}$	$\{21\}$	2	$\|22\{21\}, \frac{3}{2}, \frac{5}{2}\{21\}2\rangle = -\|2_3, 1\frac{3}{2}, \frac{5}{2})$
2	$\frac{5}{2}$	$\{21\}$	2	$\|22\{3\}, \frac{1}{2}, \frac{5}{2}\{21\}2\rangle = \frac{1}{\sqrt{2}}\big[\|2_2, 0\frac{1}{2}, \frac{5}{2}) + \|2_1, 0\frac{1}{2}, \frac{5}{2})\big]$
2	$\frac{5}{2}$	$\{21\}$	2	$\|22\{21\}, \frac{1}{2}, \frac{5}{2}\{21\}2\rangle = -\frac{1}{2}\big[\|2_2, 0\frac{1}{2}, \frac{5}{2}) - \|2_1, 0\frac{1}{2}, \frac{5}{2})\big] + \frac{1}{\sqrt{2}}\|2_3, 1\frac{1}{2}, \frac{5}{2})$

TABLE 12.3. (continued)

N	J	f	r	$|NLf',S,J,fr\rangle = |L_\beta, SS, J)$
2	$\frac{5}{2}$	$\{111\}$		$|21\{111\},\frac{3}{2},\frac{5}{2}\{111\}\rangle = |1,1\,\frac{3}{2},\frac{5}{2})$
2	$\frac{5}{2}$	$\{111\}$		$|22\{21\},\frac{1}{2},\frac{5}{2}\{111\}\rangle = \frac{1}{2}\big[|2_2,0\,\frac{1}{2},\frac{5}{2}) - |2_1,0\,\frac{1}{2},\frac{5}{2})\big] + \frac{1}{\sqrt{2}}|2_3,1\,\frac{1}{2},\frac{5}{2})$

N	J	f	r	
2	$\frac{3}{2}$	$\{3\}$		$|20\{21\},\frac{3}{2},\frac{3}{2}\{3\}\rangle = \frac{1}{\sqrt{2}}\big[|0_2,1\,\frac{3}{2},\frac{3}{2}) + |0_1,1\,\frac{3}{2},\frac{3}{2})\big]$
2	$\frac{3}{2}$	$\{3\}$		$|22\{3\},\frac{3}{2},\frac{3}{2}\{3\}\rangle = \frac{1}{\sqrt{2}}\big[|2_2,1\,\frac{3}{2},\frac{3}{2}) + |2_1,1\,\frac{3}{2},\frac{3}{2})\big]$
2	$\frac{3}{2}$	$\{3\}$		$|22\{21\},\frac{1}{2},\frac{3}{2}\{3\}\rangle = \frac{1}{2}\big[|2_2,1\,\frac{1}{2},\frac{3}{2}) - |2_1,1\,\frac{1}{2},\frac{3}{2})\big] - \frac{1}{\sqrt{2}}|2_3,0\,\frac{1}{2},\frac{3}{2})$
2	$\frac{3}{2}$	$\{21\}$	1	$|22\{21\},\frac{3}{2},\frac{3}{2}\{21\}1\rangle = \frac{1}{\sqrt{2}}\big[|2_2,1\,\frac{3}{2},\frac{3}{2}) - |2_1,1\,\frac{3}{2},\frac{3}{2})\big]$
2	$\frac{3}{2}$	$\{21\}$	1	$|20\{21\},\frac{3}{2},\frac{3}{2}\{21\}1\rangle = \frac{1}{\sqrt{2}}\big[|0_2,1\,\frac{3}{2},\frac{3}{2}) - |0_1,1\,\frac{3}{2},\frac{3}{2})\big]$
2	$\frac{3}{2}$	$\{21\}$	1	$|22\{3\},\frac{1}{2},\frac{3}{2}\{21\}1\rangle = \frac{1}{\sqrt{2}}\big[|2_2,1\,\frac{1}{2},\frac{3}{2}) + |2_1,1\,\frac{1}{2},\frac{3}{2})\big]$
2	$\frac{3}{2}$	$\{21\}$	1	$|21\{111\},\frac{1}{2},\frac{3}{2}\{21\}1\rangle = |1,0\,\frac{1}{2},\frac{3}{2})$
2	$\frac{3}{2}$	$\{21\}$	1	$|22\{21\},\frac{1}{2},\frac{3}{2}\{21\}1\rangle = \frac{1}{2}\big[|2_2,1\,\frac{1}{2},\frac{3}{2}) - |2_1,1\,\frac{1}{2},\frac{3}{2})\big] + \frac{1}{\sqrt{2}}|2_3,0\,\frac{1}{2},\frac{3}{2})$
2	$\frac{3}{2}$	$\{21\}$	2	$|22\{21\},\frac{3}{2},\frac{3}{2}\{21\}2\rangle = |2_3,1\,\frac{3}{2},\frac{3}{2})$
2	$\frac{3}{2}$	$\{21\}$	2	$|20\{21\},\frac{3}{2},\frac{3}{2}\{21\}2\rangle = -|0_3,1\,\frac{3}{2},\frac{3}{2})$
2	$\frac{3}{2}$	$\{21\}$	2	$|22\{3\},\frac{1}{2},\frac{3}{2}\{21\}2\rangle = \frac{1}{\sqrt{2}}\big[|2_2,0\,\frac{1}{2},\frac{3}{2}) + |2_1,0\,\frac{1}{2},\frac{3}{2})\big]$
2	$\frac{3}{2}$	$\{21\}$	2	$|21\{111\},\frac{1}{2},\frac{3}{2}\{21\}2\rangle = |1,1\,\frac{1}{2},\frac{3}{2})$
2	$\frac{3}{2}$	$\{21\}$	2	$|22\{21\},\frac{1}{2},\frac{3}{2}\{21\}2\rangle = \frac{1}{2}[-|2_2,0\,\frac{1}{2},\frac{3}{2})) + |2_1,0\,\frac{1}{2},\frac{3}{2})] + \frac{1}{\sqrt{2}}|2_3,1\,\frac{1}{2},\frac{3}{2})$
2	$\frac{3}{2}$	$\{111\}$		$|21\{111\},\frac{3}{2},\frac{3}{2}\{111\}\rangle = |1,1\,\frac{3}{2},\frac{3}{2})$
2	$\frac{3}{2}$	$\{111\}$		$|22\{21\},\frac{1}{2},\frac{3}{2}\{111\}\rangle = \frac{1}{2}\big[|2_2,0\,\frac{1}{2},\frac{3}{2}) - |2_1,0\,\frac{1}{2},\frac{3}{2})\big] + \frac{1}{\sqrt{2}}|2_3,1\,\frac{1}{2},\frac{3}{2})$

N	J	f	r	
2	$\frac{1}{2}$	$\{3\}$		$|22\{3\},\frac{3}{2},\frac{1}{2}\{3\}\rangle = \frac{1}{\sqrt{2}}\big[|2_2,1\,\frac{3}{2},\frac{1}{2}) + |2_1,1\,\frac{3}{2},\frac{1}{2})\big]$
2	$\frac{1}{2}$	$\{3\}$		$|20\{21\},\frac{1}{2},\frac{1}{2}\{3\}\rangle = \frac{1}{2}\big[|0_2,1\,\frac{1}{2},\frac{1}{2}) - |0_1,1\,\frac{1}{2},\frac{1}{2})\big] - \frac{1}{\sqrt{2}}|0_3,0\,\frac{1}{2},\frac{1}{2})$
2	$\frac{1}{2}$	$\{21\}$	1	$|22\{21\},\frac{3}{2},\frac{1}{2}\{21\}1\rangle = \frac{1}{\sqrt{2}}\big[|2_2,1\,\frac{3}{2},\frac{1}{2}) - |2_1,1\,\frac{3}{2},\frac{1}{2})\big]$
2	$\frac{1}{2}$	$\{21\}$	1	$|20\{3\},\frac{1}{2},\frac{1}{2}\{21\}1\rangle = \frac{1}{\sqrt{2}}\big[|0_2,1\,\frac{1}{2},\frac{1}{2}) + |0_1,1\,\frac{1}{2},\frac{1}{2})\big]$
2	$\frac{1}{2}$	$\{21\}$	1	$|21\{111\},\frac{1}{2},\frac{1}{2}\{21\}1\rangle = -|1,0\,\frac{1}{2},\frac{1}{2})$
2	$\frac{1}{2}$	$\{21\}$	1	$|20\{21\},\frac{1}{2},\frac{1}{2}\{21\}1\rangle = \frac{1}{2}\big[|0_2,1\,\frac{1}{2},\frac{1}{2}) - |0_1,1\,\frac{1}{2},\frac{1}{2})\big] + \frac{1}{\sqrt{2}}|0_3,0\,\frac{1}{2},\frac{1}{2})$
2	$\frac{1}{2}$	$\{21\}$	2	$|22\{21\},\frac{3}{2},\frac{1}{2}\{21\}2\rangle = |2_3,1\,\frac{3}{2},\frac{1}{2})$
2	$\frac{1}{2}$	$\{21\}$	2	$|20\{3\},\frac{1}{2},\frac{1}{2}\{21\}2\rangle = \frac{1}{\sqrt{2}}\big[|0_2,0\,\frac{1}{2},\frac{1}{2}) + |0_1,0\,\frac{1}{2},\frac{1}{2})\big]$
2	$\frac{1}{2}$	$\{21\}$	2	$|21\{111\},\frac{1}{2},\frac{1}{2}\{21\}2\rangle = |1,1\,\frac{1}{2},\frac{1}{2})$
2	$\frac{1}{2}$	$\{21\}$	2	$|20\{21\},\frac{1}{2},\frac{1}{2}\{21\}2\rangle = \frac{1}{2}[-|0_2,0\,\frac{1}{2},\frac{1}{2}) + |0_1,0\,\frac{1}{2},\frac{1}{2})] + \frac{1}{\sqrt{2}}|0_3,1\,\frac{1}{2},\frac{1}{2})$
2	$\frac{1}{2}$	$\{111\}$		$|21\{111\},\frac{3}{2},\frac{1}{2}\{111\}\rangle = |1,1\,\frac{3}{2},\frac{1}{2})]$
2	$\frac{1}{2}$	$\{111\}$		$|20\{21\},\frac{1}{2},\frac{1}{2}\{111\}\rangle = \frac{1}{2}\big[|0_2,0\,\frac{1}{2},\frac{1}{2})] - |0_1,0\,\frac{1}{2},\frac{1}{2})\big] + \frac{1}{\sqrt{2}}|0_3,1\,\frac{1}{2},\frac{1}{2})$

Looking then at Fig. 12.1 we see that in the octet spinorial picture we still have a single ground level, but that the next one is quintuple degenerate and the last one we consider is triple degenerate. As in Ref. [3, Eq. (70.9a)] we can multiply the energies obtained in the scalar picture by (939/3) to get the masses in MeV, and choose the frequency ω appropriately so as to get the best fit to the masses of nonstrange octet baryons, which turned out to be $\omega = 3.33$.

In Fig. 12.2 we draw the levels for the octet in our theoretical approach by solid lines. The experimental values are taken from Ref. [1] and they appear as dashed or dotted lines depending on whether they are identified in Ref. [1] as 1 or 2 quanta states. For the ground state (i.e., the nucleon) we only indicate our theoretical result as, by construction, it coincides with the experimental one. Note that the fit is reasonable except for one state to which we appended an interrogation sign.

Turning our attention to the decuplet, we consider the state $|NLf', SJ\{3\}\rangle$ and look only for those whose orbital part corresponds to the first three states in the sequence (3.8). Thus there is only one state with $N = 0$,

$$|00\{3\}, \frac{3}{2}\frac{3}{2}\{3\}\rangle, \tag{3.14}$$

two states with $N = 1$,

$$|11\{21\}, SJ\{3\}\rangle, \tag{3.15}$$

where

$$(S, J) = \left(\tfrac{1}{2}, \tfrac{1}{2}\right), \left(\tfrac{1}{2}, \tfrac{3}{2}\right); \tag{3.16}$$

while for $N = 2$ we have three states,

$$|22\{3\}, SJ\{3\}\rangle, \tag{3.17}$$

with

$$(S, J) = \left(\tfrac{3}{2}, \tfrac{1}{2}\right), \left(\tfrac{3}{2}, \tfrac{3}{2}\right), \left(\tfrac{3}{2}, \tfrac{5}{2}\right). \tag{3.18}$$

Looking then again at Fig. 12.1, we see that in the decuplet spinorial picture we have a single ground state, but the next is doubly degenerate and the last one we consider is triply degenerate. As in Ref. [3, Eq. (70.9b)] we can multiply the energies obtained in the scalar picture by (1232/3) to get the masses in MeV, and choose a frequency ω so as to get a best fit to the masses of the decuplet baryons, which turned out to be $\omega = 1.32$.

In Fig. 12.3 we draw the levels for the decuplet in our theoretical approach by solid lines. The experimental ones are taken from Ref. [1] as dashed or dotted lines depending on whether they are identified in [1] as 1 or 2 quanta states. For the ground state we only indicate our theoretical result as, by construction, it coincides with the experimental one. In the decuplet case the fit is better than for the octet and all the experimental masses have a theoretical counterpart.

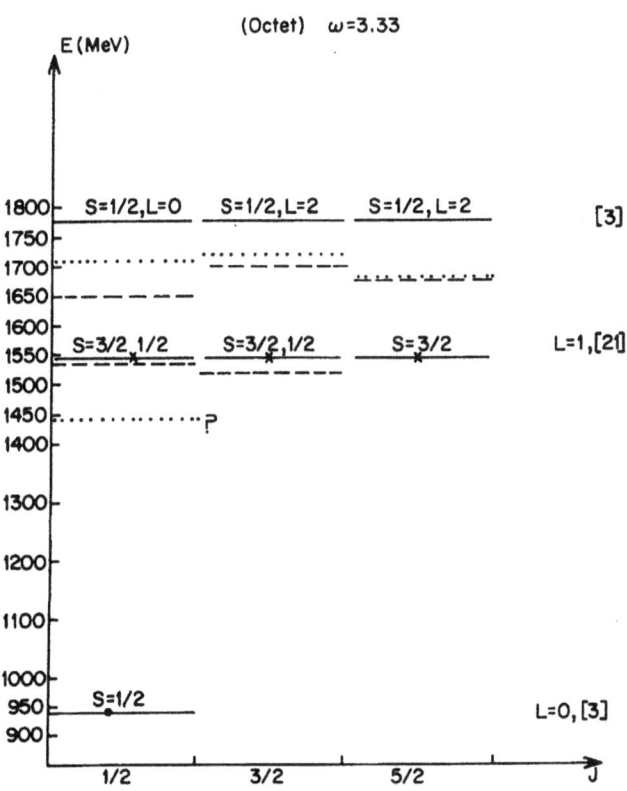

FIGURE 12.2. The experimental nonstrange octet energy levels (the masses) are indicated by dashed or dotted lines depending on whether they correspond to in Ref. [1] 1 or 2 quanta. By construction the ground state of the octet coincides with our theoretical result and thus is marked by a full line as do all our other theoretical levels. Our analysis is based on the formalism developed in Section 2, and Table 12.3. The first and second excited theoretical levels are, respectively, quintuply and triply degenerate but they coincide quite reasonably with experiment. In the abscissa we give the total angular momentum J, while the orbital angular momentum L and spin S are indicated either above or to the right of the theoretical levels. Also to the right irreps of $S(3)$ for the orbital part of the state are indicated. In our units $\hbar = m = c = 1$, with m being the mass of the u or d quarks, the frequency ω we use is, as indicated above, $\omega = 3.33$. Only one state, denoted by a question mark (?), does not seem to have a theoretical counterpart.

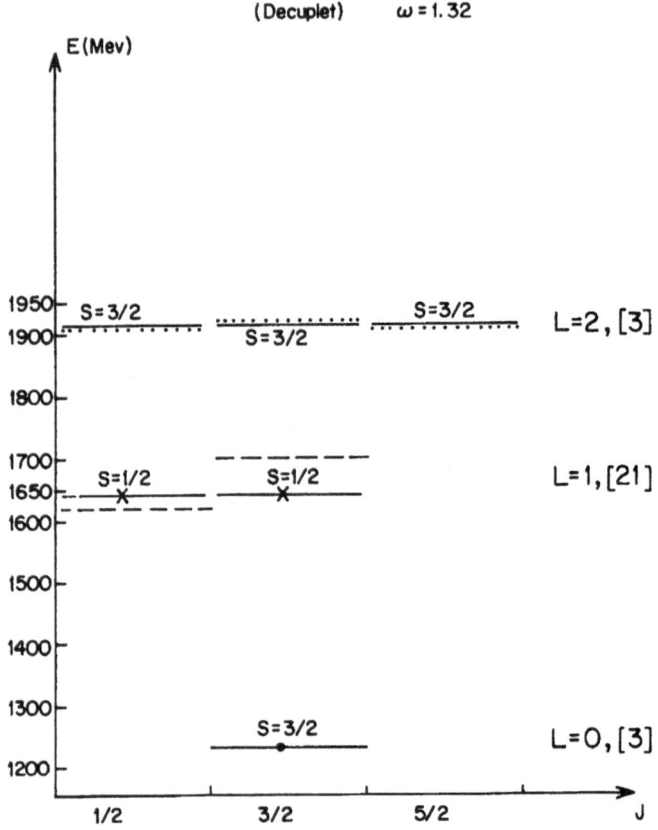

FIGURE 12.3. Most of what is mentioned in Fig. 12.2 holds for Fig. 12.3 if we replace the word *octet* by *decuplet*. We note though that now the first and second excited theoretical levels are, respectively, doubly and triply degenerate, and all lowest experimental levels seem to be fitted with $\omega = 1.32$.

We conclude by stressing that we have made a calculation using an harmonic oscillator picture with a single parameter, the frequency ω, and it is as good or as bad as many more complicated ones that start from QCD or that use many more parameters.

REFERENCES

1. Y.S. Kim and M.E. Noz, *Theory and applications of the Poincaré group*, D. Reidell Publishing Co., Boston, 1986, pp. 276–277.

2. L. I. Schiff, H. Snyder, and J. Weinberg , *On the existence of stationary states in the mesotron field*, Phys. Rev. **57** (1940), 315–318.

3. M. Moshinsky and Yu F. Smirnov, *The harmonic oscillator in modern physics*, Hardwood Academic Publishers, The Netherlands, 1996 (p. 368, formula (67.7)).

4. M. Moshinsky, *Bases for irreducible representations of the unitary groups and some applications*, J. Math. Phys. **4** (1963), 1128–1139.

5. W. Pauli and V. Weisskopf, *Uber die quantisierung der skalaren relativistichen Wellengleichung*, Helv. Phys. Acta **7** (1934), 709–731.

6. L.I. Schiff, *Quantum mechanics*, McGraw-Hill, New York, 1949, pp. 311–318.

13

Seiberg-Witten Theory Without Tears

L. O'Raifeartaigh

1 Introduction

The widespread use of the work of Jiři Patera and Paul Winternitz is evidence not only of its own intrinsic and lasting merit, of which we are all aware, but also supports the view that symmetry principles are universal in physics. Accordingly, it may not be out of place to record here yet another, quite recent, success of symmetry principles. The success I have in mind is the recent work of Seiberg and Witten [6] in which they use a combination of symmetry principles, namely, $N = 2$ supersymmetry, non-Abelian gauge-symmetry and electromagnetic duality, to connect the strong and weak coupling regimes of quantum field theory (QFT). It is true that they do this only for a specific model, indeed only for its massless (unbroken) part, but the fact that they obtain for the first time a nontrivial nonperturbative result for the strong coupling part of QFT represents a very important advance. The fact that their results are obtained almost entirely by the use of symmetry principles is, at the same time, a great triumph for the symmetry principle approach. The purpose of this talk is to give a resume of the SW results in the simplest possible mathematical terms.

2 $N = 2$ Supersymmetry

We begin by recalling the essentials [7] of the $N = 2$ supersymmetry algebra and its action. The algebra is

$$\{Q_\alpha^i, \overline{Q}_\beta^k\} = \delta_{ik}\sigma_{\alpha\beta}^\mu P_\mu, \quad \{Q_\alpha^i, Q_\beta^k\} = \varepsilon_{ik}\varepsilon_{\alpha\beta}Z, \tag{2.1}$$

plus the hermitian conjugate of the second relation, where $i, k = 1, 2$ and Z is a central charge. This algebra is realized on the simplest possible nontrivial supermultiplet, namely,

$$\Psi = \{\phi, \psi, A_\mu; F, D\}, \tag{2.2}$$

where ϕ is a complex scalar field, ψ is a Dirac spinor, A_μ is a gauge-field and F and D are complex and real dummy-fields, respectively. This $N = 2$ superfield may be decomposed into two $N = 1$ superfields, namely,

$$\Phi = \{\phi, q, F\} \quad \text{and} \quad V = \{A_\mu, \lambda, D\} \quad \text{or} \quad W_\alpha = \{F_{\mu\nu}, \lambda, D\}, \quad (2.3)$$

where Φ and V or W_α are chiral and vector multiplets, respectively, the q and λ fields being Weyl spinors of opposite chirality. Since A_μ belongs to the adjoint representation of the gauge group G and all the fields belong to the same multiplet, they must all belong to the adjoint representation of G. The simplest SW model is for $G = \text{SU}(2)$ and we shall concentrate on this case.

3 $N = 2$ Superaction

The superaction for the $N = 2$ superfield just described is

$$\mathcal{A} = \text{Im Tr} \int d^4x \, d^2\theta_\alpha \, d^2\bar\theta_\beta (\Psi)^2. \quad (3.1)$$

On expanding this in terms of the $N = 1$ superfields it becomes

$$\mathcal{A} = \text{Im} \int d^4x \, d^2\theta_\alpha \, d^2\bar\theta_\beta (\bar{A} e^{-2g_o V} A) + \text{Im} \int \tau_o \, d^4x \, d^2\theta_\alpha (W_\alpha W_\alpha), \quad (3.2)$$

where

$$\tau_o = \frac{\theta_o}{2\pi} + \frac{4\pi i}{g_o^2}, \quad (3.3)$$

the parameter g_o being the usual gauge-coupling constant and θ_o being the QCD-vacuum-angle (not to be confused with the usual supersymmetric Grassman variables). The exponential in the first term is just the supersymmetric generalization of the covariant derivative. Expanding (3.2) further in terms of conventional fields we obtain

$$\mathcal{A} = \text{tr} \int d^4x \left\{ \frac{1}{2}(\phi^\dagger D^2 \phi) + \bar\psi D\psi + g_o(\phi[\bar\lambda, q] + \phi^\dagger[\bar{q}, \lambda]) + g_o^2[\phi^\dagger, \phi]^2 \right\}$$

$$+ \text{tr} \int d^4x \left\{ \frac{1}{4g_o^2} F^{\mu\nu} F_{\mu\nu} + \frac{\theta_o}{32} \tilde{F}^{\mu\nu} F_{\mu\nu} \right\}. \quad (3.4)$$

This action will be immediately recognized as the standard action for a Quark–Gluon–Higgs system in which all the fields are in the adjoint representation and the coupling constants are reduced to g and θ by the supersymmetry. Thus it is not very exotic.

4   Textbook Properties

The action (3.4) is actually so normal that it embodies all the properties of quantum gauge theory that have surfaced over the past 30 years and could even be used as a model to teach quantum gauge theory. It might be worthwhile to list these properties:

(1) It contains a gauge-field coupled to matter.

(2) It is asymptotically free.

(3) It is scale-invariant, but with a scale-anomaly.

(4) It has spontaneous symmetry breaking.

(5) It has central charges (Z and $\bar{Z}$).

(6) It admits both instantons and monopoles.

Because of the supersymmetry it has some further special properties, whose significance will become clear later, namely—

(7) It generalizes the Montonon–Olive mass formula [5] for gauge-fields and monopoles from

$$ M = |v|\left(N_e + \frac{1}{g^2} n_m \right) \text{ to } M = |Z|, \qquad (4.1) $$

where $Z = (an_e + a_d n_m)$ and where n_e and n_m denote the gauge-field and monopole charges respectively, and the coefficients a and a_d will be explained later.

(8) It is symmetric with respect to a Z_4 symmetry that is the relic of the R-symmetry ($\theta_\alpha \to e^{i\varepsilon}\theta_\alpha$) that survives the axial anomaly breakdown.

(9) It has a holomorphic structure.

(10) It has a duality that connects the weak and strong coupling regimes.

(11) The duality generalizes to an $\mathrm{SL}(2, Z)$ symmetry.

In Section 6 we explain these last three concepts in more detail.

The one property that the model does not have (or at least does not exhibit explicitly) is confinement. However, it has been argued by SW that, if the $N = 2$ supersymmetry is broken to $N = 1$ supersymmetry in a certain explicit, but natural, manner, the resulting $N = 1$ model may admit monopole condensation and, according to the usual dual Meissner-effect argument, this is tantamount to admitting confinement.

5   Spontaneous Symmetry-Breaking

For SU(2) this concept is very simple. From the form of the Higgs potential in Eq. (3.4) we see at once that there is a Higgs vacuum for $\phi = v\sigma$ where v is any complex number and σ is any fixed generator of SU(2). Furthermore, for $v \neq 0$ this breaks the gauge-symmetry from SU(2) to U(1). (For other gauge-groups G the corresponding statement is that $v\sigma$ must lie in the Cartan subalgebra of G). On the other hand there is no spontaneous breakdown of supersymmetry. Thus the full breakdown is

$$\text{SU}(2) \rightarrow \text{U}(1) \quad N = 2 \text{ supersymmetry unbroken.} \tag{5.1}$$

Indeed it is the fact that the supersymmetry is unbroken that gives the model its nice properties, since otherwise the classical properties would not be preserved after quantization.

Since the adjoint representation of U(1) is trivial, the restriction of the classical action (3.2) to the massless U(1) sector is just the free-field action

$$\mathcal{A} = \text{Im} \int d^4x \, d^2\theta_\alpha \, d^2\bar{\theta}_\beta (\bar{A}A) + \text{Im} \int \tau_o \, d^4x \, d^2\theta_\alpha (W_\alpha W_\alpha). \tag{5.2}$$

The first great virtue of the $N = 2$ supersymmetric model is that the effective Lagrangian, obtained by integrating out the massive, SU(2)/U(1), sector takes a very similar form, namely,

$$\mathcal{A} = \frac{1}{2} \int d^4x \, d^2\theta \, d^2\bar{\theta} (\bar{A}A_d) + \text{Im} \int d^4x \, d^2\theta (\tau(A))(W_\alpha W_\alpha), \tag{5.3}$$

where
$$A_d = F'(A) \quad \text{and} \quad \tau(A) = F''(A), \tag{5.4}$$

for some function $F(A)$. Thus the effective Lagrangian is completely governed by the single function $F(A)$. As we shall see, the SW solution is nothing but a special Ansatz for the functional form of $F(A)$.

6   Holomorphy and Duality

It is now easy to quantify what is meant by holomorphy and duality. Holomorphy is simply the statement that $F(A)$ depends only on A and not on $\bar{A}$. Duality means that the physics described by the effective action (5.3) is invariant with respect to the duality transformation.

$$\begin{pmatrix} A \\ A_d \end{pmatrix} \rightarrow \begin{pmatrix} 0 & 1 \\ -1 & 0 \end{pmatrix} \begin{pmatrix} A \\ A_d \end{pmatrix}, \quad \tau(A) \rightarrow (\tau(A))^{-1}. \tag{6.1}$$

Note that the duality transformation is closely linked to the Legendre transform of $F(A)$ with respect A. By noting that in the free classical theory with $\theta_o = 0$ the transformation (6.1) reduces to

$$\vec{E} \to \vec{B} \quad \text{and} \quad g \to \frac{1}{g}, \tag{6.2}$$

we see that it is just the generalization of well-known electromagnetic (Maxwell–Dirac) duality. Thus the action (5.3) produces electrodynamic duality in the context of a genuine dynamical model. Furthermore, the duality generalizes to

$$\begin{pmatrix} A \\ A_d \end{pmatrix} \to \begin{pmatrix} p & q \\ r & s \end{pmatrix} \begin{pmatrix} A \\ A_d \end{pmatrix} \quad \text{and} \quad \tau(A) \to \frac{p\tau(A) + q}{r\tau(A) + s}, \tag{6.3}$$

where

$$\begin{pmatrix} p & q \\ r & s \end{pmatrix} \in \mathrm{SL}(2, Z).$$

The integer-valuedness of the transformation follows from the requirement that, in the perturbation theory at least, the θ angle should change only by multiples of 2π and that the mass-formula (5.1) should be form-invariant.

7 Perturbative and Nonperturbative $F(A)$

Before going on to describe the SW Ansatz for $F(A)$ let us consider the perturbative contribution $F_p(A)$ of $F(A)$. This turns out to be

$$F_p(A) = A^2 + \hbar A^2 \ln(A^2/\Lambda^2), \tag{7.1}$$

where Λ is the renormalization scale. This is evidently a classical plus a one-loop expression but it is correct to all orders in perturbation. The reason for this unusual result is that the energy momentum tensor is in the same multiplet as the axial current

$$\delta\Theta_{\mu\nu} = -\frac{\bar{\varepsilon}}{4}(\sigma^{\mu\kappa}\partial_\kappa j^\mu + \sigma^{\nu\kappa}\partial_\kappa j^\mu), \tag{7.2}$$

$$\delta j_\mu^5 = i\bar{\varepsilon}\gamma_5 j_\mu, \tag{7.3}$$

$$\delta j_\mu^\alpha = \varepsilon\left(2\gamma^\nu\Theta_{\mu\nu} - i\gamma_5\gamma^\nu\partial_\nu j_\mu^5 + \frac{i}{2}\varepsilon_{\mu\nu\kappa\lambda}\gamma^\nu\partial^\kappa j_5^\lambda\right), \tag{7.4}$$

which means that the correction to the effective potential is a partner of the axial anomaly, which, in turn, is well known to be a one-loop effect. Furthermore, for $N = 2$ supersymmetry this relationship remains unchanged to all orders in perturbation.

For the nonperturbative part of $F(A)$ the only solid a priori pieces of information are

$$\mathrm{Im}(F''(A)) \geq 0, \quad F_{np}(A) \neq 0, \quad F_{np}(iA) = F_{np}(A), \tag{7.5}$$

and the fact that it vanishes for large A. The first relation comes from the convexity of the effective potential, specifically from the fact that $\mathrm{Im}(F''(A))$, is the coefficient of the kinetic term for the gauge-field; the second relation, from one-instanton computations, and the third relation, from the residue of R-invariance that is left after spontaneous symmetry-breaking. $F(A)$ has to be guessed from this apparently meager information.

The SW-Ansatz

8   Preliminaries

Seiberg and Witten (SW) begin by reducing the problem to one in complex analysis by considering only the vacuum value $A = v$ of the chiral scalar superfield and determining the functional form of $F(v)$. Afterwards, (A) can be recovered by the simple substitution $F(v) \to F(v + \tilde{A}) \equiv F(A)$. This is analogous to the substitution

$$V(m, f, g) \to V(m + f\phi + g\phi^2, f + g\phi, g) \equiv V_{\mathrm{eff}}(\phi), \qquad (8.1)$$

which is made to obtain the effective potential from the partition function $P(m, f, g)$ of a standard renormalizable theory with a single scalar field ϕ and masses m, and coupling constants f and g.

Next, they note that asymptotic freedom allows them to identify the perturbative region as the region in which $v \to \infty$ and thus obtain the boundary value

$$\tau(v) \to \frac{i}{\pi}\ln(v) \quad \text{for} \quad v \to \infty. \qquad (8.2)$$

The fact that v is expected to be singular in the small scale (strong coupling) region leads them to postulate the existence of a more universal (complex) parameter $u \in C$ normalized so that $a(u) \to u^2$ for $u \to \infty$. Assembling all this information reduces the problem to the search for a function $\tau(u)$ such that

$$\mathrm{Im}(\tau(u)) \geq 0 \quad \text{and} \quad \tau(u)_{u \to \infty} \to \frac{i}{2\pi}\ln(u), \quad \tau(v) = \tau(-v). \qquad (8.3)$$

The procedure for choosing a $\tau(u)$ to satisfy (8.3) is actually rather similar to that used by Veneziano in choosing his formula for the S-matrix $S(s, t, u)$, where s, t, and u are the invariant squares of the momenta. That is to say, instead of computing the function directly from the underlying theory, one uses its properties (symmetries, boundary conditions, etc.), to try to guess what it should be. Indeed duality (actually triality) plays here a role that is analogous to that played by crossing symmetry (symmetry in s, t, and u) in the Veneziano case. But first one has to decide the general class of functions out of which the function $\tau(u)$ should be chosen.

The standard class of functions that map the upper part C_+ of the complex plane into itself modulo subgroups of $SL(2, Z)$ is the class of *Fuchsian* functions [2] and the choice will be made out of these. So, to put the results in perspective, we digress for a moment to consider Fuchsian functions.

9   Fuchsian Maps

The Fuchsian maps $\tau(u)$ are maps from the upper half of the complex plane C_+ to subdomains D of it given by $D = C_+/\Gamma$, where Γ is a subgroup of $SL(2, R)$. In our case, because of the discreteness in the variation of the θ-angle, the restriction is to subgroups of $SL(2, Z)$. The domains D are circular polygons (i.e., polygons whose sides may be straight lines or circles) and for $\Gamma \subset SL(2, Z)$ the corners of the polygon correspond to points on the real axis in the u-variable. Because Γ contains the element $\theta \to \theta + 2\pi$, or equivalently $\tau \to \tau + 1$, one corner of polygon may be the point at positive imaginary infinity. In general the polygons may not be of genus zero because the sides may be have to be identified in a nontrivial manner. If the genus is zero, u can be extended to cover the whole complex plane, which in turn may be compactified to the Riemann sphere. Otherwise the compactification of the u-space leads to a Riemann surface of higher genus. The essential property of all these Fuchsian maps is they guarantee $\text{Im}(\tau(u)) \geq 0$.

10   The Schwarzian Derivatives

The Fuchsian functions $\tau(u)$ just described are fairly complicated but a great simplification is achieved by considering not the functions themselves but their Schwarzian derivatives

$$S(\tau) = \frac{\tau'''}{\tau'} - \frac{3}{2}\left(\frac{\tau''}{\tau'}\right)^2. \qquad (10.1)$$

In general the main property of Schwarzian derivatives is that they are invariant with respect to the modular transformations (6.3). However, in the case of Fuchsian functions, they have the added advantage that they are simple meromorphic functions of the form

$$S(\tau(u)) = \sum_{i=1}^{n}\left\{\frac{1}{2}\frac{(1 - \alpha_i^2)}{(u - a_i)^2} + \frac{\beta_i}{(u - a_i)}\right\},$$

$$S(\tau(u)) \to \frac{1}{u^2}, \quad u \to \infty, \qquad (10.2)$$

and this is why it is convenient to use $S(\tau)$ rather than τ itself. Once $S(\tau)$ is known there is a simple and elegant way to recover τ from it, namely, to write

$$\tau = \frac{y_1}{y_2} \quad \text{where} \quad \mathbf{y}'' + \frac{1}{2}S(\tau(u))\mathbf{y} = 0, \tag{10.3}$$

and boldface denotes a 2-vector constructed from any two linearly independent solutions of the equation. Such functions τ are tailor-made for the SW model where $\tau(u)$ is defined to be a quotient, namely, a_d'/a'. The parameters n, α_i, and a_i, in Eq. (10.2) have a simple geometrical meaning. The number n is the number of corners of the polygon D (excluding the point at infinity), the α_i are the internal angles at the corners and the a_i are the images of the positions of the corners in u-space. Thus the parameters n, α_i, and a_i define the polygon. The β_i parameters are subsidiary in the sense that they are functions of the n, α_i and a_i. But they have no simple geometrical interpretation and are not known explicitly. The only simple information concerning them is given by the two linear equations that follow from the fact that $S(u) \to 1/2u^2$ as $u \to \infty$. These equations require that $n \geq 2$ and determine two of the β's in terms of the others. Only for $n = 2$ do they determine the β's uniquely. It is clear from this discussion that the choice of Fuchsian function reduces to a choice of n and the corresponding parameters α_i and a_i for $i = 1, \ldots, n$ and $n \geq 2$.

11   SW Choice

The question is, which Fuchsian function to choose? The SW choice is made by adding two further inputs to the basic conditions, namely,

Minimality: $n = 2$

Duality: $M_\infty = \left(\begin{smallmatrix} 1 & 2 \\ 0 & 1 \end{smallmatrix}\right) \leftrightarrow M_1 = \left(\begin{smallmatrix} 1 & 0 \\ 2 & 1 \end{smallmatrix}\right)$

Here M_i denotes the monodromy matrix, defined as the linear transformation induced on the vector $\mathbf{y}$ in Eq. (10.3) by looping the variable u around the point at infinity or any of the poles $u = a_i$ of the Schwarzian $S(u)$. Duality in this sense means that the monodromy matrix M_1 at one of the singularities is the dual of the monodromy matrix M_∞ at infinity. It is motivated physically by the requirement that the asymptotic freedom of g for large scales be the dual of the infrared slavery of g (asymptotic freedom of g^{-1}) for low scales. It turns out that M_∞ and M_1 generate a monodromy group Γ_2 that includes the monodromy group of the second singularity, and consists of all matrices of the form

$$I + 2 \begin{pmatrix} e & f \\ g & h \end{pmatrix}, \quad e, f, g, h \in Z. \tag{11.1}$$

According to our previous analysis this determines a unique Fuchsian map (modulo the positions of the two singularities that are normalized to be

$a_i = \pm 1$) and it turns out to be the map with $\alpha_i = 0$ and $\beta_i = \pm 1/4$. Thus with SW Ansatz equation (10.3) becomes

$$\tau(u) = \frac{y_1}{y_2} \quad \text{where} \quad \mathbf{y}'' + \frac{1}{4}\left[\frac{3+u^2}{(u^2-1)^2}\right]\mathbf{y} = 0. \tag{11.2}$$

There is actually a simplification in this case on changing to the variable $\mathbf{a}$, where $\mathbf{a}' = \mathbf{y}$ because then (11.2) becomes

$$\tau = \frac{a_1'}{a_2'} \quad \text{where} \quad \mathbf{a}'' + \frac{1}{4}\left(\frac{1}{u^2-1}\right)\mathbf{a} = 0, \tag{11.3}$$

and the differential equation in Eq. (11.3) is just a hypergeometric equation. So finally $\tau(u)$ is simply the ratio of the derivatives of two simple hypergeometric functions. In fact the function $\tau^{-1}(u)$ is a well-known automorphic function called the elliptic modular function. Thus when all the smoke has cleared away it turns out that the SW Ansatz is nothing but the assumption that $F''(A)$ is the inverse of the elliptic modular function!

One of the most interesting features of the SW choice is that it exhibits explicitly the duality used as input; in fact, it generalizes it to triality. The point is that Γ_2 is an *invariant* subgroup of $SL(2, Z)$ and

$$SL(2, Z)/\Gamma_2 = P_3, \tag{11.4}$$

where P_3 is the permutation group of order 3. Mathematically the permutation group P_3 interchanges the point at infinity and the two singularities, and physically it interchanges gauge-fields, monopoles, and dyons. The original duality thus emerges as the symmetry between the the gauge-field at infinity and the monopoles and dyons at the singularities.

12 Correctness

Since the SW proposal is only an Ansatz, one has to check whether it is the correct Ansatz. In principle this can be done by making direct computations of the nonperturbative part of the action using instanton computations. In practice this has been done only for arbitrary gauge-groups in the one-instanton configurations [3] and for SU(2) in the two-instanton configurations [1]. In these cases the SW Ansatz agrees with direct computations.

The general idea of the instanton computations is as follows: First, for any Fuchsian function satisfying the boundary conditions and R-invariance we have the asymptotic expansion

$$F(v) = v^2 + \hbar v^2 \ln\left(\frac{v}{\Lambda}\right)^2 + v^2 \sum c_m \left(\frac{\Lambda}{v}\right)^{4m}, \tag{12.1}$$

where Λ is the renormalization scale for the mth instanton sector and is defined as $\Lambda = \mu e^{-(8m\pi^2)/(g^2)}$ where μ is independent of m. Second, according to general arguments concerning the relationship between the effective potential and the partition function $F(v)$ (modified to suit the supersymmetric situation) the mth term in the expansion (12.1) should be the contribution to the partition function coming from the instanton sector of charge m. Thus we should have

$$P_m(v) = p_m v^2 \left(\frac{\Lambda}{v}\right)^{4m}, \tag{12.2}$$

where $P_m(v)$ is the contribution to the partition function from the mth instanton sector and p_m is constant. The problem therefore is to compute $P_m(v)$ and compare the coefficients p_m in Eq. (12.2) with the coefficients c_m in Eq. (12.1). To compute $P_m(v)$ one writes the kinetic term for ϕ in the action (3.4) as

$$\int d^4x \phi^\dagger D^2 \phi = \int d\Omega (\phi^\dagger, D\phi) + \int d^4x |D\phi|^2, \tag{12.3}$$

and makes a supersymmetrically invariant decomposition of the field configurations into two classes, namely, the short-range fields, which do not contribute to the surface integral in Eq. (12.3), and the long-range fields, which do. The first class of configurations are assumed to give no contribution to the partition function on the usual grounds that the fermionic and bosonic contributions to the partition function cancel for supersymmetric Lagrangians containing only first derivatives of the fields (Nicolai theorem). Thus all the contribution should come from the second class of configurations. The latter class is assumed to consist of all field configurations of the form

$$F_{\mu\nu} = F^*_{\mu\nu}, \quad \gamma^\mu D_\mu \psi = 0, \quad D^2\phi = [\bar{\psi}, \psi], \quad \phi(\infty) = v, \tag{12.4}$$

where F^* denotes the dual of F. The configurations (12.4) have very nice properties. First they make the volume term in the action (3.4) vanish, leaving only the surface term shown in Eq. (12.3). Second they allow the surface term to be expressed as a function of the $16m$ ADHMN parameters, which parametrize the self-dual gauge-fields and their zero modes as defined by the first two equations in Eq. (12.4). On converting the field-measure in the functional integral to an ADHMN-measure one finds that for these configurations the functional integral reduces to an ordinary integral of the form

$$P_m(v) = \Lambda^{4m} \int \rho^{4m-3} d\rho \, d_m(\beta, \eta) e^{\int d\vec{\Omega}.(\phi, \bar{D}\phi)}, \tag{12.5}$$

where β, η denotes all the parameters except the instanton size ρ and $d_m(\beta, \eta)$ is the ADMHN measure. Unfortunately, the ADMHN measure is known explicitly only for $m \leq 3$. On dimensional grounds one sees that the integral in Eq. (12.5) must be of the form

$$P_m(v) = \Lambda^{4m} \int \rho^{4m-3} d\rho \, dm(\beta, \eta) e^{-(v\rho)^2 f(\beta,\eta)} = v^2 \left(\frac{\Lambda}{v}\right)^{4m} p_m, \quad (12.6)$$

as anticipated in Eq. (12.2). Comparing the coefficients c_m and p_m it turns out that they agree for all the computed cases. That is, $c_1 = p_1$ for all groups and $c_2 = p_2$ for SU(2). This agreement provides strong support for the correctness of the SW Ansatz.

It should be mentioned, however, that the theoretical basis on which the above computations are made is still a little shaky. First, the computations were actually made only as an approximation, valid for low powers of vg, whereas the SW result is exact. Second, the background configurations used are not solutions of the classical field equations and are supersymmetrically invariant only in a very restricted sense. However, the agreement of the coefficients suggests that the computations could be put on a firmer basis and we hope to return to this question in a later paper.

13 Uniqueness

Further study [4] of the $N = 2$ model has shown that the minimality and duality assumptions of SW are not really necessary. More precisely, they can be replaced by the much weaker assumption that the mass be finite for finite u. This requirement is reasonable since asymptotic freedom implies that an infinity of the masses should correspond to $u \rightarrow \infty$. The finite-mass condition is implemented explicitly as follows:

First, since the mass formula is given in terms of $\mathbf{a}$ rather than $\mathbf{y}$ it is necessary to lift $\mathbf{y}$ to $\mathbf{a}$ and it turns out that the unique way to do this is to write

$$\mathbf{a}(u) = f(u)\mathbf{y}'(u) - f'(u)\mathbf{y}(u), \quad (13.1)$$

where $f(u)$ is a line bundle. At first sight the lifting would appear to decrease the possibility of uniqueness, since it adds the arbitrary line bundle $f(u)$ to the arbitrary vector-bundle $\mathbf{y}(u)$. However, if the indices r of $f(u)$ are defined as $f(u) \rightarrow (u - u_o)^r$ as $u \rightarrow u_o$ for any u_o (with $r = 0$ for logarithmic singularities), then the finiteness of the mass and the fact that $f(u)$ is a line bundle imply that r can take only certain discrete values and that for regular and singular points of $\mathbf{a}$ we must have $r = 0$ or $r \geq 1$ and $r \geq 1/4$, respectively. But then the asymptotic value $f(u) \rightarrow u$, which is obtained from perturbation theory, puts an upper bound $n = 2$ on the number of singularities. This is just the SW minimality assumption and given that $n = 2$ there are only four possible choices of Fuchsian group Γ. A closer analysis then shows that the discrete values permitted for r allow only the SW choice $\Gamma = \Gamma_2$. Thus, once the mass is required to be finite for finite u, the SW solution is unique. In particular, duality is an output rather than an input.

Acknowledgments: This chapter was compiled with the help of R. Flume, I. Sachs, and M. Magro.

REFERENCES

1. N. Dorey, Valentin V. Khoze, and Michael P. Mattis, *Multi-instanton calculus in $N = 2$ supersymmetric gauge theory*, Phys. Rev. D(3) **54** (1996), no. 4, 2921–2943.

2. A. Erdélyi, W. Magnus, F. Oberhettinger, and F. G. Tricomi, *Higher transcendental functions. III. Based*, vol. 3, McGraw-Hill, New York, 1955; Z. Nehari, *Conformal mapping*, McGraw-Hill, New York, 1952; R.C. Gunning, *Lectures on modular forms*, Princeton University Press, Princeton, NJ, 1962; L.R. Ford, *Automorphic functions*, Chelsea Publishing Company, New York, 1951.

3. D. Finnell and P. Pouliot, *Instanton calculations versus exact results in four dimensional SUSY gauge theories*, Nucl. Phys. B **453** (1995) 225; K. Ito and N. Sasakura, *One-instanton calculations in $N = 2$ supersymmetric SU(N_c) Yang-Mills theory*, Phys. Lett. B **382** (1996), Nos. 1–2, 95–103; A. Yung, *Instanton-induced effective Lagrangian in the Seiberg–Witten model*, Nucl. Phys. B **485** (1997), 38–62; F. Fucito and G. Travaglini, *Instanton calculus and nonperturbative relations in $N = 2$ supersymmetric gauge theories*, Phys. Rev. D **55** (1997), 1099–1104; T. Harano and M. Sato, *Multi-instanton calculus versus exact results in $N = 2$ supersymmetric QCD*, Nucl. Phys. B **484** (1997), 167–195.

4. R. Flume, M. Magro, L. O'Raifeartaigh, I. Sachs, and O. Schnets, *Uniqueness of the Seiberg–Witten effective Lagrangian*, Nucl. Phys. B **494** (1997), 331–345.

5. C. Montonen and D. Olive, *Magnetic monopoles as gauge particles*, Phys. Lett. B **72** (1977), 117–120.

6. N. Seiberg and E. Witten, *Electric-magnetic duality, monopole condensation, and confinement in $N = 2$ supersymmetric Yang-Mills theory*, Nuclear Phys. B **426** (1994), No. 1, 19–52.

7. J. Wess and R. Bagger, *Supersymmetry and supergravity*, Princeton Ser. Phys., Princeton University Press, Princeton, NJ, 1983; P. West, *Introduction to supersymmetry and supergravity*, World Scientific, 1990.

14

Bargmann Representation for Some Deformed Harmonic Oscillators with Non-Fock Representation

Michèle Irac-Astaud and Guy Rideau

ABSTRACT We prove that Bargmann representations exist for some deformed harmonic oscillators that admit non-Fock representations. In specific cases, we explicitly obtain the resolution of the identity in terms of a true integral on the complex plane. We prove in explicit examples that Bargmann representations cannot always be found, particularly when the coherent states do not exist in the whole complex plane.

1 Introduction

We define a deformed harmonic oscillator by the following algebraic relations between three operators a, $a^\dagger$ and N:

$$[a, N] = a, \quad [a^\dagger, N] = -a^\dagger, \tag{1.1}$$

and

$$a^\dagger a = \psi(N), \quad aa^\dagger = \psi(N+1), \tag{1.2}$$

where ψ is some real function and $a^\dagger$ has to be the adjoint of a and N is self-adjoint. The algebra defined in Eqs. (1.1), (1.2) was studied in Ref. [4]; and related algebras, under different forms, were considered in Refs. [5, 7]. In the case of the usual harmonic oscillator, $\psi(N) = N$, a particularly interesting representation of Eq. (1.1), Eq. (1.2) is the Bargmann representation, realized in a Hilbert space of the entire functions of order ≤ 1 with a reproducing kernel.

The main purpose of this paper is to study whether similar representations can be obtained when $\psi(N)$ is a strictly positive function. In this case, we are faced with non-Fock representations since in any representation the spectrum of N has no lower and upper bounds. In the next section, we give a brief description of these representations and we discuss the existence of the corresponding coherent vectors. In Section 3, we discuss in some detail the possibility of a Bargmann representation when the function ψ goes from

0 to ∞ when x goes from $-\infty$ to $+\infty$. In Section 4, we prove the existence of a Bargmann representation in the particular case of a q-oscillator. In Section 5, we investigate other examples proving that a Bargmann representation does not always exist. In Section 6, we briefly discuss few examples where $\lim_{x \to -\infty} \psi(x) > 0$ and/or $\lim_{x \to +\infty} \psi(x) < \infty$. In such cases, though coherent states exist but are associated with a bounded part of the complex plane, we can prove the impossibility of getting a Bargmann representation. It is not irrelevant to point out that the choice, although usual, of the letter N is rather misleading as apparently referring (in the Fock representation) to the number of particles. The eigenvalues of N label the levels of the energy operator that is not a priori given.

2   Representations

Let us assume N has an eigenvector $|0\rangle$, with eigenvalue μ. The inequivalent representations are labeled by the decimal part of μ. The Hilbert space $\mathcal{H}$ of the representation is spanned by the normalized eigenvectors $|n\rangle$ given by

$$|n\rangle = \begin{cases} \lambda_n a^{\dagger n} |0\rangle, & n \in \mathbb{Z}_+, \\ \lambda_n a^{-n} |0\rangle, & n \in \mathbb{Z}_-, \end{cases} \tag{2.1}$$

with

$$\lambda_n^{-2} = \psi(\mu+n)! = \begin{cases} \prod_{i=1}^{n} \psi(\mu+i), & n \in \mathbb{Z}^+, \\ \prod_{i=0}^{n+1} \psi(\mu+i), & n \in \mathbb{Z}^-. \end{cases} \tag{2.2}$$

We have

$$\begin{cases} a^{\dagger}|n\rangle &= \left(\psi(\mu+n+1)\right)^{1/2} |n+1\rangle, \\ a|n\rangle &= \left(\psi(\mu+n)\right)^{1/2} |n-1\rangle, \quad n \in \mathbb{Z}, \\ N|n\rangle &= (\mu+n) \ |n\rangle. \end{cases} \tag{2.3}$$

The first step toward a Bargmann representation requires us to study the coherent vectors, i.e, the eigenvectors of the annihilation operator a. Let us denote by $|z\rangle$ the vector

$$a|z\rangle = z|z\rangle; \tag{2.4}$$

$|z\rangle$ is given up to a constant of normalization by

$$|z\rangle = \sum_{n=-1}^{-\infty} z^n \left(\psi(\mu+n)!\right)^{1/2} |n\rangle + \sum_{n=0}^{\infty} z^n \left(\psi(\mu+n)!\right)^{-1/2} |n\rangle, \tag{2.5}$$

with the convention $\psi(\mu)! = 1$.

The vector $|z\rangle$ belongs to $\mathcal{H}$ only if the two series on the right-hand side of Eq. (2.5) are simultaneously norm convergent. This implies that

$$|z| < \lim_{p \to +\infty} \psi(p)^{1/2} \equiv r_2, \tag{2.6}$$

and

$$|z| > \lim_{p \to -\infty} \psi(p)^{1/2} \equiv r_1. \tag{2.7}$$

So the coherent states exist only if ψ is such that $r_1 < r_2$, and their domain of existence is a ring defined by $r_1 < |z| < r_2$, which can be extended to the complex plane without the origin when $r_1 = 0$ and $r_2 = \infty$.

When r_1 is larger than r_2, the annihilation operator a has no eigenstates, but then the creation operator $a^\dagger$ has eigenstates. Both situations are analogous and in the following we restrict ourselves to the case where $r_1 < r_2$.

Although μ is a significant quantity for labeling inequivalent representations, it does not play a part in the present problem. So we simplify the notation in assuming $\mu = 0$ from now on. Indeed, this is equivalent to substitute $N - \mu$ to N and $\psi_\mu(N) = \psi(\mu + N)$ to $\psi(N)$.

Moreover, we assume from now on that, unless otherwise specified, the function $\psi(N)$ is such that $r_1 = 0$ and $r_2 = \infty$; this means that the coherent vectors exist for all complex z, except $z = 0$.

Remark. $\mathcal{H}$ can be seen as the space of the functions $f(\theta)$, $0 \le \theta \le 2\pi$ on the unit circle such that

$$\int_0^{2\pi} |f(\theta)|^2 < \infty. \tag{2.8}$$

To the basis vectors $|n\rangle$ correspond the functions $\exp(in\theta)$, $n \in \mathbb{Z}$ and the operators a, $a^\dagger$ and N read

$$\begin{cases} a &= \exp(-i\theta)\psi\left(-i\frac{d}{d\theta} + \mu\right)^{1/2}, \\ a^\dagger &= \psi\left(-i\frac{d}{d\theta} + \mu\right)^{1/2}\exp(i\theta), \\ N &= -i\frac{d}{d\theta}, \end{cases} \tag{2.9}$$

which makes sense since $\psi(x)$ is strictly positive. From this point of view, we can speak of a deformed rotator, in the same sense as in Ref. [6].

3   Toward a Bargmann Representation

Let us stress that in this paper we look for a Bargmann representation in the strict sense where the integrals involved in the scalar product, or equivalently in the resolution of identity, are Riemann integrals.

So we are looking for a positive real function F, which we will call the weight function, such as we have in the following resolution of the identity:

$$\int F(z\bar{z})|\bar{z}\rangle\langle\bar{z}|\,dz\,d\bar{z} = 1, \tag{3.1}$$

where the integration is extended to the whole complex plane.

Then to a vector $|f\rangle$ in the representation space $\mathcal{H}$

$$|f\rangle = \sum_{-\infty}^{+\infty} f_n|n\rangle, \quad \sum_{-\infty}^{+\infty}|f_n|^2 < \infty, \tag{3.2}$$

should correspond the function of z:

$$f(z) \equiv \langle\bar{z}\mid f\rangle = \sum_{n\geq0} z^n f_n\big(\psi(n)!\big)^{-1/2} + \sum_{n<0} z^n f_n\big(\psi(n)!\big)^{1/2}. \tag{3.3}$$

In particular, to the basis vectors $|n\rangle$ correspond

$$\langle\bar{z}\mid n\rangle = \begin{cases} z^n\big(\psi(n)!\big)^{-1/2}, & n \geq 0, \\ z^n\big(\psi(n)!\big)^{1/2}, & n < 0. \end{cases} \tag{3.4}$$

From formula (3.1) the scalar product in $\mathcal{H}$ reads

$$\langle g\mid f\rangle = \int F(z\bar{z})\overline{g(z)}f(z)\,dz\,d\bar{z}, \tag{3.5}$$

and any operator A is represented by the kernel $A(\zeta,\bar{z}) = \langle\bar{\zeta}\mid A\mid\bar{z}\rangle$, such as

$$Af(\zeta) \equiv \langle\bar{\zeta}\mid Af\rangle = \int F(z\bar{z})A(\zeta,z)f(z)\,dz\,d\bar{z}. \tag{3.6}$$

Formula (3.3) defines a set $\mathcal{S}$ of holomorphic functions on the complex plane without the origin, the analytical properties of which are strongly depending on the function $G(x)$ defined by

$$G(x) = \sum_{n\geq0} x^n\big(\psi(n)!\big)^{-1} + \sum_{n<0} x^n\psi(n)! \tag{3.7}$$

Indeed from the Schwarz inequality, we get

$$|f(z)| \leq |f|^{1/2}G(z\bar{z})^{1/2}, \tag{3.8}$$

so that $G(x)^{1/2}$ controls the growth of the functions in $\mathcal{S}$ at infinity and near the origin.

Moreover, $G(\zeta z) = \langle\bar{z}\mid\zeta\rangle$ corresponds in $\mathcal{S}$ to the coherent state $|\zeta\rangle$; and finally, if (3.1) should be true, $G(\zeta\bar{z})$ is the reproducing kernel since we would have

$$f(\zeta) = \int F(z\bar{z})\langle\bar{\zeta}\mid\bar{z}\rangle f(z)\,dz\,d\bar{z}. \tag{3.9}$$

Taking (2.3) into account, we deduce that in our Bargmann representation, $a^\dagger$ would be the multiplication by z, a the operator $z^{-1}\psi(zd/dz)$ and N the operator zd/dz. As $G(\zeta z)$ corresponds to the coherent state $|\zeta\rangle$, we obtain for G the following functional equation:

$$xG(x) = \psi\left(x\frac{d}{dx}\right)G(x), \tag{3.10}$$

which could be obtained directly from the expansion (3.7).

From this set of inferences, we get the following necessary conditions to be verified by $F(x)$:

$$(1)\ M(n) = \int_0^\infty F(x)x^n\,dx = \begin{cases} \psi(n)!, & n \geq 0, \\ (\psi(n)!)^{-1}, & n < 0, \end{cases} \tag{3.11}$$

and

$$(2)\ \int F(z\bar z)zf(z)\overline{g(z)}\,dz\,d\bar z = \int F(z\bar z)f(z)\overline{\frac{1}{z}\psi\left(z\frac{d}{dz}\right)g(z)}\,dz\,d\bar z. \tag{3.12}$$

The first condition derives from Eq. (3.4) and the second expresses the mutual adjointness of a and $a^\dagger$. But formulas (3.11) imply that $F(x)$ decreases fast at the origin and at infinity. Then, at least if $f(z)$, $g(z)$ are finite linear combinations of powers of z, we obtain

$$\int F(z\bar z)f(z)\overline{\frac{1}{z}\psi\left(z\frac{d}{dz}\right)g(z)}\,dz\,d\bar z$$
$$= \int \left(\psi\left(-\bar z\frac{d}{d\bar z}-1\right)\frac{F(z\bar z)}{z\bar z}zf(z)\right)\overline{g(z)}\,dz\,d\bar z. \tag{3.13}$$

Using 3.12, we obtain the following functional equation for $F(x)$:

$$\psi\left(-x\frac{d}{dx}-1\right)\frac{F(x)}{x} = F(x), \tag{3.14}$$

or equivalently

$$xF(x) = \psi\left(-x\frac{d}{dx}\right)F(x), \tag{3.15}$$

for which we are looking for solutions fast decreasing at infinity and at the origin.

Let us remark that (3.11) implies the following recursive relation between the momentums $M(n)$ of $F(x)$:

$$M(n+1) = \psi(n)M(n). \tag{3.16}$$

Furthermore, according to the behavior of $F(x)$ at infinity and at the origin, we can define its Mellin transform $\widehat{F}(\rho)$:

$$\widehat{F}(\rho) = \int_0^\infty F(x)x^{\rho-1}\,dx, \tag{3.17}$$

which is meaningful for any ρ. According to Eq. (3.15), we must have

$$\widehat{F}(\rho+1) = \psi(\rho)\widehat{F}(\rho). \tag{3.18}$$

As $\widehat{F}(n+1)/\widehat{F}(1) = M(n)$, $\widehat{F}(\rho)$ is an interpolation of $M(n)$, which is univoquely determined by the condition of the mutual adjointness of a and $a^\dagger$. It is worthwhile to note the following simple formula:

$$G(x) = \widehat{F}(1) \sum_{-\infty}^{+\infty} \frac{x^n}{\widehat{F}(n+1)}. \tag{3.19}$$

According to our hypothesis, Eq. (3.15) is not a differential equation. Indeed, in this case $\psi(x)$ would be a polynomial. This is impossible, for if the higher term is an odd power of x, ψ cannot be a strictly positive function; and if the highest term is an even power of x, coherent vectors do not exist. This implies that the general solution of Eq. (3.15) may depend on an arbitrary function. The examples of the next sections will be illuminating in this respect.

4   The "q-Oscillator"

In this section, we assume $\psi_{\lambda,q}(N) = \lambda q^{-N}$, $q \le 1$, $\lambda > 0$. When $\lambda = (q^{-1} - q)^{-1}$, a and $a^\dagger$ verify the well-known q-oscillator commutation relation:

$$aa^\dagger - qa^\dagger a = q^{-N}, \tag{4.1}$$

so that the function $(q^{-1} - q)^{-1}q^{-N}$ characterizes a particular non-Fock representation of the q-oscillator. Let us remark that we also have $aa^\dagger = q^{-1}a^\dagger a$, which corresponds to the complex q-plane.

We easily get

$$G_{\lambda,q}(x) = \sum_{-\infty}^{+\infty} \left(\frac{x}{\lambda}\right)^n q^{n(n+1)/2}, \tag{4.2}$$

which obviously verifies the functional equation

$$xG_{\lambda,q}(x) = \lambda q^{-x\,d/dx}G_{\lambda,q}(x) = \lambda G_{\lambda,q}(q^{-1}x). \tag{4.3}$$

Let us note that

$$G^0_{\lambda,q}(x) = \exp\left(-\frac{\ln^2(x\lambda^{-1})}{2\ln(q)} - \frac{\ln(x\lambda^{-1})}{2}\right), \tag{4.4}$$

verifies (4.3). The general solution takes the form

$$G_{\lambda,q}(x) = q^{-1/8} G^0_{\lambda,q}(x) \sum_{-\infty}^{+\infty} \exp\left(\frac{1}{2}\ln(q)\left(n + \frac{1}{2} + \frac{\ln(x\lambda^{-1})}{\ln q}\right)^2\right), \quad (4.5)$$

where the series in the right-hand side is a strictly positive bounded periodic function of $\ln(x\lambda^{-1})/\ln q$ of period 1. So $G_{\lambda,q}(x)$ is increasing at infinity faster than any positive power of x and at the origin faster than any negative power of x. Moreover, its growth is always less than exponential of any power of x. Therefore, owing to to Eq. (3.8), this type of growth characterizes the function $f(z)$ of the set $\mathcal{S}$.

We now solve the functional equation (3.15):

$$xF_{\lambda,q}(x) = \lambda F_{\lambda,q}(qx). \quad (4.6)$$

The Mellin transform $\widehat{F}_{\lambda,q}(\rho)$ verifies

$$\widehat{F}_{\lambda,q}(\rho + 1) = \lambda q^{-\rho} \widehat{F}_{\lambda,q}(\rho), \quad (4.7)$$

with the particular solution

$$\widehat{F}^0_{\lambda,q}(\rho) = \exp\left(\rho \ln \lambda - \frac{1}{2}(\rho^2 - \rho)\ln q\right). \quad (4.8)$$

The inverse Mellin transform reads

$$F^0_{\lambda,q}(x) = \exp\left(\frac{\ln^2(x\lambda^{-1})}{2\ln(q)} - \frac{\ln(x\lambda^{-1})}{2}\right). \quad (4.9)$$

Therefore the general solution of Eq. (4.6) is given by

$$F^h_{\lambda,q}(x) = F^0_{\lambda,q}(x)h_q(x), \quad (4.10)$$

where $h_q(x)$ is a function satisfying

$$h_q(x) = h_q(qx), \quad (4.11)$$

that is, a periodic function of $\ln x/\ln q$ of period 1. We verify directly that up to a constant the momentums $M(n)$ of $F^0_{\lambda,q}(x)$ are well equal to $\lambda^{-n} q^{-n(n+1)/2}$ for $n \in \mathbb{Z}$, as desired, and this is yet true for $F^h_{\lambda,q}(x)$ given in Eq. (4.10) as implied by the definition of $h_q(x)$.

As the absolute value of a solution of Eq. (4.11) is also a solution, we can assume that $h_q(x)$ is strictly positive, so that $F^h_{\lambda,q}(x)$ is also strictly positive.

Let us now consider the integral

$$\frac{\int dz\, d\bar{z}\, F^h_{\lambda,q}(z\bar{z}) f(z) f(\bar{z})}{\int dz\, d\bar{z}\, F^h_{\lambda,q}(z\bar{z})}, \quad (4.12)$$

where $f(z)$ is defined as in Eq. (3.3). Thanks to the positivity of $F^h_{\lambda,q}(x)$, we proceed as in Bargmann's paper [1] to prove that the integral (4.12) and the series $\sum_{n\in\mathbb{Z}}|f_n|^2$ are simultaneously divergent or simultaneously convergent to the same value. The function $f(z)$ satisfies the inequality

$$|f(z)| \le C \exp$$
$$- \left(\frac{\ln^2(x\lambda^{-1})}{4\ln q} + \frac{\ln(x\lambda^{-1})}{4}\right) \sum_{-\infty}^{+\infty} \exp \frac{\ln q}{2}\left(n + \frac{1}{2} + \frac{\ln(x\lambda^{-1})}{\ln q}\right)^2, \quad (4.13)$$

which is essential to obtain pointwise convergence from norm convergence.

In particular we use 4.13 to prove the closedness of the operators z and $z^{-1}q^{zd/dz}$. Moreover, as we have

$$\left|zf(z)\right|^2 = q\left|z^{-1}q^{zd/dz}f(z)\right|^2, \quad (4.14)$$

we prove that these operators have the same domain of definition and their mutual adjointness follows easily as in Ref. [1].

So we can summarize the result of this section as follows: for $\psi_{\lambda,q}(N) = \lambda q^{-N}$ the necessary conditions (3.11) and (3.12) are sufficient conditions for having a Bargmann representation. There exist infinitely many equivalent norms defined by Eq. (4.9) and Eq. (4.10) for any strictly positive function $h(x)$ verifying (4.11).

5   Generalization of the Previous Example

In this section, we point out some directions to extend the results of the previous section when $\psi(x)$ is of the form:

$$\psi(x) = \exp\left(\sum_0^{2p+1} a_n x^n\right), \quad a_{2p+1} > 0. \quad (5.1)$$

Then the equation (3.18) can be solved and gives

$$\widehat{F}(\rho) = \exp\left(\sum_0^{2p+1} \frac{a_n}{n+1} B_{n+1}(\rho)\right), \quad (5.2)$$

where $B_{n+1}(\rho)$ are the Bernoulli polynomials. The term of highest degree in Eq. (5.2) is $a_{2p+1}\rho^{2p+2}/(2p+1)$. When ρ is pure imaginary $\rho = i\sigma$, $\sigma \in R$, this term is $a_{2p+1}(-1)^{p+1}\sigma^{2p+2}/(2p+1)$; therefore, the inverse Mellin transform of $\widehat{F}(\rho)$ exists only if p is an even number. The function $F(x)$ thus obtained is always real, but not necessarily positive. Nevertheless, in specific cases, for example, when the exponent in Eq. (5.2) contains only the term of highest degree, $F(x)$ is actually strictly positive and the Bargmann procedure works as before.

The proof of the closedness of the operators z and $z^{-1}\psi(zd/dz)$ results from Eq. (3.8) as before. But we have

$$\left|zf(z)\right|^2 = \sum_{-\infty}^{\infty} |f_n|^2 \psi(n+1),$$ (5.3)

and

$$\left|z^{-1}\psi\left(z\frac{d}{dz}\right)f(z)\right|^2 = \sum_{-\infty}^{\infty} |f_n|^2 \psi(n).$$ (5.4)

Since $\psi(n+1)/\psi(n)$ grows indefinitely as $n \to \pm\infty$, the domain $z^{-1}\psi(zd/dz)$ is included in the domain of z but cannot be identical. Nevertheless, the mutual adjointness can be proved as ensured by Eq. (3.12).

This example points out the following significant fact: Apart from its behavior at infinity, the function ψ must necessarily verify supplementary conditions for admitting a Bargmann representation. Unfortunately, we did not succeed in finding general results in this direction. At least this means that there exist functions ψ such that the Eq. (3.15) has no solution fast decreasing at infinity and at the origin simultaneously.

6   Deformed Algebra Associated to a Given Weight Function

Conversely, we can use Eq. (3.17) to deduce the function $\psi(x)$ from a given $F(x)$. As pointed out just before, we have to prove that the resulting $\psi(x)$ verifies our fundamental assumptions.

We illustrate this construction by the following example.

Let $F(x)$ be

$$F(x) = \exp\left(-\nu(\ln x)^{2n}\right), \quad \nu > 0, \ n \geq 1.$$ (6.1)

The Mellin transform is a strictly positive function given by

$$\widehat{F}(\rho) = \int_{-\infty}^{+\infty} \exp\left(-\nu t^{2n}\right)\exp(\rho t)\, dt.$$ (6.2)

For $\rho > 0$, let us write

$$\widehat{F}(\rho) = \int_{-\infty}^{+\infty} \exp\left(-\nu(t+\mu)^{2n} + \rho(t+\mu)\right) dt,$$ (6.3)

with $\mu = \left(\rho/(2n\nu)\right)^{1/(2n-1)}$ and take as new variable $u = t\mu^{n-1}$. Thus, we get

$$\widehat{F}(\rho) = e^{(2n-1)\nu(\rho/(2n\nu))^{2n/(2n-1)}} \mu^{1-n}$$

$$\times \int_{-\infty}^{+\infty} e^{-\nu n(2n-1)u^2 - \nu \sum_{p\geq 3} C_{2n}^p \frac{u^p}{\mu^{n(p-2)}}} du, \quad (6.4)$$

where C_{2n}^p are the binomial coefficients. It results in the following asymptotic expansion for $\widehat{F}(\rho)$:

$$\widehat{F}(\rho) = \exp\left((2n-1)\nu\left(\frac{\rho}{2n\nu}\right)^{2n/(2n-1)}\right)\rho^{-(n-1)/(2n-1)} \sum_{p\geq 0} a_p \rho^{-p}. \quad (6.5)$$

Therefore, thanks to Eq. (3.18), for $x \to +\infty$, $\psi(x)$ has the asymptotic expansion

$$\psi(x) = \exp\left(\frac{x}{2n\nu}\right)^{1/(2n-1)} \sum_{p\geq 0} b_p x^{-p}, \quad (6.6)$$

implying $\lim_{x\to+\infty} \psi(x) = \infty$. Now from $\widehat{F}(\rho) = \widehat{F}(-\rho)$, we deduce $\psi(-x) = \psi(x-1)^{-1}$ and thus $\lim_{x\to-\infty} \psi(x) = 0$.

Now we use the functional equation (3.10) for getting a qualitative evaluation of the asymptotic behavior of $G(x)$ defined in (3.7) and directly related to the reproducing kernel of the Bargmann representation. Indeed, starting with $\psi(xd/dx) = \widehat{F}(xd/dx + 1)/\widehat{F}(xd/dx)$ we get, after replacing $\widehat{F}(\rho)$ by the expression (6.3) and after performing the needed dilations,

$$x \int_{-\infty}^{+\infty} \exp(-\nu t^{2n} + t) G\left(\exp(t)x\right) dt$$

$$= \int_{-\infty}^{+\infty} \exp(-\nu t^{2n} + t) G\left(\exp(t)x\right) dt. \quad (6.7)$$

This implies that the integral involved in the previous formula is necessarily divergent and that $G(x)$ grows faster than $\exp\left(\nu \ln(x)^{2n} - \alpha \ln(x)\right)$ with $\alpha \geq -2$. Thus, in this specific example, for a given $F(x)$, we have found the corresponding deformed algebra for which a Bargmann representation exists.

7   Bargmann Representations Corresponding to Different ψ

In this section we exhibit a relation between the functions F associated to different functions ψ. The relatively simple results can be used to derive various functions F corresponding to various functions ψ from a given F.

Let ψ_i, $i = 1, 2$, be two positive functions for which Bargmann representations exist and let $\mathcal{H}_i$, $i = 1, 2$, be the corresponding Hilbert spaces. To any sequence f_n, $n \in \mathbb{Z}$ such that $\sum_n |f_n|^2 < \infty$ is associated two vectors $f_i(z) \in \mathcal{H}_i$, $i = 1, 2$:

$$f_i(z) = \sum_{n \geq 0} f_n z^n \left(\psi_i(n)!\right)^{-1/2} + \sum_{n < 0} f_n z^n \left(\psi_i(n)!\right)^{1/2}. \tag{7.1}$$

Thus, we can define a unitary mapping A_{12} from $\mathcal{H}_2$ on $\mathcal{H}_1$ by

$$A_{12} f_2(z) = f_1(z). \tag{7.2}$$

Specifying this equation to the basis elements, we deduce easily that A_{12} can be written

$$A_{12} = a_{12}\left(z \frac{d}{dz}\right), \tag{7.3}$$

where the function $a_{12}(\rho)$ verifies

$$a_{12}(\rho) = \left(\frac{\psi_2(\rho)}{\psi_1(\rho)}\right)^{1/2} a_{12}(\rho - 1). \tag{7.4}$$

Now, we have, for $f_1, g_1 \in \mathcal{H}_1$,

$$(f_1, g_1)_1 = \int F_1(z\bar{z}) f_1(z) \overline{g_1(z)} \, dz \, d\bar{z},$$

$$= \int F_1(z\bar{z}) a_{12}\left(z \frac{d}{dz}\right) f_2(z) \overline{a_{12}\left(z \frac{d}{dz}\right) g_2(z)} \, dz \, d\bar{z} = (f_2, g_2)_2, \tag{7.5}$$

according to Eq. (7.2). Taking into account the behavior of $F_1(z\bar{z})$ at infinity and near the origin, the last integral may be written

$$\int \left\{ a_{12}\left(-\frac{d}{dz} z\right) a_{12}\left(-\frac{d}{d\bar{z}} \bar{z}\right) F_1(z\bar{z}) \right\} f_2(z) \overline{g_2(z)} \, dz \, d\bar{z}, \tag{7.6}$$

so we get

$$F_2(x) = a_{12}^2\left(-1 - x \frac{d}{dx}\right) F_1(x). \tag{7.7}$$

Conversely, this formula can be used to determine a new function $F_2(x)$ from a known one $F_1(x)$ by giving a priori the function $a_{12}^2(\rho)$. In this case, we get F_2 by Eq. (7.7) and ψ_2 by Eq. (7.4):

$$\psi_2(\rho) = \psi_1(\rho) \frac{a_{12}^2(\rho)}{a_{12}^2(\rho - 1)}. \tag{7.8}$$

Obviously, we must be careful in choosing $a_{12}^2(\rho)$ to be sure that the resulting $\psi_2(\rho)$ fits the conditions for admitting a Bargmann representation.

For instance, we choose

$$a_{12}^2(\rho) = \sum_0^p a_n \exp(\alpha_n \rho), \quad a_n > 0, \tag{7.9}$$

where the α_n are in increasing order. Then we have

$$F_2(x) = \sum_0^p a_n \exp(\alpha_n) F_1\big(x \exp(-\alpha_n)\big), \tag{7.10}$$

and

$$\psi_2(\rho) = \psi_1(\rho) \frac{\sum_0^p a_n \exp(\alpha_n \rho)}{\sum_0^p a_n \exp(\alpha_n(\rho - 1))}, \tag{7.11}$$

on which appears clearly that the conditions for a Bargmann representation are satisfied.

8 The Case of an Annulus

Let us consider now the case where the domain of convergence of the coherent states is an annulus $\mathcal{R}$ in the complex plane defined by $r_1 < |z| < r_2$. We are looking for $F(z\bar{z})$ such that

$$\int_{\mathcal{R}} F(z\bar{z})\langle \bar{z}\rangle\langle \bar{z}| \, dz \, d\bar{z} = 1. \tag{8.1}$$

We can always suppose that $F(z\bar{z})$ is identically zero outside $\mathcal{R}$. Then, we can develop the same consideration as above and from the mutual adjointness of z and $z^{-1}\psi(zd/dz)$, we get the same functional equation (3.15) with the limit conditions that $F(x)$ is zero outside the interval $r_1^2 < x < r_2^2$.

Let us illustrate the situation with two examples:

(a) $\psi(x) = 1 + q^x, \quad q > 1.$

In this case $\mathcal{R}$ is the exterior of the disk of radius 1 and the functional equation (3.15) reads

$$(qx - 1)F(qx) = F(x). \tag{8.2}$$

So, if $F(x) \equiv 0$ for $x < 1$, then $F(x) \equiv 0$ for $x < q$. Using Eq. (8.2), we obtain

$$(q^{n+1}x - 1)F(q^{n+1}x) = F(q^n x). \tag{8.3}$$

With this equation, we prove recursively that $F(x)$ is identically zero in the complex plane.

(b) $\psi(x) = q^x/(1 + q^x), \quad q > 1.$

In this case $\mathcal{R}$ is the disk of radius 1. The functional equation (3.15) reads

$$xF(x) = (1 - q^{-1}x)F(q^{-1}x). \qquad (8.4)$$

If $F(x) \equiv 0$ for $x > 1$, then $F(x) \equiv 0$ for $x > q^{-1}$ and as previously we show recursively that $F(x)$ is identically zero in the complex plane.

Although we have not obtained a general proof, we conjecture that the results proved in the two previous examples are general: when the coherent states are restricted to a ring of the complex plane, we cannot obtain a Bargmann representation because the function involved in the resolution of the identity is identically equal to zero. We think that the phenomenon is general thanks to the fact that the nonlocal operator $\psi(-xd/dx)$ does not leave invariant the limit conditions. One can also think that this negative result is depending on our restricted definition for Bargmann representation, and that we would have advantage to generalize it, by introducing another type of integration like for instance the q-integration used in Refs. [2, 3] that precisely is the inverse of a nonlocal operator.

9   Conclusion

The deformed oscillators considered in this paper are essentially different from the well-known harmonic oscillator, since their irreducible representations never admit a fundamental vector. Nevertheless, it has been possible for some of these exotic cases to realize their representations in a Hilbert space of holomorphic functions in complete analogy with the well-known Bargmann construction, taken in the more restricted sense, i.e., by excluding any new concept of integration as in Refs. [2, 3]. We have given necessary conditions to be verified by the weight function, or equivalently, by its Mellin transform. These conditions were shown to be sufficient in several specific cases, including the so called q-oscillator. Meanwhile, we encountered cases where we proved the nonexistence of any weight function. Although our results have some degree of generality, we did not get a general statement giving a complete characterization of the deformed algebras, considered in this paper, admitting a Bargmann representation.

REFERENCES

1. V. Bargmann, *On a Hilbert space of analytic functions and an associated integral transform*, Comm. Pure Appl. Math. **14** (1961), 187–214.

2. A.J. Bracken, D.S. McAnally, R.B. Zhang, and M.D. Gould, *A q-analogue of Bargmann space and its scalar product*, J. Phys. A **24** (1991), No. 7, 1379–1391.

3. R.W. Gray and C.A. Nelson, *A completeness relation for the q-analogue coherent states by q-integration*, J. Phys. A **23** (1990) L945–L950.

4. M. Irac-Astaud and G. Rideau, *Deformed quantum harmonic oscillator*, Proc. Third Internat. Wigner Symposium (Oxford, 1993). (to appear)

5. M. Irac-Astaud and G. Rideau, *On quantum multi-Hamiltonian systems*, Lett. Math. Phys. **29** (1993), 197–203; Theoret. and Math. Phys. **99** (1994), No. 3, 658–661.

6. K. Kowalski, J. Rembielinski, and L.C. Papaloucas, *Coherent states for a particle on a circle*, J. Phys. A **29** (1996), 4149–4167.

7. C. Quesne and N. Vansteenkiste, *Representation theory of deformed oscillator*, Helv. Phys. Acta **69** (1996), 141–157.

15

The Vector-Coherent-State Inducing Construction for Clebsch–Gordan Coefficients

D.J. Rowe

ABSTRACT   The Vector-Coherent-State (VCS) theory has been demonstrated, in a wide range of applications, to be a powerful tool for inducing the explicit matrices of irreducible Lie group (and Lie algebra) representations from those of suitably defined subgroups (subalgebras). It has recently been shown that VCS theory also provides a simple algorithm for computing the Clebsch–Gordan coupling coefficients for reducing tensor products of group representations, from those of subgroups. The latter development was used to induce SU(3) coupling coefficients from those of SU(2). This chapter outlines how to derive coupling coefficients for SU(4) from those of SU(3) as an example of a general construction for inducing coefficients for SU(n) from those of SU($n - 1$).

1   Introduction

The use of mathematics in modern physics has long surpassed the traditional applications. The discovery of gauge field theory, quasisymmetry in crystals, and systems described by nonlinear equations, are just a few examples. Thus, a physicist with imagination and creativity who can speak the language of mathematics and apply sophisticated modern techniques to the solution of genuine physical problems can make enormous contributions to his subject. Our much esteemed colleagues, Jiři Patera and Pavel Winternitz, are among the most successful in this endeavor. They command the respect of both mathematicians and physicists and have provided an invaluable service in bringing the two groups together. Thus, I am honoured to participate in this conference to celebrate their birthdays and their research contributions.

My own interest in mathematical physics came with the realization that the mathematical problems I faced, in describing dynamical systems in physics, were different from those that concern mathematicians. For example, it may be enough for a mathematician to know the unitary action of a group on a Hilbert space. But, for a unitary representation of a group to be useful in physics, one needs the explicit matrices of the representation in a basis that reduces some physically relevant subgroup chain. It is not enough to have an integral expression for an inner product if one cannot do the integrals quickly and easily. One needs orthonormal bases of states, to diagonalize Hamiltonians, and compute matrix elements of physical operators.

What I want to present at this symposium is an illustration of the way vector coherent state theory [14, 22, 28–30] serves to generate effective and efficient algorithms for solving some basic problems physicists face in constructing the matrices of a Lie algebra in a unitary representation and in finding the Clebsch–Gordan coefficients for reducing tensor product representations. There is not enough time to give more than a few brief insights. Thus, to be specific, I will give an example of the way VCS theory can be used to induce SU(4) representations and Clebsch–Gordan coefficients. A fuller account of this application [25], in collaboration with J. Repka, will be given elsewhere.[1]

Coupling coefficients for su(4) have many physical applications, e.g., in Wigner supermultiplet theory [17] and in the interacting boson model [7, 18]. Since they apply to the finite-dimensional irreps of sl(4, $\mathbb{R}$), they are also of importance in relativity and field theory. Moreover, the algorithms given for su(4) apply to any su(n) [24].

VCS theory is a formal framework that can be used to tackle a variety of group theoretical problems. Fundamentally, it is an inducing construction that enables one to extend the representations of a group H to representations of a group G that contains H as a subgroup.

There are many inducing constructions. What is special about VCS theory is that it gives results in the form physicists need. In particular, it enables one to derive irreducible representations (irreps), construct orthonormal bases that reduce suitable subgroup chains, and construct the explicit matrices for the infinitesimal generators of a Lie group.

The theory makes essential use of the tensor structures of Lie algebras. This is particularly appropriate for an inducing construction because it means that one can effectively use knowledge of a subgroup $H \subset G$ in computing properties of G; one doesn't need to start over again from the beginning. Thus, VCS theory takes full advantage of the Wigner-Eckart theorem.

Consider, for example, a chain of subgroups

$$G \quad \supset \quad H \supset \quad T$$
$$\lambda \qquad \omega \qquad \tau \qquad\qquad (1.1)$$

and suppose that an irrep of each group is characterized by the symbol shown in Eq. (1.1) beneath the corresponding symbol for the group. Let

---

[1] At this point I want to acknowledge many long and stimulating discussions with Joe Repka and the essential contributions he has made to this work and to the development of VCS theory generally.

$\{|\lambda\omega\tau\rangle\}$ denote an orthonormal basis for an irrep λ of G that reduces the subgroup chain. In general, we need additional multiplicity indices for a complete labeling of basis states but, for present purposes, we suppose they are included in the given labels. Let $\{T_{\omega\tau}^\lambda\}$ denote the components of a tensor operator T^λ. Then, the Wigner–Eckart theorem states that the matrix elements of $T_{\omega_2\tau_2}^{\lambda_2}$ can be expressed

$$\langle\lambda_3\omega_3\tau_3\|T_{\omega_2\tau_2}^{\lambda_2}\|\lambda_1\omega_1\tau_1\rangle = \sum_\alpha (\omega_1\tau_1,\omega_2\tau_2 \mid \alpha\omega_3\tau_3)\langle\lambda_3\omega_3\|T_{\omega_2}^{\lambda_2}\|\lambda_1\omega_1\rangle_\alpha,$$

(1.2)

where $(\omega_1\tau_1,\omega_2\tau_2 \mid \alpha\omega_3\tau_3)$ is a Clebsch–Gordan coefficient for H and $\langle\lambda_3\omega_3\|T_{\omega_2}^{\lambda_2}\|\lambda_1\omega_1\rangle_\alpha$ is a so-called H-reduced matrix element. Thus, if one can assume prior knowledge of the CG coefficients for H, one needs only the H-reduced matrix elements of a tensor operator to have a complete knowledge of all its matrix elements.

In fact, one can go further. The H-reduced matrix elements can be expanded

$$\langle\lambda_3\omega_3\|T_{\omega_2}^{\lambda_2}\|\lambda_1\omega_1\rangle_\alpha = \sum_\rho (\lambda_1\omega_1,\lambda_2\omega_2\|\rho\lambda_3\omega_3)_\alpha \langle\lambda_3\||T^{\lambda_2}\||\lambda_1\rangle_\rho, \quad (1.3)$$

$\langle\lambda_3\||T^{\lambda_2}\||\lambda_1\rangle_\rho$ being G-reduced matrix element and $(\lambda_1\omega_1,\lambda_2\omega_2\|\rho\lambda_3\omega_3)_\alpha$ an H-reduced CG coefficient for G (sometimes called an isoscalar factor). Thus, to use the Wigner–Eckart theorem in the evaluation of matrix elements one needs only the H-reduced CG coefficients if one already knows the CG coefficients for H. Equivalently, one has an expansion of CG coefficients for G

$$(\lambda_1\omega_1\tau_1,\lambda_2\omega_2\tau_2\|\rho\lambda_3\omega_3\tau_3)$$
$$= \sum_\alpha (\omega_1\tau_1,\omega_2\tau_2 \mid \alpha\omega_3\tau_3)(\lambda_1\omega_1,\lambda_2\omega_2\|\rho\lambda_3\omega_3)_\alpha. \quad (1.4)$$

2 Induced Representations of su(4)

It is convenient to regard an su(4) irrep as a subrepresentation of u(4). We can then identify an irrep with a highest weight of the form $(\lambda, 0)$, where $\lambda \equiv (\lambda_1, \lambda_2, \lambda_3)$ is a triple of integers with $\lambda_1 \geq \lambda_2 \geq \lambda_3 \geq 0$.

The structure of the su(4) algebra can be viewed graphically as shown in Fig. 15.1.

The complex extension of su(4) is a sum

$$su(4)^C = u(3)^C + \mathbf{n}_+ + \mathbf{n}_-, \quad (2.1)$$

where $\mathbf{n}_\pm$ are nilpotent subalgebras of raising and lowering operators, respectively. The elements of $\mathbf{n}_+$ are the components of a u(3) tensor of

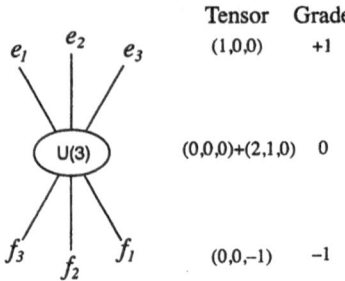

FIGURE 15.1. A schematic root diagram for u(3) showing the stability subalgebra u(3).

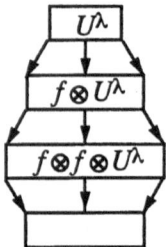

FIGURE 15.2. Graded u(3)-invariant subspaces of the carrier space for an su(4) irrep of highest weight λ. The highest-grade subspace is irreducible.

highest weight $(1,0,0)$, whereas the components of $\mathbf{n}_-$ are the components of a u(3) tensor of highest weight $(0,0,-1)$. The u(3) algebra itself is the sum of a u(1) scalar and an su(3) subalgebra whose elements are the components of a u(3) tensor of highest weight $(2,1,0)$. As the figure shows, the su(4) Lie algebra can be graded such that the raising operators (e_1, e_2, e_3) have grade 1 and are regarded as grade-raising operators while the lowering operators (f_1, f_2, f_3) have grade -1 and are regarded as grade-lowering operators.

The grade structure extends to the carrier space for a representation of su(4), as shown schematically in Fig. 15.2, such that each graded subspace is u(3) invariant.

Let V^λ denote the carrier space for an irrep of su(4) of highest weight λ and let U^λ denote its highest-grade subspace. Then U^λ is both invariant under u(3) and irreducible. Let ρ^λ denote the u(3) irrep carried by U^λ. This irrep has highest weight λ.

We suppose that all the irreps and coupling coefficients of u(3) are known. The objective is to extend this knowledge to su(4). The first step is to extend the irrep ρ^λ of u(3) with highest λ to an irrep of su(4) with the same highest weight. This is a well-defined mathematical task that has been solved by the Gel'fand–Tseitlin [13] construction. The VCS construction gives an alternative solution that makes effective use of what is known about u(3).

Let $\{\xi_\nu^\lambda\}$ denote an orthonormal basis for U^λ. And let $\{|\lambda\omega\nu\rangle\}$ denote an orthonormal basis for V^λ with the notation that ω labels the highest weight of a u(3) subrepresentation and ν indexes a basis for this subrepresentation. Thus, we have the identity

$$|\lambda\lambda\nu\rangle \equiv \xi_\nu^\lambda. \tag{2.2}$$

It will be convenient to use Dirac's bracket notation so that the orthonormality of the basis is expressed by the equation

$$\langle\lambda\omega\nu \mid \lambda\omega'\nu'\rangle = \delta_{\omega\omega'}\delta_{\nu\nu'}. \tag{2.3}$$

The basis states $\{\xi_\nu^\lambda\}$ for the u(3) irrep ρ^λ are presumed to be known. VCS theory then gives explicit wave functions $\{\psi_{\omega\nu}^\lambda\}$ for the states $\{|\lambda\omega\nu\rangle\}$ and enables us to evaluate the matrices representing elements of the su(4) Lie algebra.

The VCS wave function $\psi_{\omega\nu}^\lambda$ for a state $|\lambda\omega\nu\rangle$ is a function of three complex variables $z \equiv (z_1, z_2, z_3)$ with vector values in U^λ defined by

$$\psi_{\omega\nu}^\lambda(z) = \sum_\mu \xi_\mu^\lambda \langle\lambda\lambda\mu|e^{\Sigma_i z_i e_i}|\lambda\omega\nu\rangle. \tag{2.4}$$

VCS theory gives the following results:

(i) The VCS wave functions have explicit expressions as u(3)-coupled products

$$\psi_{\omega\nu}^\lambda(z) = K_\omega^\lambda [P^n(z) \otimes \xi^\lambda]_{\omega\nu}, \tag{2.5}$$

where

$$K_\omega^\lambda = \sqrt{\frac{(\lambda_1 + 2)!(\lambda_2 + 1)!\lambda_3!}{(\omega_1 + 2)!(\omega_2 + 1)!\omega_3!}}, \tag{2.6}$$

and $P^n(z)$ is a u(3) tensor of holomorphic polynomials in z of degree $n = \sum_i(\lambda_i - \omega_i)$ with components

$$P_\nu^n(z) = \frac{z_1^{\nu_1} z_2^{\nu_2} z_3^{\nu_3}}{\sqrt{\nu_1!\nu_2!\nu_3!}}, \quad \nu_1 + \nu_2 + \nu_3 = n. \tag{2.7}$$

(ii) U(3)-reduced matrix elements of the lowering operators are given by

$$\langle\lambda\omega'\|f\|\lambda\omega\rangle = \sqrt{n+1}\; U(\lambda\tilde{n}\omega'\tilde{1}; \widetilde{\omega n + 1})\frac{K_{\omega'}^\lambda}{K_\omega^\lambda}, \tag{2.8}$$

where $U(\lambda\tilde{n}\omega'\tilde{1}; \widetilde{\omega n + 1})$ is a u(3) Racah coefficient and $\tilde{n} = (0, 0, -n)$.

(iii) Matrix elements of the raising operators are given by

$$\langle \lambda \omega \nu \mid e_i \mid \lambda \omega' \nu' \rangle = \langle \lambda \omega' \nu' \mid f_i \mid \lambda \omega \nu \rangle^*. \tag{2.9}$$

The derivations of these results are given in Ref. [25]. Similar results were obtained [16] for the representations of su(n) in terms of u($n-1$)-coupled wave functions and u($n-1$) Racah coefficients for any $n > 1$.

3   SU(4) Clebsch–Gordan Coefficients

Biedenharn and his colleagues [2–5, 8–12, 19, 20] spent many years developing a strategy for computing CG coefficients for the special unitary groups. It turns out that their strategy is easily implemented within the framework of VCS theory [25–27].

The Biedenharn–Louck strategy

The idea is to define a model space

$$F = \sum_\lambda F^\lambda, \tag{3.1}$$

comprising one copy of each irrep and to seek a set of so-called *shift tensors* $\{T^{\lambda \sigma \tau}\}$ on F. A tensor T^λ of highest weight λ on the su(4) model space F, for example, is said to have shift weight τ if all its components $\{T^\lambda_{\omega \nu}\}$ map all states of an su(4) subspace $F^{\lambda_1} \subset F$ to the subspace $F^{\lambda_1 + \tau} \subset F$. The tensors in a BL set should be irreducible and indexed by three labels: λ denotes the highest weight (i.e., the rank) of a tensor; τ is a shift weight; and σ is a label needed to distinguish different tensors of the same λ and τ.

According to Eq. (1.3), the u(3)-reduced matrix elements of shift tensor operators are linear combinations of the desired u(3)-reduced CG coefficients; i.e.,

$$\langle \lambda_3 \omega_3 \| T^{\lambda_2 \sigma \tau}_{\omega_2} \| \lambda_1 \omega_1 \rangle_\alpha$$
$$= \delta_{\lambda_3, \lambda_1 + \tau} \sum_\rho (\lambda_1 \omega_1, \lambda_2 \omega_2 \| \rho \lambda_3 \omega_3)_\alpha \langle \lambda_3 \| | T^{\lambda_2 \sigma \tau} \| | \lambda_1 \rangle_\rho. \tag{3.2}$$

To appreciate the significance of this equation, it is important to understand the role of the multiplicity index ρ. The symbol ρ enumerates the number of copies of the irrep of highest weight λ_3 that occur in the su(4) tensor product $F^{\lambda_2} \otimes F^{\lambda_1}$. If the tensor products of su(4) irreps were multiplicity free, as they are for su(2), the sum over ρ in Eq. (3.2) would reduce to a single term. One could then evaluate the corresponding CG coefficients by multiplying the reduced matrix elements by renormalization factors such that the CG coefficients obey the standard orthonormality relations. However, in general, when there is a nontrivial multiplicity, we have to determine linear combinations of the matrix elements to obtain orthogonal as well as normalized CG coefficients. It is clear that for this to be possible the number of linearly independent tensors of given λ and τ in a BL set must not be less than the multiplicity of the $\lambda_2 \otimes \lambda_1 \rightarrow \lambda_3$ coupling.

VCS-induced shift tensors

The construction of a set of shift tensors having the required properties proves to be remarkably simple within the framework of VCS theory [27].

Let $\{t^{\omega\alpha\tau}\}$ denote a set of shift tensors for u(3). And let $\{T^{\lambda\omega\alpha\tau}\}$ be a set of su(4) tensors whose highest-grade components are the u(3)-coupled products

$$T^{\lambda\omega\alpha\tau}_{\lambda\nu} = [P^n(\nabla) \otimes t^{\omega\alpha\tau}]_{\lambda\nu}, \tag{3.3}$$

where $P^n(\nabla)$ is a u(3) tensor of homogeneous polynomials in the derivative operators $\{\nabla_i = \partial/\partial z_i\}$ with components

$$P^n_\nu(\nabla) = \frac{\nabla_1^{\nu_1} \nabla_2^{\nu_2} \nabla_3^{\nu_3}}{\sqrt{\nu_1! \nu_2! \nu_3!}}, \quad \nu_1 + \nu_2 + \nu_3 = n, \tag{3.4}$$

and $n = \sum_i (\lambda_i - w_i)$. Then, as shown in Ref. [27], the tensors $\{T^{\lambda\omega\alpha\tau}\}$ are a complete set of BL tensors for su(4). Note that the index σ of the above defined set of su(4) shift tensors $\{T^{\lambda\sigma\tau}\}$ is now interpreted as a pair; i.e., $\sigma \equiv (\omega\alpha)$.

The matrix elements of the u(3) tensors of Eq. (3.3) are easy to evaluate with the VCS wave functions of Eq. (2.5). One has

$$[T^{\lambda\omega\alpha\tau}_\lambda \otimes \psi^{\lambda_1}_{\omega_1}]_{\beta\omega_3\nu_3} = K^{\lambda_1}_{\omega_1}[[P_n(\nabla) \otimes t^{\omega\alpha\tau}]_\lambda \otimes [P^{n_1} \otimes \xi^{\lambda_1}]_{\omega_1}]_{\beta\omega_3\nu_3}, \tag{3.5}$$

where

$$n = \sum_i [\lambda_i - w_i], \quad n_1 = \sum_i [(\lambda_1)_i - (w_1)_i]. \tag{3.6}$$

Thus, the components of $t^{\omega\alpha\tau}$ map the vectors $\{\xi^{\lambda_1}_\nu\}$ to linear combinations of the vectors $\{\xi^{\lambda_1+\tau}_\nu\}$ while the components of $P^n(\nabla)$ operate on the homogeneous polynomials $P^{n_1}_\mu(z)$, of degree n_1 in (z_1, z_2, z_3), to generate homogeneous polynomials of degree $n_1 - n$. Thus, given the matrix elements of the u(3) shift tensors (which can be defined in a convenient way), it is a simple matter to evaluate the matrix elements of $T^{\lambda\omega\alpha\tau}_{\lambda\nu}$. One finds, for example, that

$$\langle\lambda_3\lambda_3\|T^{\lambda_2\omega\alpha\tau}_{\lambda_2}\|\lambda_1\omega_1\rangle_\beta = \delta_{\lambda_3,\lambda_1+\tau}\delta_{n,n_1}K^{\lambda_1}_{\omega_1}$$

$$\times \sqrt{\tfrac{1}{2}(n+1)(n+2)}\begin{bmatrix} \lambda_1 & \tilde{n} & \omega_1 & - \\ \omega & n & \lambda_2 & - \\ \lambda_3 & 0 & \lambda_3 & - \\ \alpha & - & \beta & \end{bmatrix}, \quad (3.7)$$

where the array in square brackets is a u(3) $9j$ symbol and we use the notation $n \equiv (n,0,0)$, $\tilde{n} \equiv (0,0,-n)$.

One can evaluate the other $\langle\lambda_3\lambda_3\|T^{\lambda_2\omega\alpha\tau}_{\omega_2}\|\lambda_1\omega_1\rangle_\beta$ matrix elements by stepping down ω_2 from its highest value λ_2 with the f lowering operators. We start with the identity

$$[f, T^{\lambda_2\omega\alpha\tau}_{\omega_2}]_{\omega'_2\nu_2}$$
$$= (f \otimes T^{\lambda_2\omega\alpha\tau}_{\omega_2})_{\omega'_2\nu_2} - (-1)^{\phi(\omega_2,\omega'_2)}(T^{\lambda_2\omega\alpha\tau}_{\omega_2} \otimes f)_{\omega'_2\nu_2}, \quad (3.8)$$

where $\phi(\omega_2,\omega'_2)$ is a phase factor. This gives the equation

$$\langle\lambda_3\lambda_3\|[f, T^{\lambda_2\omega\alpha\tau}_{\omega_2}]_{\omega'_2\nu_2}\|\lambda_1\omega_1\rangle_\beta$$
$$= -(-1)^{\phi(\omega_2,\omega'_2)}\langle\lambda_3\lambda_3\|(T^{\lambda_2\omega\alpha\tau}_{\omega_2} \otimes f)_{\omega'_2\nu_2}\|\lambda_1\omega_1\rangle_\beta. \quad (3.9)$$

From the basic su(4) representation theory (cf. Eq. (2.8)), the lhs is expressed

$$\langle\lambda_3\lambda_3\|[\hat{f}, \widehat{T}^{\lambda_2\omega\alpha\tau}_{\omega_2}]_{\omega'_2}\|\lambda_1\omega_1\rangle_\beta = \langle\lambda_3\lambda_3\|\widehat{T}^{\lambda_2\omega\alpha\tau}_{\omega'_2}\|\lambda_1\omega_1\rangle_\beta f^{\lambda_2}_{\omega'_2\omega_2}\frac{K^{\lambda_2}_{\omega_2}}{K^{\lambda_2}_{\omega'_2}}, \quad (3.10)$$

with

$$f^\lambda_{\omega'\omega} = \sqrt{n+1}\,\mathrm{U}(\lambda\tilde{n}\omega'\tilde{1}; \omega\widetilde{n+1})\left(\frac{K^\lambda_{\omega'}}{K^\lambda_\omega}\right)^2. \quad (3.11)$$

A Racah recoupling of the rhs gives

$$\langle\lambda_3\lambda_3\|[\widehat{T}^{\lambda_2\omega\alpha\tau}_{\omega_2} \otimes \hat{f}]_{\omega'_2}\|\lambda_1\omega_1\rangle_\beta$$
$$= \sum_{\rho\omega'_1}\mathrm{U}(\omega_1\tilde{1}\lambda_3\omega_2; \omega'_1\rho\omega'_2\beta)\langle\lambda_3\lambda_3\|\widehat{T}^{\lambda_2\omega\alpha\tau}_{\omega_2}\|\lambda_1\omega'_1\rangle_\rho f^{\lambda_1}_{\omega'_1\omega_1}\frac{K^{\lambda_1}_{\omega_1}}{K^{\lambda_1}_{\omega'_1}}. \quad (3.12)$$

Thus, if we write

$$\langle\lambda_3\lambda_3\|T^{\lambda_2\omega\alpha\tau}_{\omega_2}\|\lambda_1\omega_1\rangle_\beta = K^{\lambda_1}_{\omega_1}K^{\lambda_2}_{\omega_2}F(\lambda_1\omega_1, \lambda_2\omega_2\|\omega\alpha\lambda_3\lambda_3)_\beta, \quad (3.13)$$

and equate the above expressions for the left and right sides of Eq. (3.9), we obtain the recursion relation:

$$F\big(\lambda_1\omega_1, \lambda_2\omega_2\|\omega\alpha\lambda_3\lambda_3\big)_\beta$$
$$= -(-1)^{\phi(\omega_2,\omega_2')} \sum_{\rho\omega_1'} U(\omega_1 \tilde{1}\lambda_3\omega_2'; \omega_1'\rho\omega_2\beta) \frac{f^{\lambda_1}_{\omega_1'\omega_1}}{rf^{\lambda_2}_{\omega_2\omega_2'}}$$
$$\times F\big(\lambda_1\omega_1', \lambda_2\omega_2'\|\omega\alpha\lambda_3\lambda_3\big)_\rho. \quad (3.14)$$

CG coefficients

First the special CG coefficients $\big\{(\lambda_1\omega_1, \lambda_2\omega_2\|\sigma\lambda_3\lambda_3)_\alpha\big\}$ are derived as linear combinations of the reduced matrix elements

$$\big(\lambda_1\omega_1, \lambda_2\omega_2\|\sigma\lambda_3\lambda_3\big)_\beta = \sum_{\omega\alpha} C^{\lambda_1\lambda_2\lambda_3}_{\omega\alpha} \langle\lambda_3\lambda_3\|T^{\lambda_2\omega\alpha\tau}_{\omega_2}\|\lambda_1\omega_1\rangle_\beta, \quad (3.15)$$

which satisfy the orthogonality relations

$$\sum_{\alpha\omega_1\omega_2} \big(\lambda_1\omega_1, \lambda_2\omega_2\|\sigma\lambda_3\lambda_3\big)_\alpha \big(\lambda_1\omega_1, \lambda_2\omega_2\|\rho\lambda_3\lambda_3\big)_\alpha = \delta_{\sigma\rho}. \quad (3.16)$$

This is achieved by regarding the expansion coefficients in Eq. (3.15) as the components of vectors in a nonorthogonal basis. Thus, they can be inferred by an orthogonalization process [23].

Having determined these special coefficients, we can construct the highest-grade states

$$\Psi^{\sigma\lambda_3}_{\lambda_3\nu_3} = \sum_{\omega_1\omega_2\alpha} \big(\lambda_1\omega_1, \lambda_2\omega_2\|\sigma\lambda_3\lambda_3\big)_\alpha [\psi^{\lambda_2}_{\omega_2} \otimes \psi^{\lambda_1}_{\omega_1}]_{\alpha\lambda_3\nu_3}, \quad (3.17)$$

where $[\psi^{\lambda_2}_{\omega_2}\otimes\psi^{\lambda_1}_{\omega_1}]_{\alpha\omega_3\nu_3}$ is a u(3)-coupled wave function in the tensor product space $F^{\lambda_2}\otimes F^{\lambda_1}$. We can then apply the f lowering operators to these states, using the known representation theory of su(4), to obtain all the states $\Psi^{\sigma\lambda_3}_{\omega_3\nu_3}$. Hence, we can compute general CG coefficients as the overlaps

$$\big(\lambda_1\omega_1, \lambda_2\omega_2\|\sigma\lambda_3\omega_3\big)_\alpha = \langle[\psi^{\lambda_2}_{\omega_2} \otimes \psi^{\lambda_1}_{\omega_1}]_{\alpha\omega_3\nu_3} \mid \Psi^{\sigma\lambda_3}_{\omega_3\nu_3}\rangle. \quad (3.18)$$

In this way, we obtain the explicit expression,

$$\big(\lambda_1\omega_1, \lambda_2\omega_2\|\sigma\lambda_3\omega_3\big)_\alpha = \frac{K^{\lambda_1}_{\omega_1}K^{\lambda_2}_{\omega_2}}{K^{\lambda_3}_{\omega_3}} \sum_{\omega_1'\omega_2'\beta} \big(\lambda_1\omega_1', \lambda_2\omega_2'\|\sigma\lambda_3\lambda_3\big)_\beta$$

$$\times \frac{1}{K^{\lambda_1}_{\omega_1'}K^{\lambda_2}_{\omega_2'}} \binom{n}{n_1 - n_1'} \sqrt{\frac{n_1!n_2!}{n_1'!n_2'!n!}}$$

$$\times \mathrm{U}(\lambda \widetilde{n}_1' \omega_1 \overbrace{n_1 - n_1'}; \omega_1' \widetilde{n}_1) \, \mathrm{U}(\lambda \widetilde{n}_2' \omega_2 \overbrace{n_2 - n_2'}; \omega_2' \widetilde{n}_2)$$

$$\times \begin{bmatrix} \overbrace{\omega_1'} & \overbrace{\omega_2'} & \lambda_3 & \beta \\ n_1 - n_1' & n_2 - n_2' & \widetilde{n} & - \\ \omega_1 & \omega_2 & \omega_3 & \alpha \\ - & - & - & \end{bmatrix}. \tag{3.19}$$

Apart from the special coefficients $\left\{ \left(\lambda_1 \omega_1, \lambda_2 \omega_2 \| \sigma \lambda_3 \lambda_3 \right)_\beta \right\}$, which can be computed once and stored for future use, this expression involves only some simple K_ω^λ coefficients and u(3) coupling and recoupling coefficients, for which computer codes already exist (cf. the Draayer–Akiyama package [1, 6]).

4   Summary

Although I have focused on the representations and CG coefficients for su(4), the above algorithms work for all su(n) with $n > 1$. However, to apply them, one needs to know the coupling and recoupling coefficients of su($n - 1$). Thus, to complete the Biedenharn–Louck program of giving algorithms for all su(n) coupling coefficients, we also need algorithms for computing Racah and $9j$ symbols from the CG coefficients. This can presumably be done by the methods of Hecht [15], Draayer and Akiyama [1, 6], and Millener [21]. But it is clear that the computer time needed to evaluate the coefficients will generally increase rapidly with n. Fortunately, practical applications often require only special couplings for which simplifications arise. For example, the expressions for CG coefficients simplify enormously for multiplicity-free $\lambda_2 \otimes \lambda_1$ couplings of the type $(N_2, 0, 0) \otimes (N_1, 0, 0)$. Indeed, it is probably that such coefficients can be evaluated analytically as was shown, in Ref. [25], to be the case for su(3).

The solution of the su(n) coupling problem in an su($n - 1$) basis is important as a prototype of other situations that can be treated in a parallel way. Such situations concern subgroup chains $G \supset H$ for which the VCS representations are holomorphic and there are no inner multiplicities (i.e., no missing labels in the $G \supset H$ classification of states). The representation theory is then especially simple and VCS theory gives analytical expressions for the matrices.

However, it is known that VCS representation theory can also handle many situations in which inner multiplicities do occur. Moreover, as shown in Ref. [27], VCS theory gives shift tensors for a much larger class of subgroup chains for which the representations are holomorphic. Thus, it will be of interest to see if these tensors can be used to derive CG coefficients. Two examples of physical interest that come to mind are the groups SO(6) and Sp(3, $\mathbb{R}$) in U(3) bases.

It is also noteworthy that VCS theory has been successfully applied to the construction of nonholomorphic representations. For example, it has been applied to construct representations of SU(3) and SO(5) in SO(3) bases. It will be of considerable practical importance if corresponding shift tensors can be derived to induce CG coefficients for these groups from those of their SO(3) subgroups.

REFERENCES

1. Y. Akiyama and J.P. Draayer, *A user's guide to FORTRAN programs for Wigner and RACAH coefficients of* SU$_3$, Comput. Phys. Comm. **5** (1973), 405–406.

2. G.E. Baird and L.C. Biedenharn, *On the representations of the semisimple Lie groups.* II, J. Math. Phys. **4** (1963), 1449–1466.

3. G.E. Baird and L.C. Biedenharn, *On the representations of the semisimple Lie groups.* III. *The explicit conjugation operation for* SU$_n$, J. Math. Phys. **5** (1964), 1723–1730.

4. L.C. Biedenharn, A. Giovannini, and J.D. Louck, *Canonical definition of Wigner coefficients in* U$_n$, J. Math. Phys. **8** (1967), 691–700.

5. L.C. Biedenharn and J.D. Louck, *A pattern calculus for tensor operators in the unitary groups*, Comm. Math. Phys. **8** (1968), 89–131.

6. J.P. Draayer and Y. Akiyama, *Wigner and Racah coefficients for* SU$_3$, J. Math. Phys. **14** (1973), 1904–1912.

7. J.P. Elliott and J.A. Evans, *An intrinsic spin for interacting bosons*, Phys. Lett. B **101** (1981), 216–218.

8. D. Flath, *Coherent tensor operators*, Lie Algebras, Cohomology, and New Applications to Quantum mechanics (Springfield, 1992), Contemp. Math. Vol. 160, Amer. Math. Soc., Providence, RI, 1994, pp. 75–84.

9. D.E. Flath and J. Towber, *Tensor operators.* I. *The concept of a coherent tensor operator*, Comm. Algebra **18** (1990), 4047–4086.

10. D.E. Flath and J. Towber, *Tensor operators.* II. *The algebra of tensor operators*, Comm. Algebra **20** (1992), 2903–2917.

11. D.E. Flath and J. Towber, *Tensor operators.* I. *The concept of a coherent tensor operator*, Comm. Algebra **18** (1990), 4047–4086.

12. D.E. Flath and J. Towber, *Tensor operators.* III. *Some fundamental tensor operator identities*, J. Math. Phys. **34** (1993), 1523–1547.

13. I.M. Gel'fand and M.L. Tseitlin, *Finite-dimensional representations of the group of unimodular matrices*, Doklady Akad. Nauk SSSR (N.S.) **71**, (1950), 825–828. (Russian)

14. K.T. Hecht, *The vector coherent state method and its application to problems of higher symmetries*, Lecture Notes in Physics, Vol. 20, Springer-Verlag, Berlin-New York, 1987.

15. K.T. Hecht, SU$_3$ *recoupling and fractional parentage in the* 2s − 1d *shell*, Nuclear Phys. B **62** (1965), 1–36.

16. K.T. Hecht, R. Le Blanc, and D.J. Rowe, *New perspective on the* U(n) *Wigner–Racah calculus. I. Vector coherent state theory and construction of Gel'fand bases*, J. Phys. A **20** (1987), No. 9, 2241–2250.

17. K.T. Hecht and S.C. Pang, *On the Wigner supermultiplet scheme*, J. Math. Phys. **10** (1969), 1571–1616.

18. F. Iachello and A. Arima, *The interacting boson model*, Cambridge University Press, Cambridge, 1987.

19. J.D. Louck, *Recent progress toward a theory of tensor operators in the unitary groups*, Amer. J. Phys. **38** (1970), 3–42.

20. J.D. Louck and L.C. Biedenharn, *Canonical unit adjoint tensor operators in* U(n), J. Math. Phys. **11** (1970), 2368–2414.

21. D.J. Millener, *A note on recoupling coefficients for* SU(3), J. Math. Phys. **19** (1978), No. 7, 1513–1514.

22. D.J. Rowe, *Coherent state theory of the noncompact symplectic group*, J. Math. Phys. **25** (1984), No. 9, 2662–2671.

23. D.J. Rowe, *Properties of overcomplete and nonorthogonal basis vectors*, J. Math. Phys. **10** (1969), 1774–1777.

24. D.J. Rowe and J. Repka, *The representations and coupling coefficients of* su(n); *application to* su(4), Found. Phys. **27** (1997), No. 8, 1179–1209.

25. D.J. Rowe and J. Repka, *An algebraic algorithm for calculating Clebsch–Gordan coefficients; application to* SU(2) *and* SU(3), J. Math. Phys. **38** (1997), No. 8, 4363–4388.

26. D.J. Rowe and J. Repka, *The Racah–Wigner algebra and coherent tensors*, J. Math. Phys. **37** (1996), No. 5, 2498–2509.

27. D.J. Rowe and J. Repka, *Induced shift tensors in vector coherent state theory*, J. Math. Phys. **36** (1995), No. 4, 2008–2029.

28. D.J. Rowe, G. Rosensteel, and R. Carr, *Analytical expressions for the matrix elements of the noncompact symplectic algebra*, J. Phys. A **17** (1984), No. 8, L399–L403.

29. D.J. Rowe, G. Rosensteel, and R. Gilmore, *Vector coherent state representation theory*, J. Math. Phys. **26** (1985), No. 11, 2787–2791.

30. D.J. Rowe, B.G. Wybourne, and P.H.Butler, *Unitary representations, branching rules and matrix elements for the noncompact symplectic groups*, J. Phys. A **18** (1985), No. 6, 939–953.

16

Highest-Weight Representations of Borcherds Algebras

Richard Slansky

ABSTRACT General features of highest-weight representations of Borcherds algebras are described. These algebras provide a way to extend a Kac–Moody algebra to include the Hamiltonian and number-changing operators in a generalized symmetry structure. To show their typical features, the weight systems of several representations of a Borcherds extension of su(2) are computed. Then the extension of affine-su(2) to a Borcherds algebra is examined.

1 Introduction

This chapter examines the highest-weight representation theory of Borcherds algebras with the goal of giving some rather primitive thoughts on how these algebras might be used in quantum field theory. Borcherds algebras were proposed in a 1988 paper by Richard Borcherds [1]. They are extensions of Kac–Moody algebras, a topic to which J. Patera has contributed significantly, for example, Ref. [2]. The work I am reporting today would not have been possible without the knowledge I derived from my collaborations with Patera.

The physics setting of this work is the energy operator of a quantum field theory, which is associated with the time-translation invariance of the Lagrangian. Its role in the overall symmetry structure of the Lagrangian varies: In four-dimensional theories it belongs to the Lie algebra of the Poincaré group, which is a contraction of noncompact SO(5). In two-dimensional conformal field theories, the Hamiltonian is a member of the infinite-dimensional Virasoro algebra. If the symmetry is further extended by a finite-dimensional Lie algebra, the conformal theory is a representation of a Kac–Moody algebra (the vertex construction), and is a solution to the corresponding two-dimensional current algebra.

The usual construction of a Lie algebra and its representation theory requires a nondegenerate bilinear form that is derived from the Cartan matrix. In the affine Kac–Moody case where the determinant of the Cartan matrix is zero, this nondegenerate form can be obtained by extending the Cartan subalgebra by an independent diagonalizable operator, which is the Hamiltonian $-L_0$ in simple two-dimensional conformal field theories. L_0 does not have the full status of the other members of the Cartan subalgebra of a Kac–Moody algebra, since from one point of view its role is defined through this extension, but from another point of view, it is contained in the Virasoro algebra, and the full algebra is a semidirect product of the Virasoro algebra with the affine Kac–Moody algebra. However, this new matrix for the nondegenerate bilinear form actually defines a Lie algebra, although it is not a Kac–Moody algebra. It is a Borcherds algebra.

It is the purpose of this talk to study a few simple representations of these Lie algebras. The number operator and the Hamiltonian can be promoted to full-fledged members of the Cartan subalgebra of a Borcherds algebra [3]. There is an interesting twist in this construction when applied to affine su(2): The extended representations contain states of all numbers of particles. The original Hamiltonian becomes a member of the Cartan subalgebra, as does the operator that measures the level (or number of particles). The highest-weight representations of the Borcherds-extended affine Kac–Moody algebra contain affine representations of all levels. (The branching rules of various highest-weight representations are "sliced" or organized according to the eigenvalue of the number operator.)

A possible shortcoming of this construction is that states of all statistics are in a Borcherds representation. There is no restriction to Bose or Fermi statistics. The construction does give an example where particle-number changing operators are in a Lie algebra. However, we have not constructed a field theory from which this symmetry structure is derived.

Section 2 is a brief summary of results on Borcherds algebras, including a survey of the theory of its highest-weight representations. As extensions of Kac–Moody algebras, most of the highest-weight representation theory of Kac–Moody algebras directly applies to Borcherds algebras [1]. The review of Section 2 is completed by computing the roots and their multiplicities of a Borcherds extension of su(2) and the weights and multiplicities of several of its highest-weight representations.

The solution to the problem of constructing the extended Cartan subalgebra that includes the Hamiltonian in an affine Kac–Moody algebra [2] is summarized in Section 3. In this construction the natural choice of the linear functional of the Hamiltonian with itself is zero; if this bilinear form were used as a Cartan matrix for some new algebra, it would violate the rule for Kac–Moody algebras that the diagonal elements of the Cartan matrix all be 2. However, this nondegenerate bilinear form defines a Borcherds Lie algebra. It may be useful to analyze the representations in terms of this new algebra. The Borcherds algebras extend and generalize Kac–Moody algebras by adding an imaginary root to the set of simple roots, so a zero-valued diagonal element of the symmetrized Cartan matrix is possible.

Section 4 explores how a two-dimensional current algebra is extended to include number-changing operators. Adding an imaginary simple root to an affine algebra leads to two infinite directions, one Hamiltonian-like and the other counts the number of particles in the state. Both operators are constructed as sums of the diagonal operators that correspond to the simple roots of the extended algebra. Results and speculations are briefly restated in Section 5.

This introduction concludes with a brief review of those aspects of Kac–Moody algebras that are of greatest importance to the Borcherds generalization. Speaking roughly, the approach to Lie algebra theory that has been the most fruitful for generalization starts from an analysis of the space $\mathcal{P}$ spanned by the eigenvalues of the Cartan subalgebra, which is a set of independent simultaneously diagonalizable operators in the Lie algebra. In the finite-dimensional case $\mathcal{P}$ is a Euclidean space with a positive-definite metric and of dimension equal to the rank ℓ of the Lie algebra. The vectors in this space are the roots of the Lie algebra and the weights of its highest-weight representations.

The basis of the Cartan subalgebra is a set of ℓ linearly independent diagonalizable operators h_i, $i = 1, \ldots, \ell$, conveniently chosen. A root α is an ℓ-component vector in $\mathcal{P}$ whose components are the eigenvalues of h_i in the sense of the Lie bracket:

$$[h_i, e_\alpha] = \alpha(h_i)e_\alpha, \tag{1.1}$$

where e_α is a ladder operator that transforms a Hilbert space representation vector with weight λ to one with weight $\lambda + \alpha$ (if $\lambda + \alpha$ is a weight of the representation); the functional $\alpha(h_i)$ is the ith component of the ℓ-dimensional root α.

For a vector $|\mathbf{r}, \lambda\rangle$ in the Hilbert space of a representation $\mathbf{r}$, the ℓ-dimensional vector λ in $\mathcal{P}$ is called a weight. The components of the weight λ are defined by

$$h_i|\mathbf{r}, \lambda\rangle = \lambda(h_i)|\mathbf{r}, \lambda\rangle. \tag{1.2}$$

Additional labels of the vector $|\mathbf{r}, \lambda\rangle$ when the multiplicity of λ is greater than unity have been suppressed.

The Cartan subalgebra is dual to the space of roots, so $\alpha(h_i)$ and $\lambda(h_i)$ are linear functionals. The weight or root component $\lambda(h_i)$ is defined by

$$\lambda(h_i) = (\lambda \mid \alpha_i), \tag{1.3}$$

where this definition of $\lambda(h_i)$ differs from the usual one by a normalization factor $2/((\alpha_i \mid \alpha_i))$, see Refs. [2, 4]. The simple root α_i in $\mathcal{P}$ corresponds to the diagonal operator h_i in the Cartan subalgebra, and the component of α_i associated with h_j is $\alpha_i(h_j) = (\alpha_i \mid \alpha_j)$, where $(\alpha_i \mid \alpha_j)$ is the natural scalar product in $\mathcal{P}$ of the simple roots.

The commutation relations that define the Lie algebra are defined in terms of a set of "generators" that generate the full Lie algebra through multiple commutators. For a rank ℓ algebra there are 3ℓ generators, e_i, f_i, and h_i, $i = 1, \ldots, \ell$. The "presentation" of the Lie algebra is then

$$
= (\alpha_j \mid \alpha_i)e_j, \quad [h_i, f_j] = -(\alpha_j \mid \alpha_i)f_j,
$$
$$
[e_i, f_j] = \frac{2}{(\alpha_i \mid \alpha_i)}\delta_{ij}h_i. \tag{1.4}
$$

The values of the scalar products $(\alpha_j \mid \alpha_i)$ are specified by the Cartan matrix that defines the Kac–Moody algebra. The presentation Eq. (1.4) normalizes the Cartan subalgebra differently from the usual conventions, but is convenient for Borcherds algebras.

Equation (1.4) does not completely define the Lie algebra. The "Serre relations" impose the requirement that certain multiple commutators vanish:

$$
(\mathrm{ad}\, e_i)^{1-2(\alpha_i \mid \alpha_j)/(\alpha_i \mid \alpha_i)}e_j = 0, \quad (\mathrm{ad}\, f_i)^{1-2(\alpha_i \mid \alpha_j)/(\alpha_i \mid \alpha_i)}f_j = 0, \tag{1.5}
$$

where, for example, $(\mathrm{ad}\, e_i)^2 e_j$ means $\big[e_i, [e_i, e_j]\big]$. The presentation of the Lie algebra (Eqs. (1.4) and (1.5)) agrees with the usual one up to the normalization of the Cartan subalgebra. Equation (1.5) makes sense if $(\alpha_i \mid \alpha_i) > 0$, which is always the case for the simple roots of Kac–Moody algebras.

Kac–Moody algebras are defined by the geometrical relations (scalar products) among the eigenvalues (roots of Eq. (1.1)) of the diagonalizable operators. The first steps of the theory are dedicated to finding the best strategy for defining these relations from the axioms for Lie algebras. The method then requires proving that the construction of the eigenvalues may be lifted to a definition of the Lie algebra through its commutation relations. The geometry is defined by a matrix $A_{ij} = (\alpha_i \mid \alpha_j)$, where the ℓ-component Euclidean vectors α_i are called the simple roots, and the matrix A_{ij} is called the symmetrized Cartan matrix. The Cartan matrix itself for symmetrizable Kac–Moody algebras is defined by

$$
C_{ij} = \frac{2(\alpha_i \mid \alpha_j)}{(\alpha_j \mid \alpha_j)}. \tag{1.6}
$$

Kac–Moody algebras and the Borcherds generalization are completely defined by this matrix of scalar products of simple roots. The list of rules for C_{ij} that characterizes a Kac–Moody algebra is quite simple. The Cartan matrix is an integer matrix with diagonal elements $C_{ii} = 2$ and nonpositive integers for the off-diagonal elements with zeros matching pairwise. We consider cases where C is indecomposable, that is, it cannot be brought to block diagonal form by reordering rows and columns. If C has positive determinant, then the algebra is finite-dimensional. If C has zero determinant, then the algebra is an infinite-dimensional affine algebra; this is the case of interest in a two-dimensional current algebra. Finally, if the determinant is negative, the resulting algebra is one of the hyperbolic or other infinite-dimensional Kac–Moody algebras, which are not easily listed but are rather easily studied at the level of description of this paper.

A generalization of Kac–Moody algebras by modifying the definition of the Cartan matrix was discovered by Borcherds: For symmetrizable Cartan matrices the requirement that $C_{ii} = 2$ may be dropped by introducing a simple root α_1 of zero length, $A_{11} = (\alpha_1 \mid \alpha_1) = 0$. (The simple roots of a Kac–Moody algebra are all real.) The next section summarizes an example of Borcherds' results [1].

2   Borcherds Algebras

The extension from Kac–Moody algebras to Borcherds algebras is accomplished by relaxing the rules for forming the Cartan matrix. The simplest statement of the extension is, the set of simple roots α_i, $i = 1, \ldots, \ell$, may include imaginary roots. Our discussion is restricted to the case of just one imaginary simple root, selected to be α_1 and to have zero length. (The choice of zero length does not affect the representation theory so long as α_1 is imaginary.)

The new rules for the Cartan matrix must be supplemented with new rules for the presentation of the Lie algebra: The presentation given in Eq. (1.4) and (1.5) generalizes to Borcherds algebras except where $(\alpha_1 \mid \alpha_1) = 0$ causes nonsense. This includes the last relation of Eq. (1.4), where the right hand side of the equation is zero for $i = 1$, and the Serre relations Eq. (1.5), where the relation is simply ignored for $i = 1$. Note that this is a very strong assumption since $(\operatorname{ad} e_1)^n e_j \neq 0$ for $j \neq 1$ and any positive integer n. Thus, the Lie algebra places no constraints on the "1" direction. For all other simple roots, the presentation is unchanged, and the Cartan matrix C_{ij}, $j \neq 1$ satisfies the usual rules to be a Cartan matrix, which continues to be a symmetrizable integer matrix with nonpositive off-diagonal elements and $C_{ii} = 2$ for $i \neq 1$ [1].

The Borcherds paper summarizes the algebraic structure and representation theory of these extended algebras [1]. The focus in this section is the root and weight multiplicity formulas for Borcherds algebras and their representations. The representation theory of highest-weight representations is almost identical to the Kac–Moody case: The Weyl–Kac character formula is valid for Borcherds algebras [1]. Thus, the root and weight multiplicities can be computed in a manner identical to those of Kac–Moody algebras. In particular, the Peterson formula [4] is valid and provides a computational method for determining root and weight multiplicities.

The derivation of the Peterson formula for Kac–Moody algebras is worked out in detail in Section 22 of [2]. It is an iterative formula for the multiplicities of positive roots $\beta = \sum_i n_i \alpha_i$, in terms of the multiplicities of lower

TABLE 16.1. Positive roots of Borcherds algebra with symmetrized Cartan matrix given in Eq. (2.3) for $su(2)_B$. The imaginary simple root α_1 has zero norm. The positive root system is listed here, except at $n_1 = 0$, where one negative and two zero roots are included. (Recall that if α is a root, so is $-\alpha$.) The $su(2)$ representation is indicated by its highest-weight $\Lambda(h_2) = 2j$; for example, (3) is the spin or isospin 3/2 representation of dimension 4.

n_1	$su(2)$ content	n_1	$su(2)$ content
0	(2) + (0)	5	(3)+(1)
1	(1)	6	(4)+(2)+(0)
2	(0)	7	(5)+2(3)+2(1)
3	(1)	8	(6)+2(4)+4(2)+(0)
4	(2)	9	(7)+3(5)+5(3)+5(1)

positive roots. These are nonzero roots with components $0 \le m_i \le n_i$:

$$(\beta \mid \beta - 2\rho)c_\beta = \sum_{\beta'\beta''} (\beta' \mid \beta'')c_{\beta'}c_{\beta''}, \tag{2.1}$$

where the sum is over all positive roots β' and β'' with $\beta' + \beta'' = \beta$, and ρ is defined by $2(\rho \mid \alpha_i) = (\alpha_i \mid \alpha_i)$, $i = 1, \ldots, \ell$. The quantity c_β is defined by

$$c_\beta = \sum_{n>0} \frac{1}{n} \text{mult} \left(\frac{\beta}{n} \right), \tag{2.2}$$

where "mult" is the multiplicity of the root, which is nonzero only when β/n is a root. This sum necessarily terminates at some finite n for any root β.

The iteration formula Eq. (2.1) requires several "boundary conditions." The multiplicity of a simple root is always unity. If α_i is a real simple root, Eq. (2.1) reads $0 = 0$, so we set c_{α_i} and $\text{mult}(\alpha_i)$ to unity. The multiple of a simple root $n\alpha_i$, $n \ge 2$ is never a root. This determines $c_{n\alpha_i} = 1/n$. There may be other roots β for which $(\beta \mid \beta - 2\rho) = 0$. In these cases it always turns out that the sum in Eq. (2.1) is zero, and there is an apparent freedom in defining $\text{mult}(\beta)$. This freedom is greatly restricted by the requirement that all multiplicities must be integers. A choice that works when $(\beta \mid \beta - 2\rho) = 0$ is $\text{mult}(\beta) = 0$. Equations (2.1) and (2.2) can then be used to calculate the root multiplicities of any Kac–Moody or Borcherds algebra.

Table 16.1 gives the root multiplicities for the Borcherds algebra defined by the symmetrized Cartan matrix,

$$A(su(2)_B) = \begin{pmatrix} 0 & -1 \\ -1 & 2 \end{pmatrix}. \tag{2.3}$$

The simple root α_1 is imaginary and α_2 is real. The scalar product of roots $\alpha = n_1\alpha_1 + n_2\alpha_2$ and $\beta = m_1\alpha_1 + m_2\alpha_2$ is

$$(\alpha \mid \beta) = -n_1 m_2 - n_2 m_1 + 2n_2 m_2. \tag{2.4}$$

The root system is characterized by a set of su(2) representations for each value of n_1, and n_1 behaves like a number-operator eigenvalue. The name "su(2)$_B$" refers to the Lie algebra with symmetrized Cartan matrix Eq. (2.3).

The adjoint representation is not a highest-weight representation, since if α is a root, so is $-\alpha$. The infinite set of roots is not bounded from above or below. Nevertheless, the weight multiplicities of highest-weight representations can also be constructed from the Peterson formula for an enlarged algebra. In particular, the Freudenthal formula for Kac–Moody algebras is derived from the Peterson formula of an extended algebra with simple root α_0 appended to the simple roots, $\alpha_1, \ldots, \alpha_\ell$; the additional scalar products needed to define the extended algebra are $(\alpha_0 \mid \alpha_0) = 2$ and $(\alpha_0 \mid \alpha_i) = -(\Lambda \mid \alpha_i)$, $i = 1, \ldots, \ell$, where Λ is the highest-weight of the representation.

The results are illustrated by calculations of the $(1,0)$ representation of the algebra with symmetrized Cartan matrix Eq. (2.3). In order to calculate the weight multiplicities for the $(1,0)$ representation, one uses the Peterson formula Eqs. (2.1) and (2.2), but in conjunction with the extended Cartan matrix,

$$A^{(1,0)}\left(\mathrm{su}(2)_B\right) = \begin{pmatrix} 2 & -1 & 0 \\ -1 & 0 & -1 \\ 0 & -1 & 2 \end{pmatrix}. \tag{2.5}$$

The weight multiplicities of the highest-weight representation $(1,0)$ are the root multiplicities computed from $A^{(1,0)}\left(\mathrm{su}(2)_B\right)$ of the form,

$$\alpha = \alpha_0 + n_1\alpha_1 + n_2\alpha_2. \tag{2.6}$$

The Freudenthal formula follows from this extension, as derived in Refs. [2, 4, 5]. The weight multiplicities of the $(1,0)$ and $(0,1)$ (using a different extended Cartan matrix, $A^{(0,1)}\left(\mathrm{su}(2)_B\right)$ in this latter case) representations are listed for the first ten values of n_1 in Table 16.2. Note the infinity of weights starting from the highest-weight extending in the α_1 direction; h_1 is an operator with a semi-infinite spectrum.

It is possible to unravel the Borcherds representation in terms of Fock space operators, just as can be done for the affine Kac–Moody highest weight representations. The structure of the $(1,0)$ representation of su(2)$_B$ is extremely simple: It is possible to build this representation with an su(2) doublet of creation operators $a_{1/2,m}^{(1)\dagger}$ that carries $\Delta n_1 = 1$. All products of $a_{1/2,m}^{(1)\dagger}$ acting on the ground state are linearly independent in this construction; the lack of statistics is a problem for physical particles and is likely

TABLE 16.2. Branching rules of the $(1,0)$ and $(0,1)$ representations of the Borcherds algebra $su(2)_B$ of Table 16.1. Again, we slice the representation with α_1. The $su(2)$ representation is given by its highest-weight, $\Lambda(h_2) = 2$. The results for the $(1,0)$ are derived from the Peterson formula with Eq. (2.5) for the bilinear form.

n_1	$su(2)$ content of $(1, 0)$	$su(2)$ content of $(0, 1)$
0	(0)	(1)
1	(1)	(0)
2	(2)+(0)	(1)
3	(3)+2(1)	(2)+(0)
4	(4)+3(2)+2(0)	(3)+2(1)
5	(5)+4(3)+5(1)	(4)+3(2)+2(0)
6	(6)+5(4)+9(2)+5(0)	(5)+4(3)+5(1)
7	(7)+6(5)+14(3)+14(1)	(6)+5(4)+9(2)+5(0)
8	(8)+7(6)+20(4)+28(2)+14(0)	(7)+6(5)+14(3)+14(1)
9	(9)+8(7)+27(5)+48(3)+42(1)	(8)+7(6)+20(4)+28(2)+14(0)

owing to the assumption of no Serre relation, Eq. (1.5) for $i = 1$. The $su(2)$ structure at slice $n_1 = n$ is the tensor product of the (1) representation of $su(2)$ with itself n times.

3   Cartan Subalgebra of an Affine Kac–Moody Algebra

The group theoretical role of the Hamiltonian in conformal field theory and two-dimensional current algebra arises from the need to define a nondegenerate bilinear form for the algebra. Finite-dimensional Lie algebras are characterized by positive-definite Cartan matrices, and so the definition of the Lie algebra by its presentation Eqs. (1.4) and (1.5) has no ambiguity. However, for affine algebras the determinant of A is zero, so the Cartan matrix cannot be selected naively to be the bilinear form that defines the Lie algebra. A nondegenerate bilinear form is constructed by extending the Cartan subalgebra, and consequently extending the space of roots by adding a linearly independent vector corresponding to this new operator. The extension sketched here is worked out in more detail in Section 5 of Ref. [2].

The problem of the presentation given in Eq. (1.4) and (1.5) for an affine Kac–Moody algebra is that the functional $\alpha_i(h_j) = (\alpha_i \mid \alpha_j)$ is degenerate, so that for each root α, there is an infinite number of operators e_α and f_α with no immediate way to distinguish among them. The problem of labeling the roots is trivially solved for those affine algebras constructed as central extensions of loop algebras, as is reviewed in Section 3 and Section 16 of Ref. [2]. The general solution focuses on the geometry of $\mathcal{P}$, which is now outlined.

The solution to the labeling problem is to extend the $\ell \times \ell$ Cartan matrix to an $(\ell+1) \times (\ell+1)$ nonsingular bilinear form by introducing the operator L_0:

$$[L_0, e_\alpha] = \alpha(L_0)e_\alpha, \quad [L_0, f_\alpha] = -\alpha(L_0)f_\alpha, \quad [L_0, h_i] = 0. \qquad (3.1)$$

L_0 closes with the remaining operators of the algebra and can be added to the Cartan subalgebra. An example of Eq. (3.1) is provided by the vertex construction. Our interest in Borcherds algebras was aroused by the result that the resulting nondegenerate bilinear form is the Cartan matrix of a Borcherds algebra.

It is required that $\alpha_i(h_j) = (\alpha_i \mid \alpha_j)$ not be changed by the extension, which adds $\alpha_i(L_0)$, $L_0(h_i)$ and $\Lambda_0(L_0)$, where Λ_0 is the vector in $\mathcal{P}^{\text{ext}}$ that corresponds to L_0. The extension of $\mathcal{P}$ to $\mathcal{P}^{\text{ext}}$ is $(\ell+1)$-dimensional. The symmetry structure imposes several conditions. The bilinear form must be symmetric:

$$\alpha_i(L_0) = (\alpha_i \mid \Lambda_0) = (\Lambda_0 \mid \alpha_i) = L_0(h_i). \qquad (3.2)$$

The linear dependence in the affine Cartan matrix must be treated in a consistent fashion. This linear dependence is expressed in terms of a root δ defined by

$$\delta = \sum_i c_i \alpha_i, \qquad (3.3)$$

where the integer coefficients c_i (called "marks") depend only on the algebra. Before the extension, δ is literally zero and Eq. (3.3) simply expresses the linear dependence among the rows of the Cartan matrix:

$$\delta(h_i) = (\delta \mid \alpha_i) = 0. \qquad (3.4)$$

The whole point of extending the Cartan subalgebra by L_0 is to be able to require $\delta(L_0) \neq 0$ in $\mathcal{P}^{\text{ext}}$, and avoid the degeneracy in $\mathcal{P}$ implied by Eq. (3.4). The critical definition is

$$\delta(L_0) = (\delta \mid \Lambda_0) = \left(\sum_{i=1}^{\ell} c_i \alpha_i \,\middle|\, \Lambda_0 \right) = -1. \qquad (3.5)$$

This provides an extended bilinear form that completely labels the operators in the affine Lie algebra. In cases where the Cartan matrix is symmetrical ($c_1 = 1$), the simplest solution to Eqs. (3.2) and (3.5) is to add a zeroth row (and column) of the form $(0, -1, 0, \ldots, 0)$ to the Cartan matrix.

Finally, the value of $\Lambda_0(L_0)$ is not very important so long as it is not 2. The natural choice is to set $(\Lambda_0 \mid \Lambda_0) = 0$.

The extended Cartan subalgebra is selected to include L_0 along with h_i, $i = 1, \ldots, \ell$. The basis vectors of the extended root space corresponding to these operators are then Λ_0, and α_i, $i = 1, \ldots, \ell$. Then the bilinear form is defined as $A_{ij} = (\alpha_i \mid \alpha_j)$ for $i, j = 1, \ldots, \ell$ plus a zeroth row and column. The zeroth row is constrained by $(\delta \mid \Lambda_0) = -1$; for the algebras analyzed here, we take $(\alpha_1 \mid \Lambda_0) = -1$, $(\alpha_i \mid \Lambda_0) = 0$, $i = 2, \ldots, \ell$ and $(\Lambda_0 \mid \Lambda_0) = 0$. In making the extension of the Cartan matrix to a Borcherds algebra, it is necessary to identify the operators corresponding to the simple roots.

4   Adding Energy and Number Operators to the Cartan Subalgebra

In Section 2 it was suggested from a simple quantum-mechanical example that a number operator appears in the Cartan subalgebra of a Borcherds extended finite-dimensional Lie algebra. In this section the Hamiltonian and number operators are explicitly constructed from the simple roots of the Borcherds algebra by constructing the Borcherds extension of affine-su(2).

As noted in Section 3, the addition of an imaginary simple root to an affine algebra according to the constraints Eqs. (3.2)–(3.5) gives a symmetrized Cartan matrix of a Borcherds algebra. Thus, we obtain the Borcherds algebra, affine-su(2)$_B$. (In affine-su(2), α_1 and α_2 subtend $180°$ in $\mathcal{P}$ and have equal lengths.) The Cartan matrix is

$$A(\text{affine- su}(2)_B) = \begin{pmatrix} 0 & -1 & 0 \\ -1 & 2 & -2 \\ 0 & -2 & 2 \end{pmatrix}. \tag{4.1}$$

The root system is calculated from the Peterson formula (Eq. (2.1)) with the scalar product in $\mathcal{P}$ defined by Eq. (4.1). Roots are of the form

$$\alpha = n_1\alpha_1 + n_2\alpha_2 + n_3\alpha_3. \tag{4.2}$$

The root system can be broken up into representations of affine su(2) by computing the multiplicities of roots of the form Eq. (4.2). The roots at $n_1 = 0$ correspond precisely to the root system of affine su(2), which is a series of su(2) triplets at each integer multiple of $\delta = \alpha_2 + \alpha_3$. The positive roots for $n_1 = 1$ are those of the $(1, 0)$ representation of affine-su(2). The $n_1 = 2$ slice is a reducible sequence of affine-su(2) representations, each starting at a specific multiple of δ, n_δ:

$$(2, 0)_{n_\delta=0} + (2, 0)_{n_\delta=1} + (2, 0)_{n_\delta=2} + (2, 0)_{n_\delta=3}$$

$$+ 2(2,0)_{n_\delta=4} + 2(2,0)_{n_\delta=5} + 3(2,0)_{n_\delta=6} + 4(2,0)_{n_\delta=7}$$
$$+ 5(2,0)_{n_\delta=8} + 6(2,0)_{n_\delta=9} + \cdots , \quad (4.3)$$

where the multiplicity of the affine-su(2) representation eventually grows exponentially like a typical partition function found in the theory of Kac–Moody representations. In particular, these multiplicities are the coefficients in the expansion of the partition function of the $c = 1/2$ (c is the central charge) representation of the Virasoro algebra with highest-weight $\Lambda(L_0) = 1/2$. These numbers correspond to the dimensions of the Hilbert subspaces gotten from applying an odd number of the Neveu–Schwarz (half-odd integer moded, anticommuting) operators to the ground state.

The weight-system multiplicities of the $(1,0,0)$ representation of affine-su(2) are given by the Peterson formula with the bilinear form

$$A^{(1,0,0)}(\text{affine-}\mathrm{su}(2)_B) = \begin{pmatrix} 2 & -1 & 0 & 0 \\ -1 & 0 & -1 & 0 \\ 0 & -1 & 2 & -2 \\ 0 & 0 & -2 & 2 \end{pmatrix}. \quad (4.4)$$

The weights of the $(1,0,0)$ representation are of the form

$$\alpha = \alpha_0 + n_1\alpha_1 + n_2\alpha_2 + n_3\alpha_3. \quad (4.5)$$

A detailed discussion of affine su(2) representations is contained in Ref. [2], and the tables given there are used to unravel the weight multiplicities computed from Eq. (4.4) into affine su(2) representations.

For the $(1,0,0)$ representation, the $n_1 = 0$ weight is a singlet at $n_2 = n_3 = 0$ and corresponds to the vacuum. The $n_1 = 1$ weights are in the $(1,0)$ representation of affine-su(2). As with the roots for this algebra, the $n_1 = 2$ weights are reducible under affine-su(2), and can be decomposed into affine-su(2) irreducible representations as

$$\left[(2,0) + (0,2)\right]_{n_\delta=0} + (0,2)_{n_\delta=1} + \left[(2,0) + (0,2)\right]_{n_\delta=2}$$
$$+ \left[(2,0) + (0,2)\right]_{n_\delta=3} + 2\left[(2,0) + (0,2)\right]_{n_\delta=4}$$
$$+ 2\left[(2,0) + (0,2)\right]_{n_\delta=5} + \cdots . \quad (4.6)$$

This sequence is the set of representations in the tensor product $(1,0) \times (1,0)$ of affine-su(2), which is computed in Section 6 of Ref. [2]. The $n_1 = 3$ slice has the representations of $(1,0) \times (1,0) \times (1,0)$. It appears obvious that this structure generalizes to all larger n_1. Thus, a generalization of a two-dimensional current algebra that includes multiparticle states, where the single-particle states are in the $(1,0)$ representation of affine-su(2), is the $(1,0,0)$ representation of affine-su(2)$_B$. It includes a vacuum at $n_1 = 0$, single particle states at $n_1 = 1$, two-particle states at $n_1 = 2$, three-particle states at $n_1 = 3$, and so on. Thus, the full multiparticle space of states is included in this single representation of extended affine-su(2), and the algebra contains operators that change number of particles.

5 Conclusions

It can be interesting to survey new mathematical structures for applications in physics. In this paper we have proposed the use of Borcherds algebras and their representations to describe multiparticle states. It is an algebraic structure that extends quantum mechanics, as in the examples, from a single particle description to a structure that unifies all numbers of particles. Thus, there is an interesting Fock space structure of the simplest representations of the simplest algebra $su(2)_B$, where the representation space $(1,0)$ is the Fock space of an $su(2)$ doublet of quantum mechanical operators. Similarly the analogous extension of affine $su(2)$ as a function of n_1 is a vacuum for $n_1 = 0$, single particle for $n_1 = 1$, two particle for $n_1 = 2$, and so on. We have not discussed how to construct such a theory in detail, but it seems physically clear that multiparticle states are rather natural in the representation theory of Borcherds algebras. One might speculate that such a structure is useful for second quantization of a single particle theory. More particularly, we have shown that the representation theory of these algebras is computationally tractable, and have examined several examples.

Acknowledgments: Robert Moody explained the general structure of Borcherds algebras to me very early on. Paul Ginsparg provided a number of useful comments, especially on the possible role that Borcherds algebras might play in second quantization [3].

Work performed in part under the auspices of the U.S. Department of Energy, under contract No. W-7405-ENG-48.

References

1. R.E. Borcherds, *Generalized Kac–Moody algebras*, J. Algebra **115** (1988), 501–512.

2. S. Kass, R.V. Moody, J. Patera, and R. Slansky, *Affine Lie algebras, weight multiplicities and branching rules*, University of California Press, Los Angeles, 1990.

3. R. Slansky, *An algebraic role for energy and number operators for multiparticle states*, Nuclear Phys. B **389** (1993), 349–364.

4. V.G. Kac, *Infinite dimensional Lie algebras*, Cambridge University Press, Cambridge, 1985.

5. R. Slansky, *Group theory for unified model building*, Phys. Rep. **79** (1981), 1–128.

6. M.R. Bremner, R.V. Moody, and J. Patera, *Tables of dominant weight multiplicities for representations of simple Lie algebras*, Marcel Dekker, Inc., New York, 1985.

17

Graded Contractions of Lie Algebras of Physical Interest

J. Tolar

ABSTRACT The new notion of G-graded contraction of Lie algebras is briefly surveyed. As physical applications, our results on $Z_2 \times Z_2$-graded contractions of several Lie algebras commonly used in physics are presented, based on common papers with M. de Montigny, J. Patera, and P. Trávníček.

1 Introduction

The purpose of this chapter is to present an introduction to some physical applications of the new theory of graded contractions of Lie algebras [8, 10, 11].

Originally the notion of a contraction implicitly appeared in physics as a change of symmetry, usually connected with certain asymptotic limits of physical theories, e.g., special relativity toward nonrelativistic physics, or quantum mechanics toward classical mechanics. To our knowledge, the very first explicit treatment of a contraction can be found in [14], where a limiting procedure of this kind was performed with the infinitesimal generators of the conformal group $SO(4, 2)$. However, since the pioneering work of E.P. Wigner and E. Inönü [6] a new class of relations among symmetry algebras has been studied—Wigner–Inönü (WI) contractions of Lie algebras, and the contraction literature started to expand (e.g., [3, 5, 13]).

In the recent modifications of Wigner's approach introduced by J. Patera and his collaborators [8, 10, 11] one reduces the scope of study to *contractions preserving a fixed grading by a finite Abelian group G*. Such an approach offers a useful and practical way for the classification and description of all contractions for a fixed grading.

In order to apply the new theory of graded contractions of Lie algebras in physics, it is desirable to define the gradings on the basis of some physical principles. For instance, the gradings can be defined by the automorphisms of the given Lie algebra induced by some discrete Abelian invariance group. In our physical applications [9, 16] the gradings are defined by involutive Lie algebra automorphisms induced by discrete transformations of *space inversion* and *time reversal* be preserved. In order to preserve physical types of Lie algebra generators under contraction, we impose in addition that their transformation properties under *space rotations* be preserved. In [17] the transformation of space inversion is trivial on the Lie algebra and is replaced by a discrete canonical transformation exchanging positions and momenta. Such additional assumptions then lead to a substantially restricted but physically well justified classification of contractions.

2   Notion of Graded Contractions

Let us briefly define the notion of a grading of a Lie algebra L (over $\mathbb{R}$ or $\mathbb{C}$). A grading of a Lie algebra L by a finite *Abelian* group G means the following (the additive notation for the multiplication in G is used):

- The Lie algebra is decomposed as a linear space into a direct sum of (grading) subspaces

$$L = \bigoplus_{i \in G} L_i,$$

  which are eigenspaces of the automorphisms of L induced by the elements $g \in G$.

- The commutation relations in L have graded structure, i.e., for every choice of elements $X \in L_i$ and $Y \in L_j$, we have

$$[X, Y] = Z,$$

  where Z belongs to the grading subspace L_{i+j}, whenever the commutator differs from zero. For simplicity of notation we write

$$0 \neq [L_i, L_j] \subseteq L_{i+j}.$$

A *G-graded contraction* L^ε of L is then defined as a Lie algebra that has the same linear space structure as L, but modified commutation relations

$$0 \neq [L_i, L_j]_\varepsilon := \varepsilon_{i,j}[L_i, L_j] \subseteq \varepsilon_{i,j} L_{i+j}.$$

The parameters $\varepsilon_{i,j}$ (zero or not) are such that the Jacobi identity is never violated. For contractions of complex (or real) Lie algebras, the parameters $\varepsilon_{i,j}$ are, respectively, complex (or real). The Jacobi identity implies the (first) basic set of *contraction equations* [8]:

$$\varepsilon_{i,j}\varepsilon_{k,i+j} = \varepsilon_{j,k}\varepsilon_{i,j+k}, \quad \forall i, j, k \in G. \tag{2.1}$$

Antisymmetry of the Lie bracket implies

$$\varepsilon_{i,j} = \varepsilon_{j,i}.$$

We say that each solution of Eq. (2.1) determines a *graded contraction* of L. There is always a *trivial contraction* $\varepsilon_{i,j} = 1$, $\forall i, j \in G$, which corresponds to the original Lie algebra L. The contraction is either *continuous* or *discrete* according to whether or not the corresponding solution can be reached from the trivial solution by a continuous change such that (2.1) is never violated.

It is convenient to write ε in matrix form, although most of usual matrix operations cannot be interpreted as operations on contractions. Let us note that, in many cases, the grading of a Lie algebra is *nongeneric*; i.e., some commutators $[L_i, L_j]$ are identically zero before contraction. It is convenient to store this information in the symmetric matrix κ;

$$\kappa_{i,j} = \begin{cases} 0 & \text{if } [L_i, L_j] = 0, \\ 1 & \text{otherwise.} \end{cases}$$

It is important to notice that the *scaling transformation* φ

$$X_i \xrightarrow{\varphi} X_i' = a_i X_i, \quad a_i \in \mathbb{C} \quad \text{or} \quad \mathbb{R}, \quad X_i \in L_i,$$

of the grading subspaces of the Lie algebra changes its commutation relations into

$$[X_i, X_j] = X_{i+j} \xrightarrow{\varphi} [X_i', X_j'] = \frac{a_i a_j}{a_{i+j}} X_{i+j}'.$$

In this way the nonzero contraction parameters $\varepsilon_{i,j}$, $i, j \in G$ that differ from zero can be normalized [8] to 1 over $\mathbb{C}$ and to ± 1 over $\mathbb{R}$.

In this article we shall use special notation for $G = Z_2 \times Z_2$. Lie algebra L is $Z_2 \times Z_2$-graded, if it is possible to decompose L into the direct sum of four subspaces

$$L = L_{00} \oplus L_{01} \oplus L_{10} \oplus L_{11} = \bigoplus_{\alpha, \beta \in \{0,1\}} L_{\alpha\beta},$$

such that

$$0 \neq [L_{\alpha\beta}, L_{\gamma\delta}] \subseteq L_{\alpha+\gamma, \beta+\delta}, \quad \alpha, \beta, \gamma, \delta \in \{0, 1\},$$

with subscripts added modulo 2. We will denote the subscripts shortly by $a \equiv 00$, $b \equiv 01$, $c \equiv 10$, $d \equiv 11$.

3 Physical Applications

Kinematical Lie algebras of four-dimensional spacetime

In physics, the concept of *kinematical* (or *relativity*) *groups* is of fundamental importance, since through these symmetry groups of spacetimes the basic invariance of the laws of physics can be implemented. The paradigm is the special theory of relativity, with the ten-parameter Poincaré group containing (as transformation group of the four-dimensional Minkowski space-time) the time and space translations, space rotations, and boosts (inertial transformations).

The possible Lie algebras L of kinematical groups were classified [1] under the following natural *physical assumptions*:

(1) L is a real ten-dimensional Lie algebra and its generators correspond to time translations (H), space translations (P_i), space rotations (J_i), and inertial transformations (K_i), $i = 1, 2, 3$.

(2) Rotational invariance of space imposes special transformation properties of the generators under rotations.

(3) Space inversion and time-reversal transformations are automorphisms of L.

(4) Inertial transformations in any given direction form *noncompact* one-parameter subgroups of the kinematical group.

The results of [1] (see also [2]) show that the simple real Lie algebras of the de Sitter groups should be taken as the starting points for WI contractions yielding the Poincaré Lie algebra. It was also pointed out [7] that so$(4, 1)$ and so$(3, 2)$ are the only simple Lie algebras which can be WI-contracted into the Poincaré Lie algebra. The deformation point of view [4] explains this result by cohomology methods: the Poincaré Lie algebra admits deformations only to the de Sitter Lie algebras and these, being simple, are rigid against deformations.

Our idea in [9] was therefore to start with the simple *complex* Lie algebra B_2 and to investigate in the first step G-graded contractions of B_2 with the grading group G generated by two commuting transformations Π, Θ of the space inversion and time-reversal. In the second step we studied the G-graded contractions of the *real* forms of B_2, i.e., of so(5), so$(4, 1)$ and so$(3, 2)$. We took only postulates (1)–(3) of [1] into account, in order to see the whole contraction scheme. It turned out that it is interesting not only by itself but also from the physical point of view. (The three Lie algebras that do not satisfy postulate (4) can then be easily recognized and discarded.) These contractions can be considered physically interesting because the noncompact real forms so$(4, 1)$ and so$(3, 2)$ correspond to the de Sitter and the anti–de Sitter groups, respectively, and among the contracted Lie algebras those of the Poincaré and the Galilei group are present. As a result, the classification of the ten-dimensional kinematical groups emerged under the very natural assumptions of space isotropy and preservation of the $Z_2 \times Z_2$-grading induced by the Lie algebra automorphisms of space inversion Π and time reversal Θ.

The ten "kinematical" generators of B_2 satisfy the commutation relations

$$
\begin{aligned}
[\mathbf{J}, \mathbf{J}] &= -\mathbf{J}, \quad [\mathbf{J}, H] = 0, \quad [\mathbf{J}, \mathbf{P}] = -\mathbf{P}, \quad [\mathbf{J}, \mathbf{K}] = -\mathbf{K}, \\
&\quad\ [H, H] = 0, \quad [H, \mathbf{P}] = \ \ \mathbf{K}, \quad [H, \mathbf{K}] = -\mathbf{P}, \\
&\quad\quad\quad\quad\quad\quad [\mathbf{P}, \mathbf{P}] = -\mathbf{J}, \quad [\mathbf{P}, \mathbf{K}] = \ \ H, \\
&\quad\quad\quad\quad\quad\quad\quad\quad\quad\quad\quad\quad\ \ [\mathbf{K}, \mathbf{K}] = -\mathbf{J},
\end{aligned}
\tag{3.1}
$$

where the notation $[\mathbf{A}, \mathbf{B}] = \mathbf{C}$ for $[A_i, B_j] = \varepsilon_{ijk} C_k$, $[\mathbf{A}, \mathbf{B}] = D$ for $[A_i, B_j] = \delta_{ij} D$ and $[\mathbf{A}, D] = \mathbf{B}$ for $[A_i, D] = B_i$ is used.

The commuting involutive automorphisms of space-inversion Π and time reversal Θ induce two Z_2-gradings:

$$
\Pi: B_2 = \mathrm{span}\{\mathbf{J}, H\} \oplus \mathrm{span}\{\mathbf{P}, \mathbf{K}\},
$$

and

$$
\Theta: B_2 = \mathrm{span}\{\mathbf{J}, \mathbf{P}\} \oplus \mathrm{span}\{H, \mathbf{K}\}.
$$

Taken simultaneously, they induce the $\Pi \times \Theta$-grading with the grading group $G = Z_2 \times Z_2$:

$$
B_2 = L_a \oplus L_b \oplus L_c \oplus L_d = \mathrm{span}\{\mathbf{J}\} \oplus \mathrm{span}\{H\} \oplus \mathrm{span}\{\mathbf{P}\} \oplus \mathrm{span}\{\mathbf{K}\}.
$$

Let us summarize our restrictions imposed on the contraction:

A. The Lie algebra to be contracted is defined by the commutation relations (3.1). They determine the matrix κ which we give with rows and columns in the order a, b, c, d:

$$
\kappa^{\varepsilon} = \begin{pmatrix} 1 & 0 & 1 & 1 \\ & 0 & 1 & 1 \\ & & 1 & 1 \\ & & & 1 \end{pmatrix}.
$$

B. The brackets (3.1) involving $\mathbf{J}$ are not changed under contraction (isotropy of space).

C. The contractions preserve the $\Pi \times \Theta$-grading of B_2.

Hence, in the modified commutation relations

$$
\begin{aligned}
[\mathbf{J}, \mathbf{J}] &= -\varepsilon_{a,a} \mathbf{J}, \quad [\mathbf{J}, H] = \varepsilon_{a,b} 0, \quad [\mathbf{J}, \mathbf{P}] = -\varepsilon_{a,c} \mathbf{P}, \quad [\mathbf{J}, \mathbf{K}] = -\varepsilon_{a,d} \mathbf{K}, \\
&\quad\ [H, H] = \varepsilon_{b,b} 0, \quad [H, \mathbf{P}] = \ \ \varepsilon_{b,c} \mathbf{K}, \quad [H, \mathbf{K}] = -\varepsilon_{b,d} \mathbf{P}, \\
&\quad\quad\quad\quad\quad\quad [\mathbf{P}, \mathbf{P}] = -\varepsilon_{c,c} \mathbf{J}, \quad [\mathbf{P}, \mathbf{K}] = \ \ \varepsilon_{c,d} H, \\
&\quad\quad\quad\quad\quad\quad\quad\quad\quad\quad\quad\quad\ \ [\mathbf{K}, \mathbf{K}] = -\varepsilon_{d,d} \mathbf{J},
\end{aligned}
$$

TABLE 17.1.

$$\varepsilon^{R1} = \begin{pmatrix} 1 & \square & 1 & 1 \\ & \square & 1 & 1 \\ & & 1 & 1 \\ & & & 1 \end{pmatrix}$$ complex simple Lie algebra so$(5, \mathbf{C})$ B_2

$$\varepsilon^{R2} = \begin{pmatrix} 1 & \square & 1 & 1 \\ & \square & 0 & 1 \\ & & 0 & 1 \\ & & & 1 \end{pmatrix}$$ complex Poincaré Lie algebra $P^{\mathbf{C}}$

$$\varepsilon^{R3} = \begin{pmatrix} 1 & \square & 1 & 1 \\ & \square & 1 & 0 \\ & & 1 & 1 \\ & & & 0 \end{pmatrix}$$ complex para-Poincaré Lie algebra $P'^{\mathbf{C}}$

$$\varepsilon^{R4} = \begin{pmatrix} 1 & \square & 1 & 1 \\ & \square & 0 & 0 \\ & & 0 & 1 \\ & & & 0 \end{pmatrix}$$ complex Carroll Lie algebra $C^{\mathbf{C}}$

$$\varepsilon^{A1} = \begin{pmatrix} 1 & \square & 1 & 1 \\ & \square & 1 & 1 \\ & & 0 & 0 \\ & & & 0 \end{pmatrix}$$ complex Newton–Hooke Lie algebra $N^{\mathbf{C}}$

$$\varepsilon^{A2} = \begin{pmatrix} 1 & \square & 1 & 1 \\ & \square & 0 & 1 \\ & & 0 & 0 \\ & & & 0 \end{pmatrix}$$ complex Galilei Lie algebra $G^{\mathbf{C}}$

$$\varepsilon^{A3} = \begin{pmatrix} 1 & \square & 1 & 1 \\ & \square & 1 & 0 \\ & & 0 & 0 \\ & & & 0 \end{pmatrix}$$ complex para-Galilei Lie algebra $G'^{\mathbf{C}}$

$$\varepsilon^{A4} = \begin{pmatrix} 1 & \square & 1 & 1 \\ & \square & 0 & 0 \\ & & 0 & 0 \\ & & & 0 \end{pmatrix}$$ complex static Lie algebra $\mathrm{St}^{\mathbf{C}}$

the contraction parameters $\varepsilon_{a,b}$ and $\varepsilon_{b,b}$ need not be defined, and $\varepsilon_{a,a} = \varepsilon_{a,c} = \varepsilon_{a,d} = 1$. The remaining contraction equations (2.1) are then

$$\varepsilon_{c,c} = \varepsilon_{b,c}\varepsilon_{c,d}, \quad \varepsilon_{d,d} = \varepsilon_{b,d}\varepsilon_{c,d}, \quad \varepsilon_{b,d}\varepsilon_{c,c} = \varepsilon_{b,c}\varepsilon_{d,d}.$$

Using the same index ordering for matrix ε as for κ, we can cast solutions into matrix form. At this point let us recall the results of [8, Tables I and II], where *all* 40 solutions of the full set (2.1) of $Z_2 \times Z_2$ contraction equations were found over complex numbers, i.e., for the case of B_2. Their inspection shows that only eight of them satisfy the restrictions A and B, and each of these eight belongs to the *continuous* type. We display them in Table 17.1 together with the complexified kinematical Lie algebras. Their one-to-one correspondence with the four types of "relative-time" and four types of "absolute-time" Lie algebras—as described in Ref. [1] and denoted there by $R1, \ldots, R4$, and $A1, \ldots, A4$, respectively—is used in the superscripts of the contraction matrices ε. We follow closely the notation and the names used in [1]. We remind the reader that the entries corresponding to zero entries of κ are denoted by $\square$; they may take any value. Turning to the three real forms of the complex Lie algebra B_2, i.e., to so(5), so(4,1) and so(3,2), we have to solve the contraction equations (2.1) over the reals. Solutions with nonnegative entries are exactly the already given eight solutions of *continuous type* (normalized to 1 by a scale change of basis). Over the reals, unlike in the complex domain, contractions that admit some *negative entries* have to be taken into account as distinct, even if they lead to isomorphic Lie algebras.

By solving (2.1) one finds six additional contractions—all of *discrete* type:

$$\varepsilon^{R1'} = \begin{pmatrix} 1 & \square & 1 & 1 \\ & \square & -1 & 1 & 1 \\ & & 1 & -1 \\ & & & -1 \end{pmatrix}, \quad \varepsilon^{R1''} = \begin{pmatrix} 1 & \square & 1 & 1 \\ & \square & 1 & 1 \\ & & -1 & -1 \\ & & & -1 \end{pmatrix}, \quad \varepsilon^{R1'''} = \begin{pmatrix} 1 & \square & 1 & 1 \\ & \square & -1 & 1 \\ & & -1 & 1 \\ & & & 1 \end{pmatrix},$$

$$\varepsilon^{R2'} = \begin{pmatrix} 1 & \square & 1 & 1 \\ & \square & 0 & -1 \\ & & 0 & 1 \\ & & & -1 \end{pmatrix}, \quad \varepsilon^{R3'} = \begin{pmatrix} 1 & \square & 1 & 1 \\ & \square & -1 & 0 \\ & & -1 & 1 \\ & & & 0 \end{pmatrix}, \quad \varepsilon^{A1'} = \begin{pmatrix} 1 & \square & 1 & 1 \\ & \square & -1 & 1 \\ & & 0 & 0 \\ & & & 0 \end{pmatrix}.$$

We denoted them as primed variants of the continuous contractions $R1, R2, R3, A1$. They mutually relate pairs of real forms of B_2, where—owing to restrictions A and B—two isomorphic Lie algebras so($4_{JK}, 1$) and so($4_{JP}, 1$) (with subalgebras so(4) generated by J, K and J, P, respectively) should be taken as distinct from the physical point of view.

In this way we recovered and generalized the results of [1] with the new method of graded contractions. We found that all kinematical Lie algebras of [1] were contained within the set of $\Pi \times \Theta$-graded contractions of so(5). This obviously justifies the choice of preserved grading in accordance with the physical postulate 3. Moreover, discrete contractions provide new relations among kinematical groups. The Lie algebras so(5), so($4_{JK}, 1$) and iso(4)$_{JK}$, for which K_i are compact generators, were discarded by [1] from their classification, because they did not satisfy the postulate 4. Thus the method of graded contractions not only enlarges the range of possibilities, compared to traditional WI-contractions, but also yields a new unifying view of classical results.

Conformal group of Minkowski spacetime

We found it worthwhile and also interesting from the physical point of view to extend the scope of our investigation to the *conformal group of the Minkowski spacetime* [16] because of its overall importance, especially in quantum field theories as the symmetry of theories of massless particles [15]. As a matter of fact, already I.E. Segal [14] gave the very first definition of a contraction yielding an approximate (asymptotic) symmetry from the initial exact symmetry on an example of the conformal Lie algebra so(4, 2).

In order that the results be directly comparable with the preceding ones [9], we dwelled on the assumptions of the space isotropy and of preservation of the $\Pi \times \Theta$-grading. There are two natural commuting involutive automorphisms of the conformal Lie algebra: The space inversion Π and the time reversal Θ, which induce the following $Z_2 \times Z_2$ grading of so(4, 2):

$$so(4, 2) = \bigoplus_{\alpha\beta \in \{0,1\}} L_{\alpha\beta} = L_a \oplus L_b \oplus L_c \oplus L_d$$

$$= \text{span}\{\mathbf{J}, D\} \oplus \text{span}\{p_4, q_4\} \oplus \text{span}\{\mathbf{p}, \mathbf{q}\} \oplus \text{span}\{\mathbf{M}\}.$$

It is understood that the Lie algebra of the conformal group of the Minkowski spacetime is formed by the physical differential operators

Lorentz transformations	$M_{\mu\nu} = i(x_\mu \partial_\nu - x_\nu \partial_\mu)$,
translations	$P_\mu = i\partial_\mu$,
special conformal transformations	$K_\mu = i(2x_\mu x^\nu \partial_\nu - x^2 \partial_\mu)$,
dilatations	$D = i; x^\nu \partial_\nu$,

where $\partial_\mu \equiv \partial/\partial x^\mu$. It should be noted that the generators $\mathbf{J}, H, \mathbf{P}$ and $\mathbf{K}$ of Section 3.1 have similar physical meaning as the generators $\mathbf{J}, P_4 = p_4 + q_4$, $\mathbf{P} = \mathbf{p} + \mathbf{q}$ and $\mathbf{M}$ of so(4, 2), and $K_\mu = p_\mu - q_\mu$.

The inspection of graded commutation relations of so(4, 2) [16] shows that they determine the generic κ without zeros:

$$\kappa = \begin{pmatrix} 1 & 1 & 1 & 1 \\ & 1 & 1 & 1 \\ & & 1 & 1 \\ & & & 1 \end{pmatrix}.$$

In [8] the case of the generic $Z_2 \times Z_2$-graded structure was also considered and 40 inequivalent solutions of the contraction equations (2.1) were found.

In order to preserve the usual transformation properties of the physical generators also in the contracted Lie algebras, the $\Pi \times \Theta$-graded contractions of so(4, 2) were investigated again under the *physical assumptions* (2) and (3) stated in Section 3.1.

In accordance with assumption (2) the contraction parameters $\varepsilon_{a,a}$, $\varepsilon_{a,c}$ and $\varepsilon_{a,d}$ are kept equal to 1. This assumption restricts the contraction equations to the set of eight independent equations. From them the $\Pi \times \Theta$-graded contractions of so(4, 2) were computed [16]. The results turned out to be almost the same as in Section 3.1. There is, however, one additional case of a *discrete contraction*:

$$
\varepsilon^D = \begin{pmatrix} 1 & 0 & 1 & 1 \\ & 0 & 0 & 0 \\ & & 0 & 0 \\ & & & 0 \end{pmatrix},
$$

because of the generic κ.

For further details we refer to [16]. Nevertheless, we think it is appropriate to comment on I.E. Segal's [14] treatment of contraction of $so(4, 2)$. His example was constructed with the aim to obtain the ten-dimensional Poincaré Lie algebra as subalgebra of the fifteen-dimensional contracted Lie algebra. In our treatment only one contraction has this property, namely, ε^{R2}. One of the remaining five generators of so(4, 2)—the dilatation generator D—if taken together with the Poincaré generators, forms the basis of the Lie algebra of the Weyl group. This is the largest transformation group of the Minkowski spacetime which preserves the causal order [18].

Algebraic theory of collective nuclear models

Next we turned our attention [17] to the dynamical symmetries found in the algebraic formulation [12] of nuclear collective models. They were obtained by extending the algebraic Bohr–Mottelson model of nuclear rotations and quadrupole vibrations. The most comprehensive dynamical Lie algebra is the *noncompact symplectic algebra* sp(3, $\mathbb{R}$) of dimension 21. Its generators in the microscopic Hilbert space of an N-particle system correspond to the physical quantities

$Q_{ij} = \sum_{\alpha=1}^{N-1} \varrho_{\alpha i} \varrho_{\alpha j}$, monopole and quadrupole tensor,

$K_{ij} = \sum_{\alpha=1}^{N-1} \pi_{\alpha i} \pi_{\alpha j}$, kinetic energy and quadrupole momentum tensor,

$L_{ij} = \sum_{\alpha=1}^{N-1} (\varrho_{\alpha i} \pi_{\alpha j} - \varrho_{\alpha j} \pi_{\alpha i})$, angular momentum,

$S_{ij} = \sum_{\alpha=1}^{N-1} (\varrho_{\alpha i} \pi_{\alpha j} + \varrho_{\alpha j} \pi_{\alpha i} - i\hbar \delta_{ij})$, monopole and quadrupole vibrational momentum.

These quantum collective observables are one-body bilinear products constructed from the relative position and momentum operators $\varrho_{\alpha i}$, $\pi_{\alpha j}$, i, $j \in \{1,2,3\}$, such that all particles can be considered on the same footing interchangeably.

From the adopted physical point of view, $sp(3,\mathbb{R})$ is the Lie algebra of the subgroup of linear canonical transformations of six-dimensional phase space $\mathbb{R}^6$ (with coordinates (x_i, p_j), i, $j \in \{1,2,3\}$). The set of its infinitesimal generators is given by all quadratic polynomials in x_i, p_j.

We are going to use two convenient transformations of finite order that generate $Z_2 \times Z_2$–graded structure on $sp(3,\mathbb{R})$ and have concrete physical meaning as physically motivated particular discrete transformations of the phase space.

The *first* one is the linear canonical transformation K,

$$x_i \rightarrow p_i, \quad p_j \rightarrow -x_j, \qquad i,j \in \{1,2,3\},$$

exchanging the coordinates and momenta.

The *second* involutive automorphism Θ (commuting with the first one), which provides another Z_2-graded structure on $sp(3,\mathbb{R})$ is generated by the linear operation of time reversal:

$$x_i \rightarrow x_i, \quad p_j \rightarrow -p_j, \qquad i,j \in \{1,2,3\}.$$

By combining the time-reversal Θ and the discrete canonical transformation K, we get the following $Z_2 \times Z_2$-graded structure of $sp(3,\mathbb{R})$:

$$sp(3,\mathbb{R}) = \bigoplus_{\alpha\beta\in\{0,1\}} L_{\alpha\beta} = L_a \oplus L_b \oplus L_c \oplus L_d,$$

$$= \text{span}_{\mathbb{R}}\{-\mathbf{L}\} \oplus \text{span}_{\mathbb{R}}\{-\mathbf{S}\} \oplus \text{span}_{\mathbb{R}}\{(\mathbf{K} + \{\mathbf{Q}\})\}$$
$$\oplus \text{span}_{\mathbb{R}}\{-(\mathbf{K} - \mathbf{Q})\},$$

where $\mathbf{L} \equiv \{L_{12}, L_{23}, L_{31}\}$, $\mathbf{S} \equiv \{\frac{1}{2}S_{11}, \frac{1}{2}S_{22}, \frac{1}{2}S_{33}, S_{12}, S_{23}, S_{31}\}$, $\mathbf{K} \equiv \{\frac{1}{2}K_{11}, \frac{1}{2}K_{22}, \frac{1}{2}K_{33}, K_{12}, K_{23}, K_{31}\}$, $\mathbf{Q} \equiv \{\frac{1}{2}Q_{11}, \frac{1}{2}Q_{22}, \frac{1}{2}Q_{33}, Q_{12}, Q_{23}, Q_{31}\}$.

In order to preserve the usual $SO(3)$-transformation properties of the generators as physical quantities also in the contracted Lie algebras (e.g., vectors remain vectors, pseudovectors remain pseudovectors after contraction), we have studied the $\Theta \times K$-graded contractions of $sp(3,\mathbb{R})$ under the following *physical assumptions*:

(i) Space isotropy: The rotational invariance of space means that the form of all commutation relations involving $\mathbf{L}$ remains unchanged after contraction.

(ii) The behavior of generators under time-reversal Θ is preserved through contraction. (Parity $+1$ is preserved automatically, because the generators are quadratic forms on phase space.)

(iii) The extra discrete symmetry K between coordinates and momenta is added since it is characteristic of the harmonic oscillator Hamiltonian that appears in the numerous treatments of nuclear collective models.

Solved over $\{-1, 0, 1\}$ (under the assumptions (i)–(iii)), Eqs. (2.1) give 27 different normalized contraction matrices. Under the scaling transformation involving the simultaneous sign changes of $\varepsilon_{b,c}$, $\varepsilon_{b,d}$ and $\varepsilon_{c,d}$, one finds that among 27 solutions, only 14 contraction matrices are independent. Among them are eight (continuous) contraction matrices ε^{R1}, ε^{R2}, ε^{R3}, ε^{R4}, ε^{A1}, ε^{A2}, ε^{A3}, ε^{A4}. The remaining six contractions are additional *discrete* $\Theta \times K$-graded contractions given by the contraction matrices

$$\varepsilon^{R1d1} = \begin{pmatrix} 1 & 1 & 1 & 1 \\ -1 & 1 & -1 \\ & 1 & 1 \\ & & & -1 \end{pmatrix}, \qquad \varepsilon^{R1d2} = \begin{pmatrix} 1 & 1 & 1 & 1 \\ 1 & -1 & -1 \\ & -1 & 1 \\ & & & -1 \end{pmatrix},$$

$$\varepsilon^{R1d3} = \begin{pmatrix} 1 & 1 & 1 & 1 \\ -1 & -1 & 1 \\ & -1 & 1 \\ & & & 1 \end{pmatrix}, \qquad \varepsilon^{A1d} = \begin{pmatrix} 1 & 1 & 1 & 1 \\ -1 & -1 & 1 \\ & \cdot & \cdot \\ & & & \cdot \end{pmatrix},$$

$$\varepsilon^{R2d} = \begin{pmatrix} 1 & 1 & 1 & 1 \\ \cdot & \cdot & -1 \\ & \cdot & 1 \\ & & & -1 \end{pmatrix}, \qquad \varepsilon^{R3d} = \begin{pmatrix} 1 & 1 & 1 & 1 \\ \cdot & -1 & \cdot \\ & -1 & 1 \\ & & & \cdot \end{pmatrix}.$$

The real Lie algebras $sp(3)$, $sp(3, \mathbb{R})$ are connected by discrete contractions. We are especially interested in the classification of all physically inequivalent $\Theta \times K$-graded contractions of $sp(3, \mathbb{R})$. Below, in the first and second columns, we list its *eight continuous contractions*:

$$sp(3, \mathbb{R}) \xrightarrow{\varepsilon^{R1}} \qquad sp(3, \mathbb{R}) \qquad \xleftarrow{\varepsilon^{R1d1}} sp(3),$$

$$sp(3, \mathbb{R}) \xrightarrow{\varepsilon^{R2}} \qquad gl(3, \mathbb{R})_{L, (K-Q)} \rhd T_{12; S, (K+Q)} \qquad \xleftarrow{\varepsilon^{R1d1} \bullet \varepsilon^{R2}} sp(3),$$

$$sp(3, \mathbb{R}) \xrightarrow{\varepsilon^{R3}} \qquad u(3)_{L, (K+Q)} \rhd T_{12; S, (K-Q)} \qquad \xleftarrow{\varepsilon^{R1d1} \bullet \varepsilon^{R3}} sp(3),$$

$$sp(3, \mathbb{R}) \xrightarrow{\varepsilon^{A1}} \qquad gl(3, \mathbb{R})_{L, S} \rhd T_{12; K, Q} \qquad \xleftarrow{\varepsilon^{R1d1} \bullet \varepsilon^{A1}} sp(3),$$

$$sp(3, \mathbb{R}) \xrightarrow{\varepsilon^{R4}} \qquad so(3)_L \rhd \mathcal{L}_{S, K, Q}, \ T_{6; S} \subset \mathcal{C}_{\mathcal{L}} \qquad \xleftarrow{\varepsilon^{R1d1} \bullet \varepsilon^{R4}} sp(3),$$

$$sp(3, \mathbb{R}) \xrightarrow{\varepsilon^{A2}} \qquad so(3)_L \rhd \mathcal{L}_{S, K, Q}, \ T_{6; (K+Q)} \subset \mathcal{C}_{\mathcal{L}} \qquad \xleftarrow{\varepsilon^{R1d1} \bullet \varepsilon^{A2}} sp(3),$$

$$sp(3, \mathbb{R}) \xrightarrow{\varepsilon^{A3}} \qquad so(3)_L \rhd \mathcal{L}_{S, K, Q}, \ T_{6; (K-Q)} \subset \mathcal{C}_{\mathcal{L}} \qquad \xleftarrow{\varepsilon^{R1d1} \bullet \varepsilon^{A3}} sp(3),$$

$$sp(3, \mathbb{R}) \xrightarrow{\varepsilon^{A4}} \qquad so(3)_L \rhd T_{18; S, K, Q} \qquad \xleftarrow{\varepsilon^{R1d1} \bullet \varepsilon^{A4}} sp(3).$$

Here by $\mathcal{C}_{\mathcal{L}}$ the center of the radical $\mathcal{L}_{S, K, Q}$ is denoted. The third column involving the corresponding contractions of $sp(3)$ is presented for comparison. The operation $\bullet$ is the commutative composition of matrices defined in Ref. [8] by the multiplication of equally placed elements.

There are still *three independent discrete contractions*:

$$\mathrm{sp}(3,\mathbb{R}) \xrightarrow{\varepsilon^{R2}} \mathrm{u}(3)_{\mathbf{L},i(\mathbf{K}-\mathbf{Q})} \triangleright T_{12;\mathbf{S},(\mathbf{K}+e\mathbf{Q})} \xleftarrow{\varepsilon^{R1d1}\bullet\varepsilon^{R1d2}\bullet\varepsilon^{R2}} \mathrm{sp}(3),$$

$$\mathrm{sp}(3,\mathbb{R}) \xrightarrow{\varepsilon^{R3}} \mathrm{gl}(3,\mathbb{R})_{\mathbf{L},i(\mathbf{K}+\mathbf{Q})} \triangleright T_{12;\mathbf{S},(\mathbf{K}-\mathbf{Q})} \xleftarrow{\varepsilon^{R1d1}\bullet\varepsilon^{R1d3}\bullet\varepsilon^{R3}} \mathrm{sp}(3),$$

$$\mathrm{sp}(3,\mathbb{R}) \xrightarrow{\varepsilon^{A1}} \mathrm{u}(3)_{\mathbf{L},i\mathbf{S}} \triangleright T_{12;\mathbf{K},\mathbf{Q}} \xleftarrow{\varepsilon^{R1d1}\bullet\varepsilon^{R1d3}\bullet\varepsilon^{A1}} \mathrm{sp}(3).$$

Acknowledgments: This contribution is dedicated to Professors J. Patera and P. Winternitz on the occasion of their 60th birthdays. I wish them all the best in the years to come. Besides their numerous and important contributions to several fields of mathematical physics, they actively promoted scientific exchanges and collaborations with scientists from various countries. I would like to express how indebted I am especially to Jiří Patera for his support and inspiring discussions in Prague and Montréal during our scientific collaboration.

The results summarized in this article grew out of a fruitful collaboration with J. Patera and also with M. de Montigny and P. Trávníček. I would find it a proper tribute to J. Patera if the review succeeded in turning the attention of the scientific community to this important part of his scientific achievements.

The reviewed works were supported by NSERC of Canada (1992) and by the Grant Agency of Czech Republic under the contracts 202/93/1314 and 202/96/0218.

For the kind hospitality at the symposium I am grateful to Prof. L. Vinet, Director of Centre de recherches mathématiques, Université de Montréal.

REFERENCES

1. H. Bacry and J.-M. Levy-Leblond, *Possible kinematics*, J. Math. Phys. **9** (1968), 1605–1614.

2. H. Bacry and J. Nuyts, *Classification of ten-dimensional kinematical groups with space isotropy*, J. Math. Phys. **27** (1986), 2455–2457.

3. H.D. Doebner and O. Melsheimer, *On a class of generalized group contractions*, N. Cimento A **49** (1967), 306–311.

4. M. Gerstenhaber, *On the deformation of rings and algebras*, Ann. Math. **79** (1964), 59–103.

5. R. Gilmore, *Lie groups, Lie algebras and some of their applications*, Chapter 10, J. Wiley & Sons, New York, 1974.

6. E. Inönü and E.P. Wigner, *On the contraction of groups and their representations*, Proc. Nat. Acad. Sci. US **9** (1953), 510–524; *On a*

particular type of convergence to a singular matrix, ibid. **40** (1954), 119–121.

7. M. Lévy-Nahas, *Monique deformation and contraction of Lie algebras*, J. Math. Phys. **8** (1967), 1211–1222.

8. M. de Montigny and J. Patera, *Discrete and continuous graded contractions of Lie algebras and superalgebras*, J. Phys. A **24** (1991), 525–549.

9. M. de Montigny, J. Patera, and J. Tolar, *Graded contractions and kinematical groups of spacetime*, J. Math. Phys. **35** (1994), 405–425.

10. R.V. Moody and J. Patera, *Discrete and continuous graded contractions of representations of Lie algebras*, J. Phys. A **24** (1991), 2227–2258.

11. J. Patera, *Graded contractions of Lie algebras, their representations, and tensor products*, Group Theory in Physics (Cocoyoc, 1991), AIP Conf. Proc., Amer. Inst. Phys., Vol. 266, New York, 1992, pp. 46–54.

12. G. Rosensteel and D. J. Rowe, *On the algebraic formulation of collective models. III. The symplectic shell model of collective motion*, Ann. Phys. **2** (1980), 343–369.

13. E.J. Saletan, *Contraction of Lie groups*, J. Math. Phys. **2** (1961), 1–21.

14. I.E. Segal, *A class of operator algebras which are determined by groups*, Duke Math. J. **18** (1951), 221–265.

15. I.T. Todorov, *Infinite-dimensional Lie algebras in conformal* QFT *models*, Lecture Notes in Physics, vol. 261, Springer-Verlag, Berlin, 1986, pp. 387–443.

16. J. Tolar and P. Trávníček, *Graded contractions and the conformal group of Minkowski spacetime*, J. Math. Phys. **36** (1995), 4489–4506.

17. J. Tolar and P. Trávníček, *Graded contractions of symplectic Lie algebras in collective models*, J. Math. Phys. **38** (1997), 49–56.

18. E.C. Zeeman, *Causality implies Lorentz group*, J. Math. Phys. **5** (1964), 490–493.